Vehicle Propulsion Systems

T0215926

Lino Guzzella and Antonio Sciarretta

Vehicle Propulsion Systems

Introduction to Modeling and Optimization

Third Edition

 Springer

Authors

Prof. Dr. Lino Guzzella
ETH Zurich
8092 Zurich
Switzerland

Dr. Antonio Sciarretta
IFP Energies Nouvelles
92852 Rueil-Malmaison
France

ISBN 978-3-642-43847-9   ISBN 978-3-642-35913-2
DOI 10.1007/978-3-642-35913-2
Springer Heidelberg New York Dordrecht London

Printed on acid-free paper

Springer is part of Springer Science+Business Media (www.springer.com)

Preface

Who should read this text?

This text is intended for persons interested in the analysis and optimization of vehicle propulsion systems. Its focus lies on the control-oriented mathematical description of the physical processes and on the model-based optimization of the system structure and of the supervisory control algorithms.

This text has evolved from a lecture series held during the last years in the mechanical engineering department at the Swiss Federal Institute of Technology (ETH), Zurich. The presumed audience is graduate mechanical or electrical engineering students. The prerequisites are general engineering topics and a first course in optimal control theory. Readers with little preparation in that area are referred to [41]. The most important results of parameter optimization and optimal control theory are summarized in Appendix II.

Why has this text been written?

Individual mobility relies to a large extent on passenger cars. These vehicles are responsible for a large part of the world's consumption of primary energy carriers, mostly fossil liquid hydrocarbons. The specific application profiles of these vehicles, combined with the inexorably increasing demand for mobility, have led to a situation where the reduction of fuel consumption has become a top priority for the society and the economy.

Many approaches that permit to reduce the fuel consumption of passenger cars have been presented so far and new ideas emerge on a regular basis. In most – if not all – cases these new systems are more complex than the traditional approaches. Additional electric motors, storage devices, torque converters, etc. are added with the intention to improve the system behavior. For such complex systems the traditional heuristic design approaches fail.

The only way to deal with such a high complexity is to employ mathematical models of the relevant processes and to use these models in a systematic ("model-based") way. This text focuses on such approaches and provides an introduction to the modeling and optimization problems typically encountered by designers of new propulsion systems for passenger cars.

What can be learned from this text?

This book analyzes the longitudinal behavior of road vehicles only. Its main emphasis is on the analysis and minimization of the energy consumption. Other aspects that are discussed are drivability and performance.

The starting point for all subsequent steps is the derivation of simple yet realistic mathematical models that describe the behavior of vehicles, prime movers, energy converters, and energy storage systems. Typically, these models are used in a subsequent optimization step to synthesize optimal vehicle configurations and energy management strategies.

Examples of modeling and optimization problems are included in Appendix I. These *case studies* are intended to familiarize the reader with the methods and tools used in powertrain optimization projects.

What cannot be learned from this text?

This text does not consider the pollutant emissions of the various powertrain systems because the relevant mechanisms of the pollutant formation are described on much shorter time scales than those of the fuel consumption. Moreover, the pollutant emissions of some prime movers are virtually zero or can be brought to that level with the help of appropriate exhaust gas purification systems. Readers interested in these aspects can find more information in [128].

Comfort issues (noise, harshness, and vibrations) are neglected as well. Only those aspects of the lateral and horizontal vehicle dynamics that influence the energy consumption are briefly mentioned. All other aspects of the horizontal and lateral vehicle dynamics, such as vehicle stability, roll-over dynamics, etc. are not discussed.

Acknowledgments

Many people have implicitly helped us to prepare this manuscript. Specifically our teachers, colleagues, and students have contributed to bring us to the point where we felt ready to write this text. Several people have helped us more explicitly in preparing this manuscript: Hansueli Hörler, who taught us the basic laws of engine thermodynamics, Alois Amstutz and Chris Onder who contributed to the development of the lecture series behind this text, those of our doctoral students whose dissertations have been used as the nucleus of several sections (we reference their work at the appropriate places), and Brigitte Rohrbach, who translated our manuscripts from "Italish" to English.

June 2005   Lino Guzzella and Antonio Sciarretta

Preface

Why a second edition?

The discussions about fuel economy of passenger cars have become even more intense since the first edition of this book appeared. Concerns about the limited resources of fossil fuels and the detrimental effects of greenhouse gases have spurred the interest of many people in industry and academia to work towards reduced fuel consumption of automobiles. Not surprisingly, the first edition of this monograph sold out rather rapidly. When the publisher asked us about a second edition, we decided to use this opportunity to revise the text, to correct several errors, and to add new material.

The following list includes the most important changes and additions we made:

- The section describing battery models has been expanded.
- A new section on power split devices has been added.
- A new section on pneumatic hybrid systems has been added.
- The chapter introducing supervisory control algorithm has been rewritten and expanded.
- Two new case studies have been added.
- A new appendix that introduces the main ideas of dynamic programming has been added.

Acknowledgements

We want to express our gratitude to the many colleagues and students who reported to us errors and omissions in the first edition of this text. Several people have helped us improving this monograph, in particular Christopher Onder who actively participated in the revisions.

June 2007   Lino Guzzella and Antonio Sciarretta

Preface

Why a third edition?

Over the last five years, vehicle propulsion systems have become an even more active research area. Increasing concerns over fossil fuel consumption and the associated environmental impacts have motivated many groups in industry and academia to propose new propulsion systems and to explore new optimization methodologies. This third edition has been prepared to include some of these developments. Of course, all errors and omissions that were discovered in the meantime have been corrected as well.

In particular:

- New concepts to understand vehicle energy consumption have been introduced.
- The section describing driving cycles has been updated.
- The section describing battery-electric vehicles has been expanded, while several now-common definitions have been added to classify hybrid-electric vehicles.
- The sections describing electric motor and battery models and controls have been expanded.
- The section on pneumatic hybrid systems has been rewritten and expanded; the corresponding case study has been replaced by a more up-to-date one.
- The chapter introducing supervisory control algorithm has been rewritten and expanded.

The most important additions, however, are the exercises that are included at the end of each chapter. These exercises were developed over the last years to support the vehicle propulsion systems course at ETH. The authors hope that some readers will find these exercises useful as well. Complete solutions will be provided at the URL http://www.idsc.ethz.ch.

Acknowledgements

We want to express our gratitude to the many colleagues and students who reported to us errors and omissions in the first edition of this text. Several people have helped us improving this monograph, in in particular Philipp Elbert, Christopher Onder, Tobias Ott and Christoph Voser who actively participated in the revisions.

July 2012   Lino Guzzella and Antonio Sciarretta

Contents

1

Introduction

This introductory chapter shows how the problems discussed in this text are embedded in a broader setting. First a motivation for and the objective of the subsequent analysis is introduced. After that the complete energy conversion chain is described, starting from the available primary energy sources and ending with the distance driven. Using average energy conversion efficiency values, some of the available options are compared. This analysis shows the importance of the "upstream" processes. The importance of the selected on-board energy carrier ("fuel") is stressed as well. In particular its energy density and the safety issues connected with the refueling process are emphasized. The last section of this first chapter lists the main options available for reducing the energy consumption of passenger vehicles.

1.1 Motivation

The main motivation to write this book is the inexorably increasing number of passenger cars worldwide. As Fig. 1.1 shows, some 800 million passenger cars are operated today. More interesting than this figure is the trend that is illustrated in this figure for the example of the United States of America (the same trend is observed in Japan and Europe): in wealthy societies the car density saturates at a ratio of approximately 400 to 800 cars per 1000 inhabitants.

It is corroborated empirically that the demand for personal transportation increases with the economic possibilities of a society [285]. Therefore, if the car density mentioned above is taken as the likely future value for other regions of the world, serious problems are to be expected. Countries such as China (1.3 billion inhabitants) or India (1.1 billion inhabitants) in the year 2007 have car densities of around 30 cars per 1000 inhabitants. Accordingly, in the next 20 years the car density in these countries will increase substantially, which will further increase the pressure on fuel prices and cause serious problems to the environment.

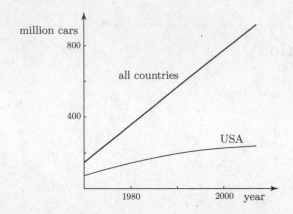

Fig. 1.1 *Schematic* representation of the development of the number of passenger cars operated worldwide

In the face of these trends, it is clear that new fuel sources must be developed and that the fuel consumption of passenger cars must be reduced substantially. This text focuses on the second approach.

1.2 Objectives

The main objectives of this text are to introduce mathematical models and optimization methods that permit a systematic minimization of the *energy consumption* of vehicle propulsion systems. The objects of this analysis are *passenger cars*, i.e., vehicles that

- are autonomous and do not depend on fixed energy-providing grids;
- have a refueling time that is negligible compared to the driving time between two refueling events;
- can transport two to six persons and some payload; and
- accelerate in approximately 10 to 15 seconds from 0 to 100 km/h, or can drive uphill a 5% ramp at the legal top speed, respectively.[1]

These requirements, which over the last one hundred years have evolved to a quasi-standard profile, substantially reduce the available options. Particularly the first and the second requirement can only be satisfied by few *on-board* energy storage systems, and the performance requirements can only be satisfied by propulsion systems able to produce a maximum power that is substantially larger than the power needed for most driving conditions.

A key element in all considerations is the on-board energy carrier system. This element must:

---

[1] These numerical values are only indicative. It goes without saying that the performance range is very wide.

- provide the highest possible energy density[2];
- allow for the shortest possible refueling time; and
- be safe and cause no environmental hazards in production, operation, and recycling.

The number of components that are necessary to realize modern and in particular future propulsion systems is inexorably increasing. Improved performance and fuel economy can only be obtained with complex devices. Of course, these subsystems influence each other. The best possible results are thus not obtained by an isolated optimization of each single component. Optimizing the entire system, however, is not possible with heuristic methods due to the "curse of exponential growth." The only viable approach to cope with this dilemma is to develop mathematical models of the components and to use model-based numerical methods to optimize the system structure and the necessary control algorithms. These models must be able to *extrapolate* the system behavior. In fact, such an optimization usually takes place before the actual components are available or requires the devices to operate in unexpected conditions. For these reasons, only first-principle models, i.e., models that are based on physical laws, will be used in this text.

Of course, some of the mathematical models and methods introduced in this text may be useful for the design of other classes of vehicles (trains, heavy-duty trucks, etc.). However, there are clear differences[3] that render the passenger car optimization problem particularly interesting.

At least three energy conversion steps are relevant for a comprehensive analysis of the energy consumption of passenger cars. As illustrated in Fig. 1.2, the actual energy source is one of the available primary energy carriers (chemical energy in fossil hydrocarbons, solar radiation used to produce bio mass or electric energy, nuclear energy, etc). In a first step, this energy is converted to an energy carrier that is suitable for on-board storage, i.e., to a "fuel" (examples are gasoline, hydrogen, etc.). This "fuel" is then converted by the propulsion system to mechanical energy that, in part, may be stored as kinetic or potential energy in the vehicle. The third energy transformation is determined by the vehicle parameters and the driving profile. In this step, the mechanical energy produced in the second conversion step is ultimately dissipated to thermal energy that is deposited to the ambient. The terms "well-to-tank," "tank-to-vehicle," and "vehicle-to-miles" are used in this text to refer to these three conversion steps. Unfortunately, all of these conversion processes cause substantial energy losses.

---

[2] The energy density here is defined as the amount of *net* energy available for propulsion purposes divided by the mass of the energy carrier necessary to generate that propulsion energy, including all containment elements but not the on-board energy transformation devices.

[3] For instance, the autonomy requirement and the dominance of part-load operation will be relevant for the optimization problems.

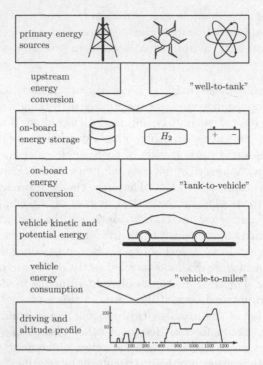

Fig. 1.2 The main elements of the energy conversion scheme

This text will not address any control problems pertaining to the "well-to-tank" energy conversion. The systems used for that conversion are very large power plants, refineries, or other process engineering systems. Of course, their average efficiency values and pollutant emission have an important impact on the economy and ecology. However, the problems arising in that area and the methods required to solve those problems belong to a different class.

In the next section an overview of the most important energy conversion approaches is presented. With this information, a preliminary estimation of the total energy consumption is possible. Note that a correct comparison is not easy, if at all possible.[4] Readers interested in a broader discussion are referred to [44].

The main physical phenomena influencing the "vehicle-to-miles" energy conversion will be discussed in Chap. 2. That chapter will mainly introduce descriptions that are *quasistatic* (this term will be precisely defined below),

---

[4] For instance, the total "well-to-miles" carbon dioxide emissions are often used to compare two competing approaches. However, such a discussion is not complete unless the "gray" energy invested in the vehicles, refineries and plants is considered. Even more difficult: how to take into account the problems associated with nuclear waste repositories, landscape degradation caused by windmills, or nitric oxide emission of coal-fired power plants?

but also *dynamic* models will be presented. In this context, it is important to understand the impact of the driving profile that the vehicle is assumed to follow. As mentioned above, only those effects are considered that have a substantial influence on the energy consumption.

The main emphasis of this text is on the modeling and optimization of the "tank-to-vehicle" energy conversion systems. For this problem suitable mathematical models of the most important devices will be introduced in Chaps. 3 through 6. Chapter 7 presents methods with which the energy consumption can be minimized. All of these methods are model-based, i.e., they rely on the mathematical models derived in the previous chapters and on systematic optimization procedures to find (local) minima of precisely cast optimization problems. Eight case studies are included in Appendix I. Appendix II then summarizes the most important facts of parameter optimization and optimal control theory and Appendix III introduces the main ideas of dynamic programming.

1.3   Upstream Processes

As mentioned above, a detailed analysis of the "well-to-tank" energy conversion processes is not in the scope of this text. However, the efficiency and the economy of these systems are important aspects of a comprehensive analysis. For this reason a rather preliminary but nevertheless instructive overview of the main energy conversion systems is given in this section.

Figure 1.3 shows a part of that complex network. The efficiency numbers given in that figure are approximate and are valid for available technology. The CO_2 factors relate the amount of carbon dioxide emitted by using one energy unit of natural gas or coal to the amount emitted when using one energy unit of oil.[5] Solar and nuclear primary energy sources are assumed to emit no CO_2, i.e., the gray energy and the associated CO_2 emission are not shown in that figure.

Only three systems are considered in Fig. 1.3 for the conversion of "fuel" to mechanical energy: a spark-ignited (SI) or gasoline internal combustion engine (ICE), a compression-ignited (CI) or Diesel ICE, and an electric motor. Average "tank-to-vehicle" efficiencies of these prime movers are shown in Fig. 1.3 as well.[6] The mechanical energy consumption (the "vehicle-to-miles" efficiency) is approximated by an equation that is valid for the European test cycle (this expression will be introduced in Sect. 2.2). With the information shown in Fig. 1.3 it is easy to make some preliminary, back-of-the-envelope-style calculations that, despite the many uncertainties, are quite instructive.

---

[5] The CO_2 factors reflect the different chemical composition *and* the different heating values. The base line is defined in Table 1.1.

[6] The peak efficiencies of all of these devices are (substantially) higher. However, the relevant data are the cycle-averaged efficiencies, which are close to the values shown in Fig. 1.3.

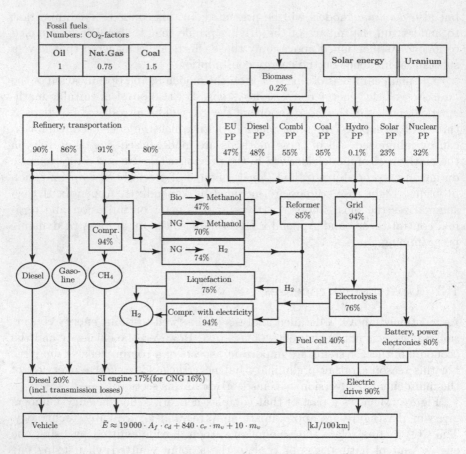

Fig. 1.3 Different paths to convert a primary energy source to mechanical energy needed to drive a car in the MVEG-95 test cycle. Source: [84] and own data.

For that purpose Table 1.1 summarizes some of the most important parameters of the fuels considered below.

Table 1.1 Main parameters of some important energy carriers (lower heating value H_l, hydrogen-to-carbon ration H/C, and mass of CO_2 emitted per mass fuel burned ν); CNG = compressed natural gas

	H_l (MJ/kg)	H/C	ν
oil	43	≈ 2	3.2
CNG ($\approx$ methane)	50	4	2.75
coal ($\approx$ carbon)	34	0	3.7
hydrogen	121	∞	0

Figure 1.4 shows the "well-to-miles" carbon dioxide emissions of three ICE-based powertrains. The vehicle assumed in these considerations is a standard mid-size passenger car. The efficiency values of the gasoline and Diesel engines are standard values as well. The efficiency of CNG engines is usually slightly smaller than the one of gasoline SI engines [20].

Of course this analysis neglects several important factors, for instance the greenhouse potential of methane losses in the fueling infrastructure. Nevertheless, the results obtained indicate that increasing the numbers of CNG engines could be one option to reduce CO_2 emissions with relatively small changes in the design of the propulsion system. Unfortunately, as mentioned before, the "well-to-miles" CO_2 emission levels are just one element of the problem space. In this case, the reduced energy density of CNG as on-board energy carrier has, so far, inhibited a broader market penetration of this vehicle class. The next section will show more details on this aspect.

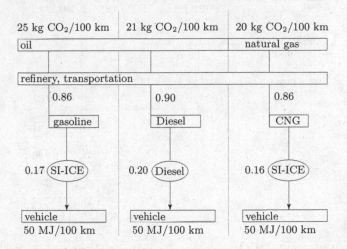

Fig. 1.4 "Well-to-miles" CO_2 emission of three conventional powertrains. The vehicle is described by the parameters $m = 1600\,\text{kg}$, $c_d \cdot A_f = 0.86\,\text{m}^2$, and $c_r = 0.013$ (see Chap. 2). The fuel properties are defined in Table 1.1.

Figure 1.5 shows what amount of CO_2 emissions can be expected when a battery-electric propulsion system is employed. The base vehicle is assumed to have the same[7] parameters as the one used to compute the values shown in Fig. 1.4. Several primary energy sources are compared in this analysis. The two CO_2-neutral[8] energy sources (solar and nuclear energy) produce no

---

[7] Of course the batteries substantially increase the vehicle mass. Here the (optimistic) assumption is adopted that the recuperation capabilities of the battery electric system compensate for the losses that are caused by this additional mass.

[8] As mentioned, only the CO_2 emission caused by the operation of the power plants are considered.

carbon dioxide emission. However, if the electric energy required to charge the batteries is generated using fossil primary energy sources, surprisingly different CO_2 emission levels result.

In the case of a natural-gas-fired combined-cycle power plant (PP) the CO_2 emission levels are substantially lower than those of traditional ICE-based propulsion systems. However, if the other limit case (coal-fired steam turbines) is taken into consideration, the "well-to-miles" carbon dioxide emission levels of a battery-electric car become even worse than those of the worst ICE-based propulsion system.[9] Moreover, in the next section it will be shown that the energy density of batteries is so small that battery electric vehicles cannot satisfy the specifications of a passenger car as defined in Sect. 1.2.

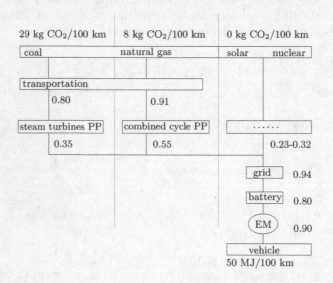

Fig. 1.5 "Well-to-miles" CO_2 emission of a battery electric vehicle. Vehicle parameters as in Fig. 1.4. Battery efficiency includes charging, discharging, and power electronic losses. The fuel properties are defined in Table 1.1.

As a last example, the estimated "well-to-miles" CO_2 emission levels of a fuel cell electric vehicle are shown in Fig. 1.6. Again, the vehicle parameters have been chosen to be the same as in the conventional case. The efficiency of the fuel cell *system* has been assumed to be around 0.40. Despite many more optimistic claims, experimental evidence, as the one published in

---

[9] Of course, low CO_2 primary energy sources should first be used to replace the worst polluting power plants that are part of the corresponding grid. In this sense, each unit of *additional* electric energy used must be considered to have been produced by the power plant in the grid that has the *worst* efficiency. Accordingly, in the example shown in Fig. 1.5 the relevant CO_2 emission number is the one valid for coal-fired power plants.

[274], has shown that the net efficiency of a fuel cell *system* will probably be close to that figure.[10] Even more uncertain are the efficiencies of on-board gasoline-to-hydrogen reformers. Including all auxiliary devices, a net efficiency of approximately 60–70% may be expected.

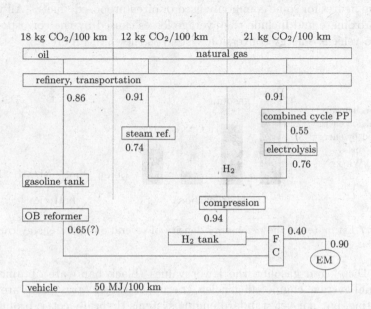

Fig. 1.6 "Well-to-miles" CO_2 emission of a fuel cell electric vehicle. Vehicle parameters as in Fig. 1.4. The efficiency of the on-board gasoline-to-hydrogen reformer is not experimentally verified. The fuel properties are defined in Table 1.1.

The main insight that can be gained from Fig. 1.6 is that as long as fossil primary energy sources are used fuel cell electric vehicles have a potential to reduce the "well-to-miles" CO_2 emission only if the hydrogen is produced in a steam reforming process using natural gas as primary energy source. As shown in Fig. 1.6, fuel-cell-based powertrains have excellent "tank-to-vehicle" but rather poor "well-to-tank" efficiencies. This fact will become very important once *renewable* primary energy sources are available on a large scale. If this comes true, then the "upstream" CO_2 emission levels are zero and the only concern will be to utilize the available on-board energy as efficiently as possible. In this situation fuel-cell-based propulsion systems might prove to be the best choice.

---

[10] Fuel cells must be supercharged to achieve sufficient power densities and to exploit in the best possible way the expensive electrochemical converters. The compressors that are necessary for that consume in the order of 20–25 percent of the electric power produced by the fuel cell [274].

1.4   Energy Density of On-Board Energy Carriers

As mentioned above, the energy density of the on-board energy carrier is one of the most important factors that influence all choices of propulsion systems for individual mobility purposes. Figure 1.7 shows *estimations* of the corresponding figures for some commonly used or often proposed "fuels." All values are approximate and include the average losses caused by the corresponding "tank-to-vehicle" energy conversion system as shown in Figure 1.3.

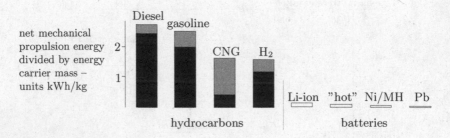

Fig. 1.7 Estimates of the *net* energy density of several on-board energy carriers

For Diesel and gasoline the lower values (black bars) are obtained using actual average engine efficiencies, the upper values (gray bars) are valid for existing, but not yet standard engine systems (hybrid-electric propulsion, downsized-supercharged gasoline engine systems, etc.).

Compressed natural gas (CNG) is stored in gaseous form. Standard engine systems have 16% efficiency and the corresponding storage systems[11] are rated at 200 bar yielding the energy density indicated in Fig. 1.7 by a black bar. Advanced engine systems can reach more than 20% efficiency and advanced storage systems can go up to 350 bar,[12] yielding the energy density indicated by the gray bar.

Hydrogen may be stored under high pressure (black bar valid for 350 bar and carbon fibre bottles) or liquefied (gray bar).[13] The energy densities shown in Figure 1.7 are based on an cycle-averaged conversion efficiency of fuel cells of 40%. The main reason for the relatively low energy density of

---

[11] Conventional steel bottles have a mass to volume ratio of approximately 1 kg/l; bottles made of carbon fibres have substantially lower mass to volume ratios of approximately 0.3-0.4 kg/l.

[12] For CNG higher pressures are not foreseeable: the compression losses become too large and outweigh the gains in energy density.

[13] Of course this approach induces an additional substantial penalty of approximately 30% in the "well-to-tank" efficiency and causes problems for long-term on-board storage since no insulation is perfect. The heat transfer to the liquid fuel will therefore either lead to some fuel evaporation and a subsequent venting or require a constant energy supply to avoid such evaporation losses.

gaseous hydrogen is the unfavorable ratio of the heating value divided by the gas constant.

The electrochemical on-board energy carriers have the lowest energy density of those "fuels" shown in Fig. 1.7, despite the fact that the electric "tank-to-vehicle" conversion systems have a very high conversion efficiency. State-of-the-art batteries optimized for high energy density achieve approximately 40 Wh/kg for lead–acid, 70 Wh/kg for nickel–metal hydrides, 150 Wh/kg for "hot" batteries such as the sodium–nickel chloride "zebra" battery, and 180 Wh/kg for lithium-ion cells. Taking the average "tank-to-vehicle" conversion efficiency to be around 80%,[14] the values illustrated in Fig. 1.7 result. Not surprisingly, none of the several battery-electric vehicle prototypes that have been developed in the past has ever evolved to become a mass-produced alternative to the existing solutions.

Batteries are, however, very interesting "medium-term" energy storage devices, which can help to improve the efficiency of other propulsion systems. Such *hybrid* powertrains will be analyzed in detail in Chap. 4 and in several case studies in Appendix I.

Note that there are several other ways to store energy on-board, notably supercapacitors (electrostatic energy), hydraulic and pneumatic reservoirs (potential energy), and flywheels (kinetic energy). All of these systems have similar or even smaller net energy densities than lead acid batteries.[15] However, supercapacitors and similar devices have relatively high *power* densities, i.e., they may be charged and discharged with high power.[16] Such "short-term" energy storage devices can be useful for the recuperation of the kinetic energy stored in the vehicle that would otherwise be lost in braking maneuvers. More details on these aspects can be found in Chaps. 2 and 4.

The main point illustrated in Figure 1.7 is that with respect to energy density, liquid hydrocarbons are unquestionably the best fuels for passenger car applications. These fuels have several other advantages:

- the refueling process is fast (several MW of power), safe, and does not require any expensive equipment;
- their long-term storage is possible at relatively low costs; and
- there were and still are large and easily exploitable reserves of crude oil.[17]

---

[14] This figure includes the losses in the transmission, in the electric motor, in the power electronics and in the battery discharging process.

[15] Supercapacitors can reach net energy densities in the order of 5 Wh/kg. Systems based on pneumatic air stored at 300 bars in bottles made of carbon fibre can reach net energy densities in the order of 20 Wh/kg.

[16] Batteries can also be optimized for high power density. However, in this case their energy densities are lower than those figures indicated in Fig. 1.7.

[17] Note that liquid hydrocarbons need not originate from crude oil sources. Many approaches that use fossil (natural gas, coal, etc.) and renewable (bio Diesel, ethanol, etc.) primary energy sources are known with which liquid hydrocarbons can be synthesized.

Particularly the last point is, of course, the topic of many vivid debates. This text does not attempt to contribute to these discussions. However, following the paradigm of a "least-regret policy," the standing assumption adopted here is that the improvement of the "tank-to-miles" efficiency, while satisfying the performance and cost requirements, is worth the efforts.

1.5   Pathways to Better Fuel Economy

As illustrated in Fig. 1.2, there are essentially three possible approaches to reducing the total energy consumption of passenger cars:

- Improve the "well-to-tank" efficiency by optimizing the upstream processes and by utilizing alternative primary energy sources.
- Improve the "tank-to-vehicle" efficiency, as discussed below.
- Improve the "vehicle-to-miles" efficiency by reducing the vehicle mass and its aerodynamic and rolling friction losses.

As mentioned above, the optimization of the "well-to-tank" efficiency is an important area in itself. These problems are not addressed in this text. The phenomena which define the "vehicle-to-miles" energy losses will be analyzed in this text, but no attempt is made to suggest concrete approaches to reduce these losses. In fact, the disciplines that are important for that are material science, aerodynamics, etc. which are not in the scope of this monograph.

This text focuses on improving the "tank-to-vehicle" efficiency. Three different approaches are discernible on the component level and two on the system level:

1. Improve the *peak* efficiency of the powertrain components.
2. Improve the *part-load* efficiency of the powertrain components.
3. Add the capability to *recuperate* the kinetic and potential energy stored in the vehicle.
4. Optimize the structure and the parameters of the propulsion system, assuming that the fuel(s) used and the vehicle parameters are fixed.
5. Realize appropriate supervisory control algorithms that take advantage of the opportunities offered by the chosen propulsion system configuration.

Items 2–5 will be discussed in detail in this text, while item 1 is not within its scope. On one hand that optimization requires completely different methods and tools and, on the other hand, the potential for improvements in that area, in most cases, is rather limited. Compared to the peak efficiency of Diesel ($\approx$ 0.40) or gasoline ($\approx$ 0.37) engines, the part-load efficiency, which determines the actual fuel consumption in regular driving conditions, is much smaller (on average $\approx$ 0.20 for Diesel and $\approx$ 0.17 for gasoline engines). Therefore, the potential offered by improving these figures is much larger. Optimized powertrain systems and appropriate control algorithms are instrumental to achieve that objective.

2

Vehicle Energy and Fuel Consumption – Basic Concepts

This chapter contains three main sections: First the dynamics of the longitudinal motion of a road vehicle are analyzed. This part contains a discussion of the main energy-consuming effects occurring in the "vehicle-to-miles" part and some elementary models that describe the longitudinal dynamics and, hence, the drivability of the vehicle.

The influence of the driving pattern on the fuel consumption is analyzed in the second section. The main result of this analysis is an approximation of the mechanical energy required to make a road vehicle follow a given driving cycle. The sensitivity of the energy consumption to various vehicle parameters and the potential for the recuperation of kinetic energy when braking are derived from that result.

The third section briefly introduces the most important approaches used to predict the fuel economy of road vehicles, the main optimization problems that are relevant in this context, and the software tools available for the solution of these problems.

2.1 Vehicle Energy Losses and Performance Analysis

2.1.1 Energy Losses

Introduction

The propulsion system produces mechanical energy that is assumed to be momentarily stored in the vehicle. The driving resistances are assumed to drain energy from this reservoir. This separation might seem somewhat awkward at first glance. However, it is rather useful when one has to distinguish between the individual effects taking place.

The energy in the vehicle is stored:

- in the form of kinetic energy when the vehicle is accelerated; and
- in the form of potential energy when the vehicle reaches higher altitudes.

The amount of mechanical energy "consumed" by a vehicle[1] when driving a pre-specified driving pattern mainly depends on three effects:

- the aerodynamic friction losses;
- the rolling friction losses; and
- the energy dissipated in the brakes.

The elementary equation that describes the longitudinal dynamics of a road vehicle has the following form

$$m_v \frac{d}{dt} v(t) = F_t(t) - (F_a(t) + F_r(t) + F_g(t) + F_d(t)), \qquad (2.1)$$

where F_a is the aerodynamic friction, F_r the rolling friction, F_g the force caused by gravity when driving on non-horizontal roads, and F_d the disturbance force that summarizes all the other not yet specified effects. The traction force F_t is the force generated by the prime mover[2] minus the force that is used to accelerate the rotating parts inside the vehicle and minus all friction losses in the powertrain. Figure 2.1 shows a schematic representation of this relationship. The following sections contain more information about all of these forces.

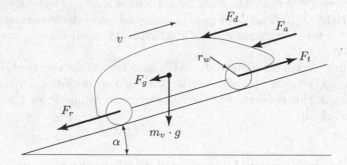

Fig. 2.1 Schematic representation of the forces acting on a vehicle in motion

Aerodynamic Friction Losses

The aerodynamic resistance F_a acting on a vehicle in motion is caused on one hand by the viscous friction of the surrounding air on the vehicle surface. On the other hand, the losses are caused by the pressure difference between the front and the rear of the vehicle, generated by a separation of the air flow. For idealized vehicle shapes, the calculation of an approximate pressure field and

---

[1] To be thermodynamically correct, the first part of this sentence should read: The amount of *exergy* transformed to useless *anergy* ...

[2] This force can be negative, for instance during braking phases.

the resulting force is possible with the aid of numerical methods. A detailed analysis of particular effects (engine ventilation, turbulence in the wheel housings, cross-wind sensitivity, etc.) is only possible with specific measurements in a wind tunnel.

For a standard passenger car, the car body causes approximately 65% of the aerodynamic resistance. The rest is due to the wheel housings (20%), the exterior mirrors, eave gutters, window housings, antennas, etc. (approximately 10%), and the engine ventilation (approximately 5%) [143].

Usually, the aerodynamic resistance force is approximated by simplifying the vehicle to be a prismatic body with a frontal area A_f. The force caused by the stagnation pressure is multiplied by an aerodynamic drag coefficient $c_d(v, \ldots)$ that models the actual flow conditions

$$F_a(v) = \frac{1}{2} \cdot \rho_a \cdot A_f \cdot c_d(v, \ldots) \cdot v^2. \qquad (2.2)$$

Here, v is the vehicle speed and ρ_a the density of the ambient air. The parameter $c_d(v, \ldots)$ must be estimated using CFD programs or experiments in wind tunnels. For the estimation of the mechanical energy required to drive a typical test cycle this parameter may be assumed to be constant.

Rolling Friction Losses

The rolling friction force is often modeled as

$$F_r(v, p, \ldots) = c_r(v, p, \ldots) \cdot m_v \cdot g \cdot \cos(\alpha), \quad v > 0, \qquad (2.3)$$

where m_v is the vehicle mass and g the acceleration due to gravity. The term $\cos(\alpha)$ models the influence of a non-horizontal road. However, the situation in which the angle α has a substantial influence is not often encountered in practice.

The rolling friction coefficient c_r depends on many variables. The most important influencing quantities are vehicle speed v, tire pressure p, and road surface conditions. The influence of the tire pressure is approximately proportional to $1/\sqrt{p}$. A wet road can increase c_r by 20%, and driving in extreme conditions (sand instead of concrete) can easily double that value. The vehicle speed has a small influence at lower values, but its influence substantially increases when it approaches a critical value where resonance phenomena start.

A typical example of these relationships is shown in Fig. 2.2. Note that the tires reach their thermal equilibrium only after a relatively long period (a few ten minutes). Figure 2.2 includes examples of typical equilibrium temperatures for three speed values. For many applications, particularly when the vehicle speed remains moderate, the rolling friction coefficient c_r may be assumed to be constant. This simplification will be adopted in the rest of this text.

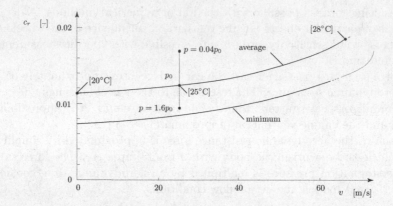

Fig. 2.2 Tire friction coefficient as a function of the vehicle speed v and variations of the tire pressure p

Uphill Driving Force

The force induced by gravity when driving on a non-horizontal road is conservative and considerably influences the vehicle behavior. In this text this force will be modeled by the relationship

$$F_g(\alpha) = m_v \cdot g \cdot \sin(\alpha), \tag{2.4}$$

which, for small inclinations α, may be approximated by

$$F_g(\alpha) \approx m_v \cdot g \cdot \alpha \tag{2.5}$$

when α is expressed in radians.

Inertial Forces

The inertia of the vehicle and of all rotating parts inside the vehicle causes fictitious (d'Alembert) forces. The inertia force induced by the vehicle mass is included in (2.1) by the term on the left side. The inertia of the rotating masses of the powertrain can be taken into account in the respective submodels. Nevertheless, sometimes for rapid calculation, it may be convenient to add the inertia of the rotating masses to the vehicle mass. Such an analysis usually considers a prime mover and a transmission with a total transmission ratio γ.

The total[3] inertia torque of the wheels is given by

$$T_{m,w}(t) = \Theta_w \cdot \frac{d}{dt}\omega_w(t) \tag{2.6}$$

and it acts on the vehicle as an additional inertia force $F_{m,w} = T_{m,w}/r_w$, where r_w is the wheel radius. Usually, the wheel slip is not considered in a first approximation, i.e., $v = r_w \cdot \omega_w$. In this case

$$F_{m,w}(t) = \frac{\Theta_w}{r_w^2} \cdot \frac{d}{dt}v(t). \tag{2.7}$$

Consequently, the contribution of the wheels to the overall inertia of the vehicle is given by the following term

$$m_{r,w} = \frac{\Theta_w}{r_w^2}. \tag{2.8}$$

Similarly, the inertia torque of the engine is given by

$$T_{m,e}(t) = \Theta_e \cdot \frac{d}{dt}\omega_e(t) = \Theta_e \cdot \frac{d}{dt}(\gamma \cdot \omega_w(t)) = \Theta_e \cdot \frac{\gamma}{r_w} \cdot \frac{d}{dt}v(t), \tag{2.9}$$

where Θ_e is the total moment of inertia of the powertrain[4] and ω_m its rotational speed. Again, assuming no wheel slip, this torque is transferred via the wheels to the vehicle body as a force

$$F_{m,e}(t) = \frac{\gamma}{r_w} \cdot T_{m,e}(t) = \Theta_e \cdot \frac{\gamma^2}{r_w^2} \cdot \frac{d}{dt}v(t). \tag{2.10}$$

Note that this expression assumes an ideal gearbox with zero losses and an engaged clutch. Since the average efficiency of realistic gearboxes is above 95%, and the clutch is closed most of the time, the errors caused by these simplifications are usually very small.

Assuming a constant gear ratio γ, the force (2.10) corresponds to an additional vehicle mass of

$$m_{r,e} = \frac{\gamma^2}{r_w^2} \cdot \Theta_e. \tag{2.11}$$

In summary, the equivalent mass of the rotating parts is approximated as follows

$$m_r = m_{r,w} + m_{r,e} = \frac{1}{r_w^2} \cdot \Theta_w + \frac{\gamma^2}{r_w^2} \cdot \Theta_e, \tag{2.12}$$

which needs to be added to the vehicle mass m_v in (2.1). The *total gear ratio* γ/r_w appears quadratically in this expression. Accordingly, for high gear ratios (lowest gears in a standard manual transmission), the influence of the rotating parts on the vehicle dynamics can be substantial and, in general, may not be neglected.

---

[3] The inertia Θ_w includes all wheels and all rotating parts that are present on the wheel side of the gear box. The speed of all wheels ω_w is assumed to be identical.

[4] The inertia Θ_e includes the engine inertia and the inertia of all rotating parts that are present on the engine side of the gear box.

2.1.2 *Performance and Drivability*

General Remarks

Performance and drivability are very important factors that, unfortunately, are not easy to precisely define and measure. For passenger cars, three main quantifiers are often used:

- top speed;
- maximum grade at which the fully loaded vehicle reaches the legal top-speed limit; and
- acceleration time from standstill to a reference speed (100 km/h or 60 mph are often used).

These three quantifiers are discussed below.

Top Speed Performance

For passenger cars, the top speed is often not relevant because that limit is substantially higher than most legal speed limits. However, in some regions and for specific types of cars that information is still provided. This limit is mainly determined by the available power and the aerodynamic resistance.[5] Neglecting all other losses, the maximum speed is obtained by solving the following power balance

$$P_{max} \approx \frac{1}{2} \cdot \rho_a \cdot A_f \cdot c_d \cdot v_{max}^3, \qquad (2.13)$$

where P_{max} is the maximum traction power available at the wheels. The relevant information contained in this equation is the fact that the power demand depends on the *cube* of the vehicle speed. In other words, the engine power must be doubled in order to increase the top speed by 25%.

Uphill Driving

For trucks and any other vehicles that carry heavy loads, a relevant performance metric is the uphill driving capability. The relationship between P_{max} and the maximum gradient angle α_{max} is obtained by neglecting in (2.1) all the resistance forces but F_g

$$P_{max} \approx m_v \cdot v_{min} \cdot g \cdot \sin(\alpha_{max}), \qquad (2.14)$$

where v_{min} is the desired uphill speed. For a given rated power and a minimum speed, this equation yields the maximum uphill driving angle. This discussion is not complete without an analysis of the influence of the gear box. However, since that analysis deserves to be treated in some detail this discussion is deferred to Chap. 3.

---

[5] At that speed, the power consumed to overcome rolling friction is typically one order of magnitude smaller than the power dissipated by the aerodynamic friction.

Acceleration Performance

The most important drivability quantifier is the acceleration performance. Several metrics are used to quantify that parameter. In this text the time necessary to accelerate the vehicle from 0 to 100 km/h is taken as the only relevant information. This value is not easy to compute exactly because it depends on many uncertain factors and includes highly dynamic effects. An approximation of this parameter can be obtained as shown below.

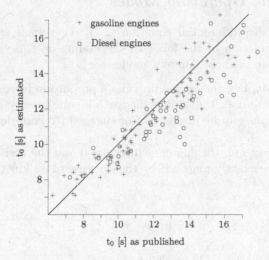

Fig. 2.3 Comparison between acceleration times as published by the manufacturers and values calculated with (2.16) for $v_0 = 100$ km/h

When accelerating a vehicle on a flat road, a large portion of the traction energy is converted to kinetic energy as follows

$$E_0 = \frac{1}{2} \cdot m_v \cdot v_0^2. \tag{2.15}$$

If all the resistance forces in (2.1) are neglected, the energy E_0 is also the energy that has to be provided by the powertrain to produce the acceleration required. Accordingly, the mean power $\bar{P}$ that has to be provided by the engine is $\bar{P} = E_0/t_0$, where t_0 is the time available for the acceleration. An approximation, which takes into account the varying engine speed and the losses neglected, is obtained by choosing $\bar{P} \approx P_{max}/2$, where P_{max} is the maximum rated power. Consequently, a simple relation between the acceleration time and the maximum power of the engine is given by the following expression

$$t_0 \approx \frac{v_0^2 \cdot m_v}{P_{max}}. \tag{2.16}$$

Figure 2.3 shows the comparison between the acceleration times published by various manufacturers and those estimated using the approximation (2.16) for $v_0 = 100\,\text{km/h}$. Although this relationship is based on many simplifying assumptions, its predictions of the acceleration times agree well with measured data of gasoline engines. Since for Diesel engines the maximum power is typically not encountered at top speed, the agreement is not as good, but the approximation (2.16) still yields acceptable estimates.

2.1.3 Vehicle Operating Modes

From the first-order differential equation (2.1), the vehicle speed v can be calculated as a function of the force F_t. Depending on the value of F_t, the vehicle can operate in three different modes:

- $F_t > 0$, *traction*, i.e., the engine provides a propulsion force to the vehicle;
- $F_t < 0$, *braking*, i.e., the brakes dissipate kinetic energy of the vehicle, where the engine can be disengaged or engaged (to consider fuel cut-off); and
- $F_t = 0$, *coasting*, i.e., the engine is disengaged and the resistance losses of the vehicle are exactly matched by the decrease of its kinetic energy.

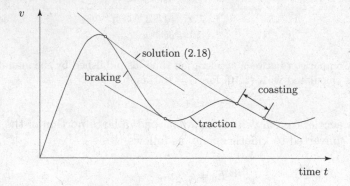

Fig. 2.4 Modes of vehicle motion

For the limit case $F_t = 0$ on a horizontal road and without disturbances, the coasting velocity $v_c(t)$ of the vehicle can be computed by solving the following ordinary differential equation derived from (2.1)

$$\frac{d}{dt} v_c(t) = \frac{-1}{2 \cdot m_v} \cdot \rho_a \cdot A_f \cdot c_d \cdot v_c^2(t) - g \cdot c_r \qquad (2.17)$$

$$= -\alpha^2 \cdot v_c^2(t) - \beta^2$$

For $v_c > 0$, this equation can be integrated in closed form yielding the result

$$v_c(t) = \frac{\beta}{\alpha} \cdot \tan\left\{\arctan\left(\frac{\alpha}{\beta} \cdot v_c(0)\right) - \alpha \cdot \beta \cdot t\right\}. \tag{2.18}$$

These equations are important because they can be used to define the three main operating modes of the vehicle. As shown in Fig. 2.4, the vehicle is in:

- *traction mode* if the speed decreases less than the coasting velocity $v_c(t)$ (2.18) would decrease when starting at the same initial speed;
- *braking mode* if the speed decreases more than the coasting velocity $v_c(t)$ would decrease when starting at the same initial speed; and
- *coasting mode* if the vehicle speed and the coasting speed $v_c(t)$ coincide for a *finite* time interval.

Therefore, once a test cycle (see Sect. 2.2) has been defined, it is possible to decide for each time interval in what mode a vehicle is operated. Particularly for the MVEG–95 cycle this analysis is straightforward and yields the result indicated in the lowest bar (index "trac" for traction mode) in Fig. 2.6.

2.2 Mechanical Energy Demand in Driving Cycles

2.2.1 Test Cycles

Test cycles consisting of standardized speed and elevation profiles have been introduced to compare the pollutant emissions of different vehicles on the same basis. After that first application, the same cycles have been found to be useful for the comparison of the fuel economy as well. In practice, these cycles are often used on chassis dynamometers where the force at the wheels is chosen to emulate the vehicle energy losses while driving that specific cycle (see Sect. 2.1.1). These tests are carried out in controlled environments (temperature, humidity, etc.), with strict procedures being followed to reach precisely defined thermal initial conditions for the vehicle (hot soak, cold soak, etc.). The scheduled speed profile is displayed on a monitor[6] while a test driver controls the gas and brake pedals such that the vehicle speed follows these reference values within pre-specified error bands.

There are several commonly used test cycles. In the United States, the federal urban driving cycle (FUDS) represents a typical city driving cycle, while the federal highway driving cyle (FHDS) reflects the extra-urban driving conditions. The federal test procedure (FTP–75) is approximately equal to one and a half FUDS cycle, but it also includes a typical warm-up phase. The first FUDS cycle is driven in cold-soak conditions. The second half of the FUDS cycle (hot soak) is driven after a 10-minute engine-off period. The cold start

---

[6] For vehicles with manual transmission, the requested gear shifts are also signalled on the driver's monitor.

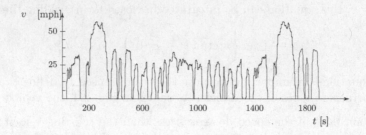

Fig. 2.5 US test cycle FTP–75 (Federal Test Procedure), length: 11.12 miles (17.8 km), duration: 1890 s, average speed: 21 mph (9.43 m/s)

is important to assess the pollutant emissions (engine and catalyst warming up) but the fuel consumption is also influenced by the cold-start conditions (higher engine friction), though not as dramatically as the pollutant emissions.

In Europe, the urban driving cycle (ECE) consists of three start-and-stop maneuvers. The combined cycle proposed by the motor vehicle emissions group in 1995 (MVEG–95, also NEDC) runs through the ECE four times (the first with a cold start) and adds an extra-urban portion referred to as the EUDC. The Japanese combined cycle is the 10–15 mode, consisting of three repetitions of an urban driving cycle plus an extra-urban portion. Figures 2.5 and 2.6 show the speed profiles for the American and the European cycles. Note that for manual transmissions, the MVEG–95 prescribes the gear to be engaged at each time instant.[7]

Critics claim that fuel consumption values measured using the MVEG–95 test procedure are optimistic and are rarely reached in practice [223]. Furthermore, the MVEG–95 test cycle does not correctly depict realistic driving scenarios (speed-acceleration distribution, trip length, cold/hot phase ratio, robotized gear shifts, etc.). Seeking to overcome these issues and to harmonize the test procedure among different countries the WLTP (Worldwide Harmonized Light Duty Test Procedure) was developed. The test procedure not only contains a driving cycle (DHC) but also rules the related test procedure (DTP). Here, when speaking of the driving cycle, we will refer to WLTP for simplicity. The speed trajectory is depicted in Fig. 2.7.

Of course, real driving patterns are often much more complex and demanding (speeds, accelerations, etc.) than these test cycles. All automotive companies have their in-house standard cycles, which better reflect the average real driving patterns. For the sake of simplicity, in this text the FTP and MVEG–95 cycles will be used in most cases. The methods introduced below, however, are applicable to more complex driving cycles as well.

---

[7] This restriction is not particularly reasonable. Using the additional degrees of freedom associated with a flexible gear selection can considerably improve the fuel economy. This will become clearer in the subsequent chapters.

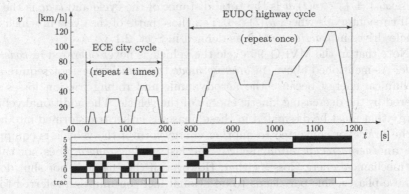

Fig. 2.6 European test cycle·MVEG–95, gears 1–5, "c": clutch disengaged, "trac": traction time intervals. Total length: 11.4 km, duration: 1200 s, average speed: urban 5.12 m/s, extra-urban 18.14 m/s, overall 9.5 m/s. The cycle includes a total of 34 gear shifts.

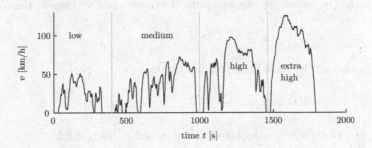

Fig. 2.7 WLTP test cycle, length: 23.1 km, duration, 1800 s, average speed: 46.1 km/h (12.8 m/s)

2.2.2 Mechanical Energy Demand

Introduction

In the following subsection the mechanical energy that is required to make a vehicle follow the MVEG–95 cycle is determined. A key role in these considerations is played by the mean tractive force $\bar{F}_{trac}$. The concept of a mean tractive force is particularly useful for the evaluation of a first tentative value for the fuel consumed by a propulsion system and to discuss various aspects of the "vehicle-to-miles" energy conversion step.

The mean tractive force $\bar{F}_{trac}$ is defined as follows

$$\bar{F}_{trac} = \frac{1}{x_{tot}} \cdot \int_{t \in trac} F_t(t) \cdot v(t)\, dt, \qquad (2.19)$$

where $x_{tot} = \int_0^{t_{\max}} v(t)\, dt$ is the total distance of the cycle and *trac* is the set of all time intervals where $F_t(t) > 0$, i.e., those parts of the cycle in which the vehicle drives in *traction mode*[8] as defined in Sect. 2.1.3.

Note that in the MVEG–95 cycle the vehicle is never operated in *coasting mode*. As mentioned above, in *braking mode* the vehicle does not require any mechanical energy because the aerodynamic and rolling friction losses are covered by the decreasing kinetic energy of the vehicle. The additional vehicle energy that must be dissipated in these phases is either transformed into heat by the brakes or converted to another energy form if the vehicle is equipped with an *energy recuperation* device. The last mode, *stopped*, does not cause any mechanical energy losses either. However, if the engine is not shut down in these phases, this part leads to additional fuel consumption referred to as *idling losses*.

The evaluation of the integral used to define the mean tractive force is accomplished by a *discretization* in time of the drive cycle as follows (more on this approach will be said in the next section). The velocity profile of a test cycle is defined for the given time instants $v(t_i) = v_i$, $t_i = i \cdot h$, $i = 0, \ldots, n$. Accordingly, the speed used to compute the mean tractive force is the average speed

$$v(t) = \bar{v}_i = \frac{v_i + v_{i-1}}{2}, \quad \forall \quad t \in [t_{i-1}, t_i) \,. \tag{2.20}$$

Similarly, the acceleration is approximated by

$$a(t) = \bar{a}_i = \frac{v_i - v_{i-1}}{h}, \quad \forall \quad t \in [t_{i-1}, t_i) \,, \tag{2.21}$$

and the traction force is calculated as described in Sect. 2.1.1

$$F_t(t) = \bar{F}_i = \bar{F}_{a,i}(\bar{v}_i) + \bar{F}_{r,i}(\bar{v}_i, \ldots) + \bar{F}_{m,i}(\bar{v}_i, \bar{a}_i), \quad \forall \quad t \in [t_{i-1}, t_i) \,. \tag{2.22}$$

Note that the uphill driving force was neglected for simplicity. Including the uphill driving force into the following analysis is a straightforward process.

An approximation of the tractive force (2.19) can be found using the expression

$$\bar{F}_{trac} \approx \frac{1}{x_{tot}} \cdot \sum_{i \in trac} \bar{F}_i \cdot \bar{v}_i \cdot h, \tag{2.23}$$

where the partial summation $i \in trac$ only covers the time intervals during which the vehicle is in traction mode. Using the fact that the aerodynamic and the rolling friction forces are always dissipative, i.e.

$$\bar{F}_{a,i} + \bar{F}_{r,i} \geq 0, \quad \forall \quad i = 0 \ldots n, \tag{2.24}$$

and the fact that kinetic energy is conserved $(v(0) = v(t_{\max}))$, i.e.

---

[8] In the MVEG–95 cycle the vehicle is in traction mode approximately 60% of the total cycle time.

$$\frac{1}{x_{tot}} \cdot \sum_{i=0}^{n} \bar{F}_{m,i} \cdot \bar{v}_i \cdot h = 0, \tag{2.25}$$

we can derive from (2.23) that the mean traction force over a driving cycle can be split up in a dissipative and a circulative part, respectively

$$\bar{F}_{trac} \approx \frac{1}{x_{tot}} \cdot \sum_{i=0}^{n} \bar{F}_i \cdot \bar{v}_i \cdot h - \frac{1}{x_{tot}} \cdot \sum_{i \notin trac} \bar{F}_i \cdot \bar{v}_i \cdot h$$

$$= \underbrace{\frac{1}{x_{tot}} \cdot \sum_{i=0}^{n} (\bar{F}_{a,i} + \bar{F}_{r,i}) \cdot \bar{v}_i \cdot h}_{\triangleq \bar{F}_{diss} \geq 0} + \underbrace{\frac{1}{x_{tot}} \cdot \sum_{i \notin trac} -\bar{F}_i \cdot \bar{v}_i \cdot h}_{\triangleq \bar{F}_{circ} \geq 0}, \tag{2.26}$$

The mean tractive force is equal to $\bar{E}$, the average energy consumed per distance travelled. When the latter is expressed using the units $kJ/100\,km$, the relationship between the two quantities is $\bar{E} = 100 \cdot \bar{F}_{trac}$. Therefore, the mean traction energy per distance is

$$\bar{E} = \bar{E}_{diss} + \bar{E}_{circ}, \tag{2.27}$$

where the first term $\bar{E}_{diss} \geq 0$ stands for the energy dissipated via the vehicle body and the second term $\bar{E}_{circ} \geq 0$ stands for the energy that is circulated through the vehicle body (temporarily stored as kinetic and potential energy) and is then available at the wheels during the braking periods. The circulated energy E_{circ} can be either dissipated via the friction brakes or transformed into some other energy form via a recuperation device.

Simulation results in [240] have shown that for a full-size passenger vehicle the ratio of recuperable energy and dissipation energy $\frac{\bar{E}_{circ}}{\bar{E}_{diss}}$ is around 15% for overall driving conditions and increases to around 30% for urban driving.

Case 1: No Recuperation

In the case of no recuperation, (2.23) can be directly used to calculate the total tractive force $\bar{F}_{trac}$ that includes contributions from three different effects

$$\bar{F}_{trac} = \bar{F}_{trac,a} + \bar{F}_{trac,r} + \bar{F}_{trac,m}, \tag{2.28}$$

with the three forces caused by aerodynamic and rolling friction and acceleration resistance, defined by

$$\bar{F}_{trac,a} \approx \frac{1}{x_{tot}} \cdot \frac{1}{2} \cdot \rho_a \cdot A_f \cdot c_d \cdot \sum_{i \in trac} \bar{v}_i^3 \cdot h,$$

$$\bar{F}_{trac,r} \approx \frac{1}{x_{tot}} \cdot m_v \cdot g \cdot c_r \cdot \sum_{i \in trac} \bar{v}_i \cdot h, \tag{2.29}$$

$$\bar{F}_{trac,m} \approx \frac{1}{x_{tot}} \cdot m_v \cdot \sum_{i \in trac} \bar{a}_i \cdot \bar{v}_i \cdot h.$$

In reasonable approximation, the sums and the distance x_0 in these equations depend only on the driving cycle and not on the vehicle parameters. Especially for artificial cycles like the MVEG–95, the ECE cycle, and the EUDC, the following numerical values are valid for a wide range of vehicle parameters

$$\frac{1}{x_{tot}} \cdot \sum_{i \in trac} \bar{v}_i^3 \cdot h \approx \{319, 82.9, 455\},$$

$$\frac{1}{x_{tot}} \cdot \sum_{i \in trac} \bar{v}_i \cdot h \approx \{0.856, 0.81, 0.88\}, \qquad (2.30)$$

$$\frac{1}{x_{tot}} \cdot \sum_{i \in trac} \bar{a}_i \cdot \bar{v}_i \cdot h \approx \{0.101, 0.126, 0.086\}.$$

Note that measured driving cycles typically include coasting phases. Therefore, the time intervals defined by $trac$ are subject to change when the vehicle parameters are varied[9], and thus, the approximation in (2.30) is only valid for the mentioned (artificial) test cycles.

Once the driving cycle is chosen, (2.23) allows for a simple estimation of the mean tractive force as a function of the vehicle parameters m_v, A_f, c_d, and c_r. Using again the relationship $\bar{E} = 100 \cdot \bar{F}_{trac}$, the average energy used for traction per distance in the MVEG–95 cycle is given by

$$\bar{E} = \bar{E}_{diss} + \bar{E}_{circ}$$
$$\approx 19\,000 \cdot A_f c_d + 840 \cdot m_v c_r + 10 \cdot m_v \quad [\text{kJ}/100\,\text{km}], \qquad (2.31)$$

where, to simplify matters, the physical parameters air density ρ_a (at sea level) and acceleration g have been integrated in the constants. Considering typical values for the vehicle parameters, the three contributions in this sum are of the same order of magnitude.

Case 2: Perfect Recuperation

In the case of *perfect recuperation* (a recuperation device of zero mass and 100% efficiency), the energy spent to accelerate the vehicle is completely recuperated during the braking phases. As a consequence, the mean force $\bar{F}$ does not have any contributions $\bar{F}_{circ}$ caused by braking the vehicle using friction brakes, i.e.,

$$\bar{F} = \bar{F}_{diss} = \bar{F}_a + \bar{F}_r. \qquad (2.32)$$

---

[9] Consider a coasting velocity profile measured on vehicle A. If a different vehicle B is intended to follow that same velocity profile, depending on its mass and friction parameters, it might be in traction or braking mode.

However, in the full recuperation case, the mean force must include the losses caused by aerodynamic and rolling resistances during the braking phases as well. In this case, (2.29) must be replaced by

$$\bar{F}_a \approx \frac{1}{x_{tot}} \cdot \frac{1}{2} \cdot \rho_a \cdot A_f \cdot c_d \cdot \sum_{i=0}^{n} \bar{v}_i^3 \cdot h,$$

$$\bar{F}_r \approx \frac{1}{x_{tot}} \cdot m_v \cdot g \cdot c_r \cdot \sum_{i=0}^{n} \bar{v}_i \cdot h,$$

(2.33)

with the following numerical values valid for the MVEG–95, the ECE, and the EUDC cycles

$$\frac{1}{x_{tot}} \cdot \sum_{i=0}^{n} \bar{v}_i^3 \cdot h \approx \{363, 100, 515\},$$

(2.34)

$$\frac{1}{x_{tot}} \cdot \sum_{i=0}^{n} \bar{v}_i \cdot h = \{1, 1, 1\}.$$

With these results, the mechanical energy required to drive the MVEG–95 cycle for 100 km can be approximated (similarly to (2.31)). In the case of full recuperation this quantity is given by

$$\bar{E} = \bar{E}_{diss} \approx 22\,000 \cdot A_f c_d + 980 \cdot m_v c_r \quad [\text{kJ}/100\,\text{km}]. \qquad (2.35)$$

Examples

Figure 2.8 shows the mean tractive force and the corresponding specific energy demand in the MVEG–95 cycle for two different vehicles. On the basis of the fuel's lower heating value, a full-size passenger car (without recuperation) requires per 100 km distance travelled an amount of mechanical energy that corresponds to 1.16 l of Diesel fuel.[10] This result is obtained by inserting the vehicle parameters shown in Fig. 2.8 into (2.31). In the case of perfect recuperation (a 100% efficient recuperation device with no additional mass), the corresponding value, obtained by using (2.35), is 0.89 l of Diesel fuel. In the best case, i.e., when equipped with an ideal recuperation device, a hypothetical advanced light car would require the equivalent of about 0.39 l Diesel fuel per 100 km. Without recuperation, the corresponding value would be 0.54 l Diesel fuel per 100 km.

---

[10] Of course, due to the losses in the engine and in the other components of the powertrain, the actual fuel consumption is substantially higher.

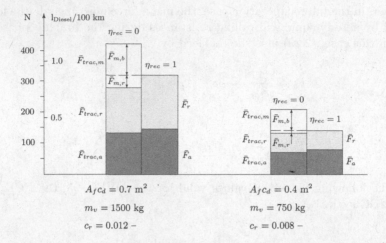

Fig. 2.8 Comparison of the energy demand in the MVEG–95 cycle for a full-size car (left) and a light-weight car (right). A mean force of 1 N is equivalent to 27.78 Wh mechanical energy per 100 km. The lower heating value for Diesel fuel is approximately 10 kWh/l.

It is worthwhile repeating that even in the case with no additional recuperation device the vehicle utilizes some of its kinetic energy to drive parts of the cycle. As illustrated in Fig. 2.8, out of the mean tractive force necessary for the acceleration, $\bar{F}_{trac,m}$ of (2.29), only the quantity $\bar{F}_{m,b}$ is later dissipated as heat by the brakes. The remaining portion $\bar{F}_{m,r}$ is used to overcome the driving resistances in the non-traction phases.

2.2.3 Some Remarks on the Energy Consumption

Comparison of Different Vehicle Classes

Using the result of (2.31), some typical numerical values of the mechanical power needed to drive the tractive parts of the MVEG–95 cycle can be estimated. As shown in Table 2.1 for a selection of vehicle classes, the *average* tractive power $\bar{P}_{MVEG--95}$ necessary to drive the MVEG–95 cycle is much smaller than the power P_{max} necessary to satisfy the acceleration requirements. For instance, a standard full-size car only needs an average of 7.1 kW to follow the MVEG–95 cycle.[11] However, using (2.16), the power necessary to accelerate the same vehicle from 0 to 100 km/h in 10 seconds is found to be around 115 kW. For a light-weight car, the numbers change in magnitude, but the ratio remains approximately the same.

---

[11] The last acceleration in the highway part of the MVEG–95 cycle requires the greatest amount of power. For the full-size vehicle this requirement is around 34 kW.

Table 2.1 Numerical values of the average and peak tractive powers for different vehicle classes

	SUV	full-size	compact	light-weight
$A_f \cdot c_d$	$1.2\,m^2$	$0.7\,m^2$	$0.6\,m^2$	$0.4\,m^2$
c_r	0.017	0.013	0.012	0.008
m_v	2000 kg	1500 kg	1000 kg	750 kg
$\bar{P}_{MVEG--95}$	11.3 kW	7.1 kW	5.0 kW	3.2 kW
P_{max}	155 kW	115 kW	77 kW	57 kW

This discrepancy between the small average power and the large power necessary to satisfy the drivability requirements is one of the main causes for the relatively low fuel economy of the current propulsion systems. In fact, as shown in the next section in Fig. 2.14, the efficiency of IC engines strongly decreases when these engines are operated at low torque. Several solutions to this *part-load problem* will be discussed in the subsequent chapters.

Sensitivity Analysis

As discussed in the previous sections, the mean tractive force necessary to drive a given cycle depends on the vehicle parameters $A_f \cdot c_d$, c_r, and m_v. Their relative influence can be evaluated with a sensitivity analysis. In the context of this vehicle energy consumption analysis, a meaningful definition of the sensitivity is

$$S_p = \lim_{\delta p \to 0} \frac{[\bar{E}(p + \delta p) - \bar{E}(p)]/\bar{E}(p)}{\delta p / p}, \tag{2.36}$$

where $\bar{E}$ is the cycle energy as defined by (2.31). The variable p stands for any of the three parameters $A_f \cdot c_d$, c_r, or m_v. Its variation is denoted by δp. The sensitivity (2.36) can also be written as

$$S_p = \frac{\partial \bar{E}}{\partial p}(p) \cdot \frac{p}{\bar{E}(p)}, \tag{2.37}$$

with the three partial derivatives

$$\frac{\partial \bar{E}}{\partial (A_f \cdot c_d)} = 19\,000$$

$$\frac{\partial \bar{E}}{\partial c_r} = 840 \cdot m_v \tag{2.38}$$

$$\frac{\partial \bar{E}}{\partial m_v} = (840 \cdot c_r + 10)$$

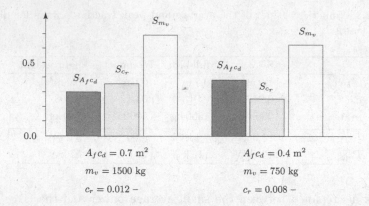

Fig. 2.9 Sensitivities (2.36) of the mechanical energy consumption in the MVEG–95 cycle with respect to the three main vehicle parameters. Two cases: full-size car (left); advanced light-weight vehicle (right).

Figure 2.9 shows the sensitivities of a typical full-size car and of an advanced light-weight vehicle. Two facts are worth mentioning:

- In the case of a standard vehicle, the sensitivity S_{m_v} is by far the most important. Accordingly, the most promising approach to reduce the mechanical energy consumed in a cycle is to reduce the vehicle's mass.[12]
- The relative dominance of the vehicle's mass on the energy consumption is not reduced when an advanced vehicle concept is analyzed. For this reason, the idea of recuperating the vehicle's kinetic energy will remain interesting for this vehicle class as well.

As shown in the sensitivity analysis, the vehicle mass m_v plays a very important role for the estimation of the energy demand of a vehicle. Therefore, Fig. 2.10 shows the mean tractive force and the corresponding equivalent of Diesel fuel as a function of the vehicle mass. Contradicting the measured fuel consumption data, the mechanical energy consumption in the test cycle is smaller than the corresponding value for the vehicle driving at constant highway speeds. As will become clear in the subsequent chapters, the reason for this fact is again to be found in the low efficiency of standard IC engines at part-load conditions.

---

[12] Of course this assertion neglects several aspects (economy, safety, ...). The full picture, which unfortunately is very difficult to obtain, would include the cost sensitivity as well. This figure, which indicates how expensive it is to reduce the parameter p, multiplied by the sensitivities (2.36), would show which approach is the most cost effective.

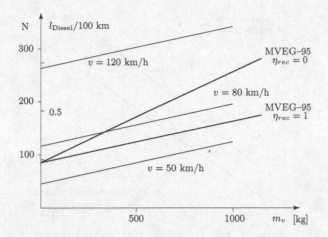

Fig. 2.10 Affine dependency of the mean force on the vehicle mass for the MVEG–95 and three values of constant speed. The example is valid for an advanced vehicle with the parameters $\{A_f \cdot c_d, c_r\} = \{0.4\,\mathrm{m}^2, 0.008\}$.

Figure 2.11 shows the influence of the vehicle mass and the *cycle-averaged* powertrain efficiency on the carbon dioxyde emissions of a passenger car.[13] A limit of 140 g/km can be reached with standard Diesel engines (cycle-averaged efficiency around 0.2) and small to medium-size vehicles (vehicle mass around 1200 kg). If a limit of 120 g/km is to be reached, either a substantial reduction of vehicle mass or a substantial increase of powertrain efficiency is needed. As mentioned above, this text focuses on the latter approach. However, it must be emphasized that very low fuel consumptions and CO_2 emissions can only be reached by combining these two approaches.

Realistic Recuperation Devices

So far it has been assumed that the device used to recuperate the vehicle's kinetic energy causes no losses (in both energy conversion directions) and has no mass. Of course, the total recuperation efficiency η_{rec} of all real recuperation devices will be smaller than 100%. Realistic values are around 60-70%.[14] Moreover, the recuperation device will increase the mass of the vehicle by m_{rec}, which in turn causes increased energy losses.

In the case of a realistic recuperation device, only a certain amount $\eta_{rec} \cdot E_{circ}$ of the circulating energy can be recuperated. Therefore, the traction

---

[13] For the sake of simplicity the aerodynamic and rolling friction parameters are kept constant.

[14] Note that this figure represents *two* energy conversions: first from kinetic to, say, electrostatic energy stored in a supercapacitor during braking and then back again into kinetic energy during acceleration.

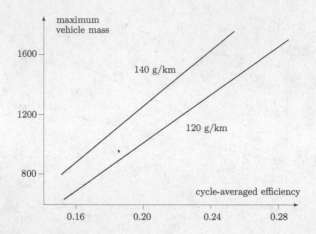

Fig. 2.11 Maximum vehicle mass permitted as a function of cycle-averaged power-train efficiency if the CO_2 emissions are to remain below $140\,\mathrm{g/km}$ and $120\,\mathrm{g/km}$, respectively. The other vehicle parameters are $\{A_f \cdot c_d, c_r\} = \{0.7\,\mathrm{m}^2, 0.012\}$.

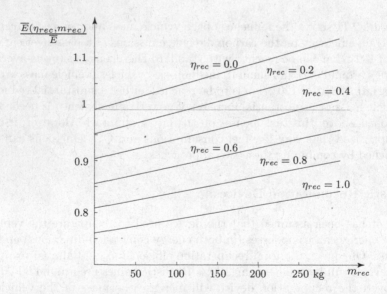

Fig. 2.12 Energy demand in the MVEG–95 cycle of a vehicle equipped with a recuperation device with mass m_{rec} and efficiency η_{rec}, normalized by the energy $\bar{E}_{MVEG-95}$ consumed by a vehicle without recuperation device. The example is valid for a vehicle with the parameters $\{A_f \cdot c_d, c_r, m_v\} = \{0.7\,\mathrm{m}^2, 0.012, 1500\,\mathrm{kg}\}$.

energy is given by

$$\bar{E}(\eta_{rec}, m_{rec}) = \bar{E}_{diss}(m_{rec}) + (1 - \eta_{rec}) \cdot \bar{E}_{circ}(m_{rec}). \tag{2.39}$$

Note that for the MVEG–95 test cycle, $\bar{E}_{diss}$ is given by (2.35) and $\bar{E}_{circ}$ is given by the difference of (2.31) and (2.35)

$$\bar{E}_{circ} \approx \bar{E} - \bar{E}_{rec}$$

$$= -3000 \cdot A_f c_d - 140 \cdot m_v c_r + 10 \cdot m_v \quad [\text{kJ}/100\,\text{km}] \tag{2.40}$$

Inserting both $\bar{E}_{diss}$ and $\bar{E}_{circ}$ into (2.39) yields an approximation for the energy needed to drive the MVEG–95 cycle in function of the recuperation efficiency and the additional mass of the recuperation device

$$\bar{E}(\eta_{rec}, m_{rec}) = [22\,000 - 3000 \cdot (1 - \eta_{rec})] \cdot A_f c_d$$

$$+ [980 - 140 \cdot (1 - \eta_{rec})] \cdot c_r \cdot (m_v + m_{rec}) \tag{2.41}$$

$$+ 10 \cdot (1 - \eta_{rec}) \cdot (m_v + m_{rec}).$$

For a typical mid-size car Fig. 2.12 shows the normalized energy recuperation potential as a function of the mass and the efficiency of the recuperation device. The maximum recuperation potential is approximately 25% in the hypothetical case $\eta_{rec} = 1$ and $m_{rec} = 0$. With the realistic values $\eta_{rec} \approx 0.6$ and $m_{rec} \approx 100\,\text{kg}$, only slightly more than 10% of the energy necessary to drive the MVEG–95 cycle can be recuperated.

Note, however, that a recuperation device has additional benefits. It can provide a certain level of power that can be used for traction during low-load or peak-load phases. On the one hand this allows shifting the operating point of the prime mover, on the other hand the prime mover can be down-sized. Both approaches will reduce the previously mentioned part-load problem of combustion-engine based vehicles and therefore will reduce their energy consumption. Furthermore, by shutting off the prime mover during stand-still, idling losses can be minimized. For these reasons, today's hybrid electric vehicles cut the fuel consumption by far more than the above-mentioned 10%.

2.3 Methods and Tools for the Prediction of Fuel Consumption

Once the mechanical energy required to drive a chosen test cycle is known, the next step is to analyze the efficiency of the propulsion system. Therefore, in this chapter the term "fuel consumption" is interpreted as the total energy consumption of the vehicle, independently of the form of the energy carried onboard. Three possible approaches are *introduced* in this section. More details on each of the three methods are discussed in the subsequent chapters and, in particular, in the case studies included in Appendix I.

2.3.1 Average Operating Point Approach

This method is often used for a first preliminary estimation of the fuel consumption of a road vehicle. The key point is to lump the full envelope of all powertrain operating points into one single representative average operating point and to compute the total fuel consumption of the propulsion system at that regime.

This approach requires a test cycle to be specified a priori. Once the driving pattern is chosen, the mean mechanical energy per distance at the wheels $\bar{E}$ can be estimated with the methods discussed in the last section. This information is then used to "work backwards" through the powertrain.

The average operating point method is able to yield reasonable estimates of the fuel consumption of *simple* powertrains (IC engine or battery electric propulsion systems). When complex propulsion systems are to be optimized, the method is less well suited. In particular it does not offer the option of including the effect of energy management strategies in these computations.

Fuel-to-Traction Efficiency

Usually the average operating point method is applied to simple powertrains as in conventional IC engine powered vehicles or pure electric vehicles. In these cases it is sufficient to define a single overall cycle efficiency that is often referred to as *tank-to-wheel* efficiency. This is correct as long as the energy flows through the powertrain in one direction only (from tank to wheel).

However, in order to be able to generalize the concept of the average operating point method it is important to distinguish between two energy flows that have opposite directions: the energy flow from fuel to traction and the energy flow associated to recuperation. The recuperation efficiency has been explained in the previous section. The *fuel-to-traction* efficiency is defined by how much of the energy stored in the fuel is converted to actual traction power. The distinction between recuperation and fuel-to-traction efficiency is very important since, as will become clear below, the definition of a single overall powertrain efficiency is not sufficient to fully describe the energy flows inside a powertrain with a recuperation device.

The fuel-to-traction efficiency is defined as the ratio of traction energy to the fuel energy consumed, i.e.

$$\eta_{ft} = \frac{\bar{E}}{\bar{E}_f} = \frac{\bar{E}_{diss} + (1 - \eta_{rec}) \cdot \bar{E}_{circ}}{\bar{E}_f}. \tag{2.42}$$

The fuel energy can thus be expressed as a function of the fuel-to-traction efficiency, the recuperation efficiency and the driving cycle specific energies $\bar{E}_{diss}$ and $\bar{E}_{circ}$

$$\bar{E}_f = \frac{1}{\eta_{ft}} \cdot \left(\bar{E}_{diss} + (1 - \eta_{rec}) \cdot \bar{E}_{circ} \right). \qquad (2.43)$$

The fuel-to-traction efficiency contains all losses that occur in the chain from the prime mover to the wheels, i.e., losses in the prime mover energy conversion step, friction losses, and losses in the clutch and the gear-box. Overall, the fuel-to-traction efficiency can be regarded as the cycle averaged conversion efficiency from the fuel to traction energy.

The average fuel-to-traction efficiency of a combustion engine based vehicle is mainly determined by the efficiency of the combustion engine and therefore is dependent on the driving scenario. In urban driving it is very low (about 20% for Diesel and 17% for gasoline engines), whereas during highway driving it can be quite high. Simulation results in [240] have shown that the average fuel-to-traction efficiency of well-designed hybrid electric vehicles (about 35% for Diesel-hybrids and 30% for gasoline-hybrids) is always closely below the peak efficiency of the prime mover. This is true for both a wide range of vehicle design parameters and driving scenarios. Note that in these simulations, the possibility of shutting off the engine during standstill and low-load phases has already been included.

Tank-to-Wheel Efficiency

One may think it might be useful to lump η_{ft} and η_{rec} into one overall cycle efficiency

$$\eta_{tw} = \frac{\bar{E}_{diss} + \bar{E}_{circ}}{\bar{E}_f}, \qquad (2.44)$$

which is often referred to as *tank-to-wheel* efficiency. However, reformulating (2.44) using the definitions of η_{ft} and η_{rec}

$$\eta_{tw} = \eta_{ft} \cdot \frac{\bar{E}_{diss} + \bar{E}_{circ}}{\bar{E}_{diss} + (1 - \eta_r) \cdot \bar{E}_{circ}}, \qquad (2.45)$$

reveals that this term can become unrealistically large if the recuperation efficiency η_{rec} is high and if the circulating energy $\bar{E}_{circ}$ is a large portion of the total energy demand of the drive cycle[15]. Therefore, the physical meaning of this definition is useless when considering vehicles with a recuperation device.

---

[15] E.g., when considering a long pure downhill driving cycle with $\frac{\bar{E}_{circ}}{\bar{E}_{diss}}=5$, $\eta_{ft}=40\%$ and $\eta_r=70\%$, (2.45) would yield a value of 96%.

Conventional Vehicle

In the case of a conventional IC engine powertrain, the average operating point method is straightforward since the energy flows through the powertrain in one direction only.

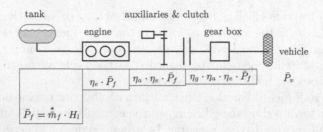

Fig. 2.13 Illustration of the components considered in the fixed operating point approach for the case of a conventional IC engine powertrain

Figure 2.13 shows the components relevant for such an analysis for the case of a conventional IC engine powertrain. Consider $\bar{P}_v$ to be the average mechanical power needed at the wheels. Then this figure illustrates:

- the losses caused by the gear box and differential (lumped into one device for the sake of simplicity) represented by the efficiency η_g;
- the losses caused by the auxiliary devices (generation of electric energy, power steering, etc.) and the clutch represented by the efficiency η_a; and
- the losses caused by the IC engine represented by the efficiency η_e.

The fuel-to-traction efficiency is thus given by $\eta_{ft} = \eta_g \cdot \eta_a \cdot \eta_e$, and the fuel energy consumed per kilometer in the cycle can be calculated using (2.43) together with $\eta_{rec} = 0$

$$\bar{E}_f = \frac{1}{\eta_g \cdot \eta_a \cdot \eta_e}(\bar{E}_{diss} + \bar{E}_{circ}). \qquad (2.46)$$

The average mass of fuel consumed per kilometer is then

$$\bar{m}_f = \frac{\bar{E}_f}{H_l}. \qquad (2.47)$$

Alternatively, one could calculate the average fuel power

$$\bar{P}_f = \frac{1}{\eta_{ft}} \cdot \bar{P}_v = \frac{1}{\eta_{ft}} \cdot \frac{(\bar{E}_{diss} + \bar{E}_{circ}) \cdot x_{tot}}{T}, \qquad (2.48)$$

which can then by used to calculate the average fuel mass flow rate

$$\overset{*}{\bar{m}}_f = \frac{\bar{P}_f}{H_l}. \tag{2.49}$$

The engine efficiency η_e can be estimated using approximations, such as the Willans rule introduced in the next chapter, or using measured *engine maps*. Figure 2.14 shows such a map. Note that η_e strongly depends on engine torque, especially in the low load regime the efficiency drops steeply, whereas the engine speed has less of an influence. In the mean operating point approach, one characteristic engine torque and one characteristic engine speed must be found. These data points follow from the average power $\bar{P}_v$ consumed at the wheels, the mean vehicle velocity $\bar{v}$, and the mean gear ratio $\bar{\gamma}$.

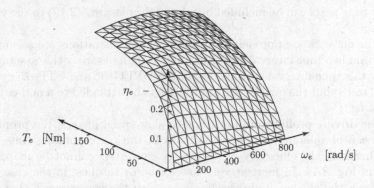

Fig. 2.14 IC engine efficiency as a function of engine torque and speed

Sect. 3.3.2 contains an example that shows in detail how to apply this approach to a standard ICE powertrain.

2.3.2 *Quasistatic Approach*

In quasistatic simulations, the input variables are the speed v and the acceleration a of the vehicle as well as the grade angle α of the road.[16] With this information, the force F_t is computed that has to be acting on the wheels to drive the chosen profile for a vehicle described by its main parameters $\{A_f \cdot c_d, c_r, m_v\}$. In this step the vehicle is assumed to run at constant speed v, acceleration a, and grade α for a (short) time period h. Once F_t is known, the losses of the powertrain are considered using operating point dependent efficiencies as in (2.46). The fuel consumption during the time interval $t \in [(i-1) \cdot h, i \cdot h)$ is calculated by simply converting the needed fuel energy into the corresponding amount of fuel.

---

[16] Regulatory test cycles or driving patterns recorded on real vehicles can be used.

Evidently, the average operating point method and the quasistatic method are very similar. The only difference is that in the latter approach the test cycle is divided into (many) intervals in which the average operating point method is applied. For each time interval, the *constant* speed and acceleration that the vehicle is required to follow are given by (2.20) and (2.21), respectively.

The basic equation used to evaluate the force required to drive the chosen profile is Newton's second law (2.1) as derived in the previous section. Using this equation, the force is calculated as follows

$$\bar{F}_{t,i} = m_v \cdot \bar{a}_i + F_{r,i} + F_{a,i} + F_{g,i} \tag{2.50}$$

$$= m_v \cdot \bar{a}_i + \frac{1}{2} \cdot \rho_a \cdot A_f \cdot c_d \cdot \bar{v}_i^2 + c_r \cdot m_v \cdot g \cos(\bar{\alpha}_i) + m_v \cdot g \cdot \sin(\bar{\alpha}_i).$$

The rotating parts can be included by adding the mass m_r (2.12) to the vehicle mass.

In the quasistatic approach the velocity and accelerations are assumed to be constant in a time interval h chosen small enough to satisfy this assumption. Usually this time interval is constant (in the MVEG–95 and FTP–75 cycles h is equal to 1 s) but the quasistatic approach can be extended to a non-constant value of h.

If the driving profile contains idling or slow-speed phases, the propulsion system can be operated at very low loads in the corresponding time intervals. Therefore, the efficiency of the energy converters cannot be mapped as shown in Fig. 2.14. In fact, at very low loads (or torques, in the case of an IC engine), the efficiency approaches zero, with the consequence that small measurement uncertainties can cause substantial errors in the estimation of the fuel consumption. Figure 2.15 illustrates a more suitable approach for the representation of the efficiency of an IC engine. Similar maps can be obtained for other energy converters.

In this approach, the fuel consumption necessary to sustain a pre-defined torque and speed combination is directly mapped to the torque-speed plane. This representation allows a visualization of the *idling losses* and the *fuel cut-off* limits. If the deceleration of the vehicle provides enough power to cover, in addition to the aerodynamic and rolling friction losses of the vehicle itself, the losses of the complete propulsion system (friction losses of the powertrain, power consumed by the auxiliaries, engine friction and pumping losses, etc.), the fuel may be completely cut off.

For the example of a standard IC engine powertrain, Fig. 2.16 shows the simplified structure of an algorithm that computes the vehicle's fuel consumption using the quasistatic method. For relatively simple powertrain structures the algorithm illustrated in Fig. 2.16 can be compactly formulated using a vector notation for all variables (speed, tractive force, etc.). This form is particularly efficient when using Matlab.[17]

---

[17] Matlab/Simulink is a registered trademark of TheMathWorks, Inc., Natick, MA.

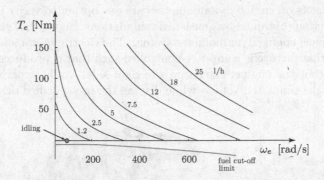

Fig. 2.15 IC engine fuel consumption in liters per hour as a function of engine torque and speed

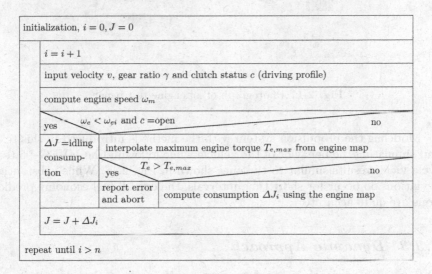

Fig. 2.16 Structure of an algorithm for quasistatic fuel consumption

The quasistatic method is well suited to the minimization of the fuel consumption of complex powertrain structures. With this approach it is possible to design *supervisory* control systems that optimize the power flows in the propulsion system. The influence of the driving pattern can be included in these calculations. Despite these capabilities, the numerical effort remains relatively low. The main drawback of the quasistatic method is its "backward" formulation, i.e., the physical causality is not respected and the driving profile that has to be followed has to be known a priori. Therefore, this method is not able to handle feedback control problems or to correctly deal with state events.

Experiments on engine dynamometers are one option to verify the quality of the predictions obtained by quasistatic simulations. Figure 2.17 shows a picture of a typical engine dynamometer system. The electric motor shown on the left side in that picture is computer-controlled such that it produces the braking torque that the engine, shown on the right side, would experience when installed in the simulated vehicle while driving the pre-specified driving cycle.

Fig. 2.17 Photograph of an engine test bench

Modeling the propulsion system with the methods introduced in this text and simulating the behavior of the vehicle while following the MVEG–95 test cycle yields results similar to the ones shown in Fig. 2.18. While substantial deviations do occur for short time intervals, the overall fuel economy predictions are quite accurate.

2.3.3 Dynamic Approach

The *dynamic approach* is based on a "correct" mathematical description of the system.[18] Usually, the model of the powertrain is formulated using sets of ordinary differential equations in the state-space form

$$\frac{d}{dt}x(t) = f(x(t), u(t)), \qquad x(t) \in I\!R^n, \, u \in I\!R^m, \qquad (2.51)$$

but other descriptions are possible, e.g., using partial differential equations or equations with differential *and* algebraic parts. The formulation (2.51) may be used to describe many dynamic effects in powertrains. While some of these effects are relevant for the estimation of the fuel consumption (engine temperature dynamics, etc.) certain others are not (inlet manifold dynamics of

---

[18] Of course, no model can exactly describe the system behavior. Modeling errors must be taken into account by appropriate robustness guarantees in the later design steps.

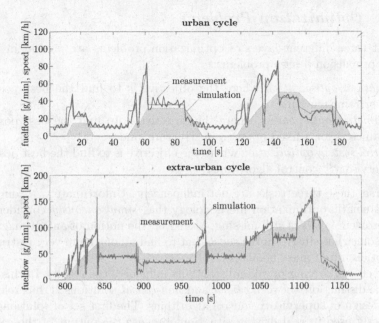

Fig. 2.18 Comparison of measured and simulated fuel consumption

SI engines, EGR rate dynamics in Diesel engines, etc). Most of the relevant effects are relatively slow, while the fast effects are significant for the optimization of comfort, drivability, and pollutant emission. Readers interested in these effects are referred to [168] and [128]. The slower effects that are important for the optimization of the fuel economy will be discussed in the subsequent chapters.

Equation (2.51) is the starting point for a plethora of feedforward and feedback system analysis and synthesis approaches. Readers interested in these approaches are referred to the standard textbooks, e.g. [160], [339], [151]. Note that in the dynamic method the inputs to the powertrain model are the same signals as those that are present in the real propulsion system. Accordingly, a module that emulates the behavior of the driver has to be included in these simulations.

The description (2.51) is very versatile, and many optimization problems can only be solved using the dynamic method. Notable examples of such problems are the design of feedback control systems or the detection and correct handling of state events in optimization problems. The drawback of using (2.51) and the associated analysis and synthesis approaches is the relatively high computational burden of these methods. For this reason, this text emphasizes the quasistatic methods. Dynamic methods are only chosen when no other option is available.

2.3.4 Optimization Problems

At least three different layers of optimization problems are present in most vehicle propulsion design problems:

- *structural optimization* where the objective is to find the best possible powertrain structure;
- *parametric optimization* where the objective is to find the best possible parameters for a fixed powertrain structure; and
- *control system optimization* where the objective is to find the best possible supervisory[19] control algorithms.

Of course, these three tasks are not independent. Unfortunately, a complete and systematic optimization methodology that *simultaneously* considers all three problem layers is still missing. Moreover, the notion of an optimal solution is somewhat elusive in the sense that to find an optimum very restricting assumptions often must be adopted.

The chosen driving profile substantially influences the results. In this context, the distinction between *causal* and *non-causal* solutions is thus relevant in the design of supervisory control algorithms. The first set of solutions can be directly used in real driving situations because the output of the control system only depends on actual and past driving profile data. Non-causal control algorithms also utilize future driving profile data to produce actual control system outputs. Clearly, such an approach is only possible in situations where the complete driving profile is known at the outset.

This situation can arise when trip planning instruments become generally available (GPS-based navigation, on-line traffic situation information, etc.). Even more important is the role of non-causal optimal solutions as benchmarks for causal solutions. A causal solution only can approach this result. Its "distance" from the non-causal optimum is a good indicator of the quality of a causal solution. In fact, if a causal solution achieves 95% of the benefits of the non-causal approach, there is little room for further improvements, such that refining the causal control system might not be worthwhile.

In summary, the minimization of the fuel consumption of a powertrain generally, is not a simple and straightforward problem that can be completely solved with systematic procedures. Many iteration loops and intuitive short-cuts are necessary in all but the simplest cases. Only well-defined partial problems can be solved using systematic optimization problems. The case studies shown in Appendix I exemplify this part of the design procedure, and the main mathematical tools used in this analysis are introduced in Appendix II and Appendix III.

---

[19] In the context of this book the emphasis is on those control algorithms that produce the set points for all low-level control loops. Such systems will be denoted by the term *supervisory controllers*.

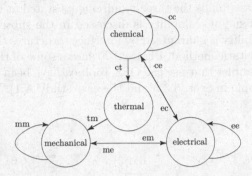

Fig. 2.19 Energy domains relevant for the modeling and optimization of vehicle propulsion systems. Also shown are the feasible energy conversion paths (reversible with double arrows, non-reversible with single arrows) [272].

2.3.5 Software Tools

General Remarks

Most non-trivial system analysis and synthesis problems can only be solved using numerical approaches. For that purpose efficient and reliable numerical computer tools must be available. The basic functions of such tools usually are provided by general-purpose software packages. Using these functions, tools specifically developed for the systems analyzed in this text can then be developed.

Such a software package must fulfill at least the following requirements:

- it allows for the interconnection of all relevant powertrain elements even if they operate in different energy domains (see Fig. 2.19);
- it allows for the scaling of all powertrain elements such that parametric optimizations can be accomplished; and
- it integrates well with the visualization and numerical optimization tools provided by the underlying general-purpose software package.

Quasistatic Simulation Tools

For quasistatic simulations, the ADVISOR (Advanced Vehicle Simulator) software package is often used. This program was developed by the National Renewable Energy Laboratory (NREL), and was later transferred to industry, see http://www.nrel.gov/vehiclesandfuels/success_advisor.html.

The examples shown in this text have been calculated with the QSS Toolbox. The QSS Toolbox is a collection of Matlab/Simulink blocks and the appropriate parameter files that can be run in any Matlab/Simulink environment. This package is available for academic purposes and can be downloaded at the URL http://www.idsc.ethz.ch/Downloads/qss/.

The QSS Toolbox fulfills the three requirements stated at the beginning of this section. The issue of scalability is discussed in the subsequent chapters. The interconnectability is guaranteed by interface structures that are compatible with the quasistatic method. Figure 2.20 shows some of the QSS Toolbox elements. Two examples that use the QSS Toolbox have been included in this text (see the example in Sect. 3.3.3 and the case study A.1).

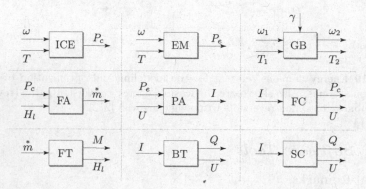

Fig. 2.20 Interfaces of main blocks of a QSS simulation environment. ICE = IC engine, FA = fuel amplifier (fuel control), FT = fuel tank, EM = electric motor, PA = power amplifier (current control), BT = battery, GB = gear box, FC = fuel cell, SC = supercapacitor.

Complex powertrain structures can be built with the basic blocks shown in Fig. 2.20. Figure 2.21 shows the example of a fuel-cell electric powertrain that includes a supercapacitor element for load-leveling purposes. A similar setup was used in [12] to analyze the potential for CO_2 reduction of fuel cell systems. The block "EMS" in Fig. 2.21 represents the supervisory control algorithm (energy management). Based on its two inputs P_e (the total required electric power) and U_2 (the supercapacitor voltage that indicates the amount of electrostatic energy stored in that device), it controls the amount of electric power that the fuel cell FC and the supercapacitor SC have to produce.[20]

Dynamic Simulation Tools

Many packages are available for the dynamic simulation of powertrain systems. The main problem encountered with the usual tools like Matlab/Simulink is the missing flexibility when the system topology is changed. Such a change usually requires a complete redesign of the mathematical model. Attempts have been made to improve this situation. Notably the approach proposed by the Dymola/Modelica software tools is cited by several authors as a very promising step in that direction (see for instance [329]).

---

[20] Of course the power balance $P_e = P_{e1} + P_{e2}$ has to be satisfied.

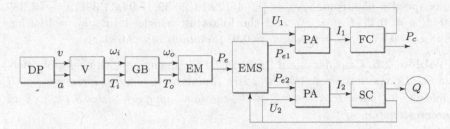

Fig. 2.21 Top level of a QSS Toolbox model of a fuel-cell electric powertrain (detailed model description in [12]). Blocks not yet introduced so far: DP = driving profile, V = vehicle, EMS = energy management strategy.

Since in this text the emphasis is on the quasistatic approach, that set of tools is not used in the subsequent parts. The few problems of dynamic system simulation and optimization are solved using classical software tools.

2.4 Problems

Vehicle Energy Losses and Performance Analysis

Problem 2.1. For a vehicle with $m_v = 1500$ kg, $A_f \cdot c_d = 0.7\,\text{m}^2$, $c_r = 0.012$, a vehicle speed $v = 120\,\text{km/h}$ and an acceleration $a = 0.027\,g$, calculate the traction torque required at the wheels and the corresponding rotational speed level (tires 195/65/15T). Calculate the road slope that is equivalent to that acceleration. *(Solution: $T_t = 330\,Nm$; $\omega_w = 101\,rpm$; $\alpha = 2.7\%$).*

Problem 2.2. Find the road slope α that is equivalent to a step of height h for a car with (a rigid) wheel radius r_w on a flat terrain. Calculate the result for $h/2r_w = \{0.01, 0.02, 0.05, 0.1, 0.2\}$. *(Solution: $\alpha = \{10, 14, 22, 31, 44\}\,(\%)$).*

Problem 2.3. Find an equation to evaluate the speed profile of an ICE vehicle under maximum engine torque. Assume a maximum torque curve of the type $T_e = a \cdot \omega_e^2 + b \cdot \omega_e + c$. Include the engine inertia and a traction efficiency η_t. *(Solution: The solution depends on the sign of $D = B^2 - 4 \cdot A \cdot C$, where*
$$A = \left(-\tfrac{1}{2} \cdot \rho_a \cdot A_f \cdot c_d + \left(\tfrac{\gamma}{r_w}\right)^3 \cdot \eta_t \cdot a\right) \cdot \tfrac{1}{m_{eq}},\ B = \left(\tfrac{\gamma}{r_w}\right)^2 \cdot \eta_t \cdot b \cdot \tfrac{1}{m_{eq}},\ C =$$
$$\left(-g \cdot m_v \cdot c_r + \tfrac{\gamma}{r_w} \cdot \eta_t \cdot c\right) \cdot \tfrac{1}{m_{eq}}.\ \textit{For (i) } D > 0,\ v(t) = \tfrac{S \cdot (1+K(t)) + B \cdot (K(t)-1)}{2 \cdot A \cdot (1-K(t))},$$
where $K(t) = \tfrac{2 \cdot A \cdot v_0 + B - S}{2 \cdot A \cdot v_0 + B + S} \cdot e^{S \cdot (t-t_0)} = K_0 \cdot e^{S \cdot (t-t_0)}$ and $S = \sqrt{D}$. For (ii)
$$D < 0,\ v(t) = \tfrac{S \cdot \tan\left(\arctan\left(\tfrac{2 \cdot A \cdot v_0 + B}{S}\right) + \tfrac{S \cdot (t-t_0)}{2}\right) - B}{2 \cdot A},\ \textit{where } S = \sqrt{-D}).$$

Problem 2.4. Evaluate the 0-100 km/h time precisely using the result of Problem 2.3. Use the following data: engine launch speed = 2500 rpm, engine max speed $= 6500$ rpm, $\gamma/r_w = \{46.48, 29.13, 20.39, 15.04, 11.39\}$, $a = -4.38 \cdot 10^{-4}$, $b = 0.3514$, $c = 80$, and the following vehicle data: $m_v = 1240$ kg, $A_f \cdot c_d = 0.65 \, m^2$, $c_r = 0.009$, $\eta_t = 0.9$. *(Solution: $t_0 = 9.6 \, s$)*.

Problem 2.5. Consider again Problem 2.4 for an engine with negligible inertia Θ_e. Compare the result with (2.16). *(Solution: $t_0 = 8.8 \, s$. Equation (2.16) yields 11.9 s, but if $\bar{P} = \frac{P_{e,max} + P_{e,min}}{2}$, the same approach yields 8.1 s, i.e., an underestimation of 9%)*.

Problem 2.6. Find an equation to calculate the takeoff time (=time to synchronize the speed before and after the clutch) as a function of engine launch speed and torque. Assume that the clutch is slipping but transmitting the whole engine torque. *(Solution: $t_{takeoff} = \frac{\ln(-K)}{S}$)*.

Problem 2.7. Evaluate the coasting speed and the roll-out time without acting on the brakes for a vehicle with an initial speed $v_0 = 50$ km/h and $m_v = 1200$ kg, $A_f \cdot c_d = 0.65 \, m^2$, $c_r = 0.009$. *(Solution: $t_{rollout} = 130.9 \, s$)*.

Mechanical Energy Demand in Driving Cycles

Problem 2.8. Evaluate the traction energy and the recuperation energy for the MVEG–95 for the vehicle example of Fig. 2.8, left and right, assuming perfect recuperation. *(Solution: For (i) the full-size car, $\bar{E} = 4.78 \, MJ$, $\bar{E} = 3.64 \, MJ$, $\Delta\bar{E} = 1.14 \, MJ$ (24% of $\bar{E}$). For (ii) the light-weight car, $\bar{E} = 2.21 \, MJ$, $\bar{E}_{rec} = 1.61 \, MJ$, $\Delta\bar{E} = 0.6 \, MJ$ (27% of $\bar{E}$))*.

Problem 2.9. Calculate the mean force and fuel consumption data shown in Fig. 2.8 left. *(Solution: In the case of (i) no recuperation, $\bar{F}_{trac,a} = 134 \, N$, $\bar{F}_{trac,r} = 151 \, N$, $\bar{F}_{trac,m} = 150 \, N$, $\overset{*}{V} = 1.2 \, l$. In the case (ii) of perfect recuperation, $\bar{F}_a = 152 \, N$, $\bar{F}_r = 177 \, N$, $\bar{F}_{trac,m} = 150 \, N$, $\overset{*}{V} = 0.91 \, l$)*.

Problem 2.10. Calculate the data in Fig. 2.9, left and right. *(Solution: In the case of (i) a full-sized car, $S(A_f \cdot c_d) = 0.31$, $S(c_r) = 0.35$, $S(m_v) = 0.70$. In the case of (ii) a light-weight vehicle, $S(A_f \cdot c_d) = 0.38$, $S(c_r) = 0.25$, $S(m_v) = 0.63$)*.

Problem 2.11. Calculate which constant vehicle speed on a flat road is responsible for the same energy demand at the wheels along a MVEG–95 cycle, in the case of no recuperation and of perfect recuperation, respectively. Assume the light-weight vehicle data of Fig. 2.8, left and right: $\{A_f \cdot c_d, m_v, c_r\} = \{0.7 \, m^2, 1500 \, kg, 0.012\}$. *(Solution: In the case of (i) no recuperation, $\bar{v} = 87 \, km/h$. In the case of (ii) perfect recuperation, $\bar{v} = 70 \, km/h$)*.

Problem 2.12. Calculate the maximum mass allowed for a recuperation system with $\eta_{rec} = 40\%$. Use the vehicle parameters of Fig. 2.12: $\{A_f \cdot c_d, m_v, c_r\} = \{0.7 \, m^2, 1500 \, kg, 0.012\}$. *(Solution: $m_{rec} = 250 \, kg$)*.

3

IC-Engine-Based Propulsion Systems

In this chapter, standard IC-engine-based propulsion systems are analyzed using the tools that will be later applied for the optimization of more complex powertrains. The main components of IC-engine-based propulsion systems are the engine and the gear box. Clutches or torque converters, which also are part of such a powertrain, are needed during the relatively short phases in which the engine must be kinematically decoupled from the vehicle. As shown in the first section of this chapter, in the context of this book the engine may be described by an engine map and two normalized engine variables. The second section shows how the gear ratios must be chosen to satisfy drivability requirements. Once these two components and the main vehicle parameters are specified, estimations of the fuel consumption can be made using any of the methods introduced in the previous chapter. The third section of this chapter includes two examples of such an analysis.

3.1 IC Engine Models

3.1.1 Introduction

As mentioned in the previous chapter, two different descriptions of the powertrain elements are used in this text: the quasistatic and the dynamic formulation. For IC engines, the corresponding input and output variables are shown in Fig. 3.1. The thermodynamic efficiency of such a device is defined by

$$\eta_e = \frac{\omega_e \cdot T_e}{P_c},\tag{3.1}$$

where ω_e is the engine angular speed, T_e the engine torque, and P_c the enthalpy flow[1] associated with the fuel mass flow

---

[1] The index c stands for "chemical" because the fuel carries the energy in the form of chemical energy.

$$\overset{*}{m}_f = P_c/H_l, \tag{3.2}$$

where H_l is the fuel's lower heating value.

(a) Quasistatic approach

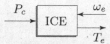

(b) Dynamic approach

Fig. 3.1 Engine input and output variables in the quasistatic and in the dynamic system description

The thermodynamic efficiency η_e of IC engines mainly depends on the engine speed and torque. The modeling of all relevant phenomena is a vast and well-documented area. A rich literature exists that describes the important points of this topic. The standard text [136] summarizes the main ideas and contains references to many other publications. In Sect. 3.1.3 below, simplified formulations of the dependency $\eta_e(\omega_e, T_e)$ are shown that are suitable for the purposes of this text.

The variables T_e and ω_e have a clear physical interpretation. Unfortunately, their range depends on the specific engine that is modeled (size, geometry, etc.). For this reason, normalized variables are introduced in the next section. Using these variables, the engine size can be used as an optimization parameter.

3.1.2 Normalized Engine Variables

When the engine runs in steady-state conditions, two normalized variables describe its operating point. These two quantities are the *mean piston speed*

$$c_m = \frac{\omega_e \cdot S}{\pi} \tag{3.3}$$

and the *mean effective pressure*

$$p_{me} = \frac{N \cdot \pi \cdot T_e}{V_d}, \tag{3.4}$$

where ω_e is the engine speed, T_e the engine torque, V_d the engine's displacement, and S its stroke. The parameter N depends on the engine type: for a four-stroke engine $N = 4$ and for a two-stroke engine $N = 2$ must be inserted in (3.4).

Obviously, the mean piston speed is the piston speed averaged over one engine revolution. It is limited at the lower end by the idling speed limit and at the upper end by aerodynamic friction in the intake part and by mechanical stresses in the valve train. Typical maximum values of c_m are below $20\,\mathrm{m/s}$.

The mean effective pressure is that amount of constant pressure that must act on the piston during one full expansion stroke to produce that amount of mechanical work that a constant engine torque T_e produces during one engine cycle. For naturally aspirated engines the maximum value of p_{me} is around $10^6\,\mathrm{Pa}$ (10 bar). Typical turbocharged Diesel engines reach maximum mean effective pressures of close to 20 bar. Even higher values are possible with special supercharging devices (twin turbochargers, pressure-wave superchargers, etc.).

The key advantages of using the normalized engine variables c_m and p_{me} are that their range is approximately the same for all[2] engines and that they are not a function of the engine size. Since for engines of similar type the speed boundaries vary less than the torque limits, engine maps are often shown in practice with c_m replaced by n_e, i.e., the engine speed in rpm.

For a fixed mean effective pressure and a mean piston speed, the equation

$$P_e = z \cdot \frac{\pi}{16} \cdot B^2 \cdot p_{me} \cdot c_m \tag{3.5}$$

describes how the mechanical power P_e produced by the engine correlates with the number of cylinders z and the cylinder bore B.

With (3.5) it is possible to estimate the necessary engine size once the desired rated engine power P_{max} has been chosen. For instance, if a lightweight vehicle with $m_v = 750\,\mathrm{kg}$ plus a payload of $100\,\mathrm{kg}$ is designed to reach $100\,\mathrm{km/h}$ in $t_0 = 15\,\mathrm{s}$, an estimation of the necessary rated power of $45\,\mathrm{kW}$ is obtained using the the approximation (2.16). Assuming the engine to be a naturally aspirated one, (3.5) indicates that the choice $z = 3$ and $B = 0.067\,\mathrm{m}$ are reasonable values.[3]

3.1.3 Engine Efficiency Representation

The engine efficiency (3.1) is often plotted in the form of an *engine map*. Figure 3.2 shows such a map of the engine specified in the last section. On top of the $p_{me} = 0$ line, this map shows the engine efficiency as calculated with a standard thermodynamic engine process simulation program. No mixture enrichment at high loads is considered in these calculations. For that reason the best efficiencies are reached at full load conditions. Also shown are the constant power curves and the estimated maximum mean effective pressure limits.

---

[2] Of course the engines have to be of the same type, e.g., naturally aspirated SI engines.

[3] A four-cylinder configuration would require a bore B which is too small to yield a satisfactory thermodynamic efficiency.

As mentioned in Sect. 2.3, engine maps similar to the one shown in Fig. 3.2 are not the only way to describe the engine efficiency. It is easy to convert this form to the "fuel-flow" description that is used in Fig. 2.15.

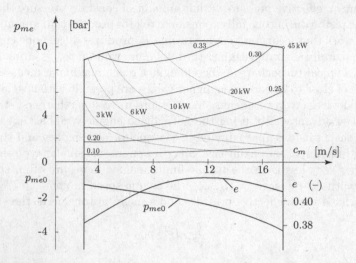

Fig. 3.2 Engine map computed with a thermodynamic process simulation program. Naturally aspirated SI engine, strictly stoichiometric air/fuel mixture. Engine parameters: displaced volume $V_d = 0.710\,l$, bore and stroke $B = S = 0.067\,m$, compression ratio $\epsilon = 12$.

Quite often it is possible to further simplify the engine model. A very simple, but nevertheless rather useful approximation is the Willans description [272], [356]. In this approach the engine mean effective pressure is approximated by

$$p_{me} \approx e(\omega_e) \cdot p_{mf} - p_{me0}(\omega_e), \tag{3.6}$$

where the input variable p_{mf} is the *fuel mean pressure*. This variable is the mean effective pressure that an engine with an efficiency of 100% would produce by burning a mass m_f of fuel with a (lower) heating value H_l

$$p_{mf} = \frac{H_l \cdot m_f}{V_d}. \tag{3.7}$$

The parameter $e(\omega_e)$ stands for the indicated engine efficiency, i.e., the efficiency of the thermodynamic energy conversion from chemical energy to pressure inside the cylinder. The parameter p_{me0} summarizes all mechanical friction and pumping losses in the engine. For a specific engine system, these two parameters significantly depend mainly on the engine speed. Figure 3.2 contains two curves in the lower part that exemplify these dependencies.[4]

---

[4] In this example, the mean friction pressure is larger than in most modern engines. This fact is a consequence of the small engine size chosen in this example.

A parametrization using low-order polynomials or spline functions is usually sufficient. For preliminary computations the variables e and p_{me0} are often assumed to be constant parameters.

3.2 Gear-Box Models

3.2.1 Introduction

Gear boxes are elements that transform the mechanical power provided by a power source at a certain speed ω_1 and torque T_1 to a different speed ω_2 and torque T_2 level. Neglecting all losses that are caused by such a device, the following relations are valid

$$\omega_1 = \gamma \cdot \omega_2, \qquad T_2 = \gamma \cdot T_1, \tag{3.8}$$

where γ is the gear ratio.

As usual, a quasistatic and a dynamic formulation may be used to describe this element. The corresponding inputs and outputs are shown in Fig. 3.3. Mathematical models of these devices are introduced in this section and later in Chap. 4.

(a) Quasistatic approach

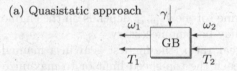

(b) Dynamic approach

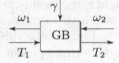

Fig. 3.3 Gear box input and output variables in the quasistatic and in the dynamic system description

In addition to gear boxes most powertrains include devices that kinematically decouple the prime mover from the vehicle. Two types are commonly used: friction clutches (dry or wet) and hydraulic torque converters. These devices will be described in Sect. 3.2.4.

3.2.2 Selection of Gear Ratios

The ratio of the maximum to the minimum speed of a standard IC engine is limited to a value of approximately 10 for SI engines and to a value of 6 for

Diesel engines. Clearly, this is not sufficient for most practical applications. Excluding the very small speed values when the clutch or the torque converter are slipping, the ratio of the maximum to the minimum speed of a standard vehicle has a value of around 30. Accordingly, a gear box must be added to the powertrain. This device must realize a minimum gear ratio of five to meet the standard driving requirements.

Three types of gear boxes are encountered in practical applications:

- manual gear boxes, which have a finite number of fixed gear ratios and are manually operated by the driver;
- automatic transmissions, which combine a fixed number of gear ratios with a gear shift mechanism and a hydrodynamic torque converter or an automated standard clutch; and
- continuously variable transmissions (CVTs), which are able to realize any desired gear ratio within the limits of this device.

Choosing the gear ratios requires the solution of a complex optimization problem [99]. In a first iteration, the gear box efficiency and its dynamic properties may be neglected. The largest gear ratio (the smallest gear in a manual gear box) is often chosen to meet the towing requirements. Using (2.4), this gear ratio is found to be

$$\gamma_1 = \frac{m_v \cdot g \cdot r_w \cdot \sin(\alpha_{max})}{T_{e,max}(\omega_e)}. \tag{3.9}$$

The maximum engine torque $T_{e,max}$ depends on the engine speed ω_e. For this reason iterations may become necessary.

The smallest gear ratio (the highest gear in a manual gear box) can be chosen either to reach the top-speed limit or to maximize the fuel economy. These two approaches can be combined by choosing the second-smallest gear to satisfy the maximum speed requirements and the smallest gear to maximize the fuel economy ("overdrive" configuration). This approach is chosen in the example illustrated in Fig. 3.4. In this case, to determine the ratio of the fourth gear, first the following equation[5] must be solved for v_{max}

$$P_{e,max} = v_{max} \cdot F_{max} = v_{max} \cdot \left(m_v g c_r(v_{max}) + \frac{1}{2}\rho_a A_f c_d v_{max}^2 \right). \tag{3.10}$$

In general, this equation must be solved numerically. If c_r may be assumed to be constant, a third-order polynomial equation results. Using Cartan's formula, it can be proven that in this case only one real solution of (3.10) exists such that no ambiguities arise.

Once the achievable top vehicle speed v_{max} and the top engine speed $c_{m,max}$ are known, the gear ratio γ_4 is found using the following equation

---

[5] This equation is similar to (2.13). However, it includes the rolling friction losses that become relevant in this context.

$$\gamma_4 = \frac{r \cdot c_{m,max} \cdot \pi}{v_{max} \cdot S}. \tag{3.11}$$

With this information, the vehicle resistance curves can be plotted in the engine map. Figure 3.4 shows the resulting resistance curves for the lightweight vehicle and the three-cylinder engine used as an example in this section.

The fifth gear can be chosen according to several fuel-economy optimization criteria. One possibility is to choose a value with which the vehicle can run at the most frequently used city speed without violating the smooth-running limits. In the case illustrated in Fig. 3.4 that speed is assumed to be 50 km/h.

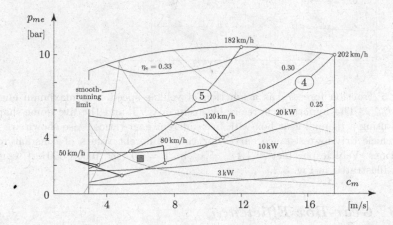

Fig. 3.4 Engine map 3.2 combined with two vehicle resistance curves (fourth and fifth gears on horizontal road) for the vehicle specified in Sect. 3.1.2. The grey square indicates the MVEG–95 average operating point.

Of course, there are many other criteria that must be observed when choosing the gear ratios. In all cases, the *gear spread*, i.e., the ratio of two neighboring gear ratios, must remain within certain boundaries. A geometric law is often chosen[6]

$$\gamma_k = \kappa \cdot \gamma_{k-1}, \quad k = 2, \dots, k_{max}, \quad \kappa \approx \frac{2}{3}. \tag{3.12}$$

Moreover, when using gear boxes that have discrete values of gear ratios, *gear-box gaps* cannot be avoided. As illustrated by the shaded areas in Fig. 3.5, such gear boxes cannot exploit the full traction potential of the engine. The condition (3.12) ensures that these regions do not become too large.

An example of a model-based numeric optimization of the gear ratios is shown in Appendix I in Case Study A.1.

---

[6] Of course, only rational gear ratios can be realized in practice.

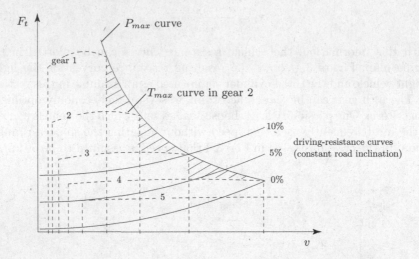

Fig. 3.5 Traction force F_t as a function of vehicle speed v at maximum engine power P_{max}. The hyperbola is realizable using a CVT and the five dome-shaped regions using a manual gear box with five different gear ratios. Also shown are the total vehicle driving-resistance curves for three different values of constant road inclinations. Vehicle parameters $\{A_f \cdot c_d, c_r, m_v\} = \{0.4\,\mathrm{m}^2, 0.008, 850\,\mathrm{kg}\}$; engine map as illustrated in Fig. 3.2.

3.2.3 Gear-Box Efficiency

The main dynamic effects caused by gear boxes have been discussed in Sect. 2.1.1. However, the result summarized in (2.11) is only valid if the efficiency η_{gb} of the gear box is 100%. Of course, this is not realistic. The losses caused by gear boxes and similar powertrain components depend on many influencing factors: speed, load, temperature, etc., just to name the most important ones. Figure 3.6 displays the structure of the system analyzed in this section and illustrates the variables that will be important in this analysis.

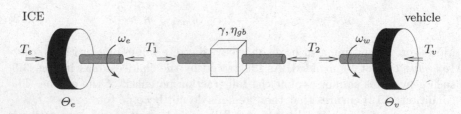

Fig. 3.6 Illustration of the definitions relevant for the modeling of the gear-box efficiency

An approximation of the losses in gear boxes can be formulated using an affine dependency between the gear box input and output power

$$T_2 \cdot \omega_w = e_{gb} \cdot T_1 \cdot \omega_e - P_{0,gb}(\omega_e), \quad T_1 \cdot \omega_e > 0, \tag{3.13}$$

where $P_{0,gb}$ is the power that the gear box needs to idle at an engine speed ω_e. Equation (3.13) is valid when the vehicle is in traction mode. If $T_1 \cdot \omega_e < 0$ a similar equation can be formulated to describe the losses in the gear box that affect the fuel cut-off torque

$$T_1 \cdot \omega_e = e_{gb} \cdot T_2 \cdot \omega_w - P_{1,gb}(\omega_e), \quad T_1 \cdot \omega_e < 0. \tag{3.14}$$

For automotive cog-wheel gear boxes, typical values for e_{gb} are between 0.95 and 0.97. Depending on the size of the gear box and on its lubrication system, the idling losses $P_{0,gb}$ can reach up to 3% of the rated power of the gear box.

The evaluation of the efficiency of CVTs is discussed in Chap. 5, where hybrid-inertial vehicles are treated.

3.2.4 Losses in Friction Clutches and Torque Converters

During those phases in which the vehicle and the engine speed are not matched, the powertrain has to be kinematically decoupled. For that purpose either friction clutches or hydrodynamic torque converters are used in practice. Figure 3.7 illustrates the structure of the powertrain and the corresponding main system variables.

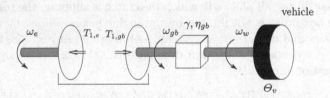

Fig. 3.7 Illustration of the definitions relevant for the modeling of the clutch and torque converter efficiency

Friction Clutches

Dry- or wet-friction clutches have no torque amplification capability, i.e., their input and output torques are identical

$$T_{1,e}(t) = T_{1,gb}(t) = T_1(t) \quad \forall \, t. \tag{3.15}$$

Friction clutches produce substantial losses only during the first acceleration phase when the vehicle starts at zero velocity. If the engine speed ω_e is assumed to be constant during this start phase, the clutch dissipates the following amount of mechanical energy

$$E_c = \frac{1}{2} \cdot \Theta_v \cdot \omega_{w,0}^2, \tag{3.16}$$

where $\omega_{w,0}$ is that wheel velocity at which the clutch input speed ω_e and the output speed ω_{gb} coincide for the first time. The inertia Θ_v includes the vehicle inertia and all inertias due to the rotating parts located after the clutch. The amount of energy dissipated does not depend on the clutch torque profile during the clutch-closing process.

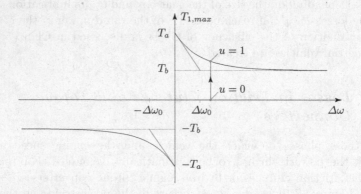

Fig. 3.8 Approximation of the maximum torque of a friction clutch

Note that during all phases in which the clutch is slipping, the torque $T_1(t)$ at the gear box input is not limited by the engine but is defined by the clutch characteristics and its actuation system. As illustrated in Fig. 3.8, the clutch torque $T_1(t)$ depends on the speed difference $\Delta\omega(t) = \omega_{1,e}(t) - \omega_{1,gb}(t)$ and on the actuation input $u(t)$

$$T_1(t) = T_{1,max}(\Delta\omega(t)) \cdot u(t), \quad 0 \le u(t) \le 1. \tag{3.17}$$

The maximum clutch torque can be approximated by

$$T_{1,max}(t) = \text{sign}(\Delta\omega(t)) \cdot \left[T_b - (T_b - T_a) \cdot e^{-|\Delta\omega(t)|/\Delta\omega_0} \right] \tag{3.18}$$

The parameters $\Delta\omega_0$, T_a and T_b must be determined experimentally. In general, they depend on the temperature and wear of the clutch.

Torque Converters

Most automatic transmissions consist of an automated cog-wheel gear system and a hydraulic torque converter. The latter device produces additional losses in those operating phases in which it is not locked up. The losses incurred in these phases can be modeled as shown below.

The torque at the input of the converter may be modeled as follows

$$T_{1,e}(t) = \xi(\phi(t)) \cdot \rho_h \cdot d_p^5 \cdot \omega_e^2(t). \tag{3.19}$$

The converter input speed $\omega_e(t)$ (the "pump speed") has a strong influence on the converter input torque, but the speed ratio

$$\phi(t) = \frac{\omega_{gb}(t)}{\omega_e(t)} \tag{3.20}$$

is important as well. The parameters ρ_h and d_p stand for the density of the converter fluid and for the pump diameter, respectively. The function $\xi(\phi)$ must be determined using experiments. Its qualitative form is illustrated in Fig. 3.9.

The converter output torque $T_{1,gb}$, which in this case may be larger than the input torque, is determined by the pump–turbine interaction

$$T_{1,gb} = \psi(\phi(t)) \cdot T_{1,e}(t). \tag{3.21}$$

The function $\psi(\phi)$ must be experimentally determined as well. Qualitatively, it will have a form similar to the one shown in Fig. 3.9. Equations (3.19) and (3.21) are valid in steady-state conditions. However, since the fluid dynamic processes inside the torque converter are substantially faster than the typical time constants of the vehicle longitudinal dynamics, the fluid dynamic effects may often be neglected.

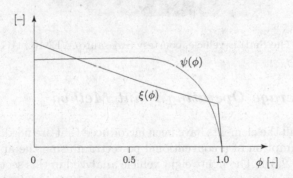

Fig. 3.9 Qualitative representation of the main parameters of a torque converter

With these preparations, the efficiency of the torque converter in traction mode is easily found to be

$$\eta_{tc} = \frac{\omega_{gb} \cdot T_{1,gb}}{\omega_e \cdot T_{1,e}} = \psi(\phi) \cdot \phi. \tag{3.22}$$

3.3 Fuel Consumption of IC Engine Powertrains

3.3.1 Introduction

In the following two sections the two methods introduced in ,the last chapter in Sects. 2.3.1 and 2.3.2 are used to predict the fuel consumption of an experimental light-weight vehicle. The powertrain consists of a downsized high-speed SI engine and a manual cog-wheel gear box. The vehicle is a standard small four-seat passenger car with improved aerodynamics and rolling friction parameters.

The data of this vehicle are similar to those of the experimental SmILE prototype that was developed and realized by Swissauto/WENKO AG, Burgdorf Switzerland.[7] A picture of that vehicle is shown in Fig. 3.10. A detailed description of that project and its results can be found in [131].

Fig. 3.10 The SmILE vehicle, courtesy Swissauto/WENKO AG, Burgdorf

3.3.2 Average Operating Point Method

At this point all the elements have been introduced that are needed to estimate the fuel consumption of a conventional powertrain using the approach introduced in Sect. 2.3.1. The lightweight vehicle analyzed in this section is characterized by its main parameters $\{A_f \cdot c_d, c_r, m_v\} = \{0{,}4\,\mathrm{m}^2, 0.008, 750\,\mathrm{kg}\}$. The engine analyzed below is the one whose main characteristics are illustrated in Fig. 3.2.

Using (2.28) and (2.29), and adding a payload mass of 100 kg to the vehicle mass, the mean traction force at the wheel $\bar{F}_{trac}$ necessary to drive the

---

[7] The engine developed in that project was a downsized supercharged SI engine. That approach proved to be very effective to improve the part-load efficiency of a stoichiometrically operated SI engine.

MVEG–95 cycle is found to be approximately 210 N, which yields an average traction power at the wheels of approximately

$$\bar{P}_{trac} = \frac{\bar{F}_{trac} \cdot \bar{v}}{trac} = \frac{210\,\mathrm{N} \cdot 9.5\,\mathrm{m/s}}{0.6} \approx 3.3\,\mathrm{kW}. \tag{3.23}$$

The parameter $trac$ denotes the time fraction in which the vehicle is in traction mode (see Fig. 2.6), and the mean power $\bar{P}_{trac}$ is the relevant information needed to compute the engine load.[8]

The powertrain is assumed to include a conventional cog-wheel gear box and friction clutch. Using (3.13) the power at the input of the gear box can be estimated to be

$$P_1 = \frac{1}{e_{gb}} \cdot \left(\bar{P}_{trac} + P_{0,gb} \right) = \frac{1}{0.97} \cdot (3.3\,\mathrm{kW} + 0.3\,\mathrm{kW}) \approx 3.7\,\mathrm{kW}. \tag{3.24}$$

The auxiliaries, including the electric power generator, and the friction clutch consume some of the power produced by the engine. In this approach the losses caused by the auxiliaries are taken into account by an additional average mechanical power $\bar{P}_{aux}$ of 0.25 kW. This value is rather low, i.e., the vehicle is assumed to have no power steering and no air conditioning.

According to (3.16), each start causes an energy loss of

$$E_c = \frac{1}{2} \cdot \Theta_v \cdot \omega_{w,0}^2 = \frac{1}{2} \cdot m_v \cdot v_0^2 = \frac{1}{2} \cdot 850\,\mathrm{kg} \cdot (3\,\mathrm{m/s})^2 \approx 3.8\,\mathrm{kJ}. \tag{3.25}$$

In the MVEG–95 on average one start from rest occurs every kilometer. Since in this cycle the average velocity is 9.5 m/s, such an event takes place every 105 s. Accordingly, the average power consumed by the starts is around $3.8\,\mathrm{kJ}/105\,\mathrm{s} \approx 35\,\mathrm{W}$.

In summary, during the traction phases the engine has to produce the average power $\bar{P}_e = (3.7 + 0.25 + 0.035)\,\mathrm{kW} \approx 4\,\mathrm{kW}$. Therefore, assuming a mean piston speed of $\bar{c}_m = 6\,\mathrm{m/s}$,[9] the engine is operated with an average mean effective pressure $\bar{p}_{me}$ of approximately 2.5 bar. This value is obtained by inserting the engine parameters into (3.5).

The efficiency of the engine at that operating point is found using (3.6), whereas the numerical values for $e(c_m)$ and $p_{me0}(c_m)$ are taken from Fig. 3.2

$$\eta_e = \frac{p_{me}}{p_{mf}} = \frac{e(c_m) \cdot p_{me}}{p_{me} + p_{me0}(c_m)} \approx \frac{0.4 \cdot 2.5\,\mathrm{bar}}{2.5\,\mathrm{bar} + 1.6\,\mathrm{bar}} \approx 0.24. \tag{3.26}$$

Therefore, the average fuel power consumed by the engine in the MVEG–95 cycle is approximately

$$\bar{P}_f = trac \cdot \bar{P}_e/\eta_e = 0.6 \cdot 4\,\mathrm{kW}/0.24 \approx 10\,\mathrm{kW}. \tag{3.27}$$

---

[8] To obtain the correct average fuel consumption, the factor $1/trac$ will be compensated later in (3.27).

[9] For the engine specified this corresponds to approximately 2700 rpm.

This corresponds to a fuel flow of

$$\overset{*}{V}_f = \bar{P}_f / (H_l \cdot \rho_f). \tag{3.28}$$

Assuming standard RON–95 gasoline and inserting the corresponding numerical values of H_l=43.5 MJ/kg for the fuel's lower heating value and ρ_f=0.75 kg/l for its density yields a fuel consumption of approximately 0.31 ml/s or, with the value of 9.5 m/s for the average speed in the MVEG–95 cycle, of approximately 3.3 l/100 km.

So far it has been assumed that in all braking *and* idling phases the engine is shut down. While it is easy to cut off fuel in the braking phases, automatic starters that avoid idling losses are more expensive and, thus, most engines have non-zero idling losses.

The idling fuel mean pressure can be estimated from (3.6) by setting $p_{me} = 0$

$$p_{mf,0} = p_{me0}(\omega_{e,idle})/e(\omega_{e,idle}) \tag{3.29}$$

from which the fuel flow follows to be

$$\overset{*}{V}_{f,idle} = p_{mf,0} \cdot \frac{V_d}{H_l \cdot \rho_f} \cdot \frac{c_{m,idle}}{N \cdot S}. \tag{3.30}$$

Choosing $c_{m,idle}$=2.5 m/s as the idling mean piston speed[10] and assuming a four-stroke engine yields the following numerical values

$$\overset{*}{V}_{f,idle} = 4 \cdot 10^5 \,\mathrm{Pa} \cdot \frac{710 \cdot 10^{-6} \,\mathrm{m}^3}{43.5 \cdot 10^6 \,\mathrm{J/kg} \cdot 0.75 \,\mathrm{kg/l}} \cdot \frac{2.5 \,\mathrm{m/s}}{4 \cdot 0.067 \,\mathrm{m}} \approx 8.3 \cdot 10^{-5} \,\mathrm{l/s}. \tag{3.31}$$

In the MVEG–95 cycle the engine is idling for approximately 300 s. Accordingly, in that cycle the fuel spent for idling is approximately

$$300 \,\mathrm{s} \cdot 8.3 \cdot 10^{-5} \,\mathrm{l/s} \cdot 100/11.4 \,\mathrm{km} \approx 0.2 \,\mathrm{l/100 \,km}. \tag{3.32}$$

This figure has to be added to the 3.3 l/100 km used for vehicle propulsion. The sum of 3.5 l/100 km is the estimated total fuel consumption for the chosen example of a lightweight vehicle and downsized engine system. Cold-start losses and other detrimental effects not considered so far are likely to increase that figure somewhat.

3.3.3 Quasistatic Method

In this section a quasistatic approach is used to predict the fuel consumption of the vehicle and powertrain described in the previous section. The QSS toolbox serves as the computational platform for the powertrain modeling and simulation. Figure 3.11 shows the top layer of the resulting model description. Readers familiar with Matlab/Simulink will immediately recognize the characteristic elements of that software tool.

---

[10] For the engine chosen in this example this corresponds to approximately 1100 rpm. This is a reasonable value for such a small engine.

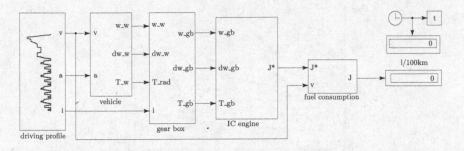

Fig. 3.11 Top layer of the QSS model of the powertrain analyzed in this section

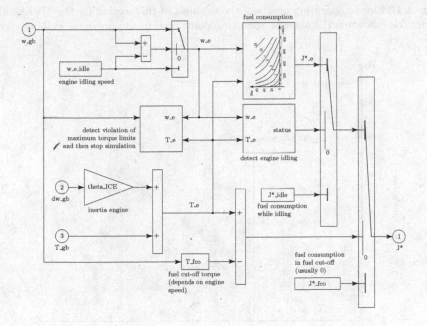

Fig. 3.12 Structure of the block "IC engine" of Fig. 3.11

As an example, the contents of the block "IC engine" are shown in Fig. 3.12. The complete model, including all necessary system parameters, is part of the QSS toolbox package.

Simulating the behavior of this powertrain yields a total fuel consumption of 3.6 l/100 km in the MVEG–95 cycle. This value correlates well with the value of 3.5 l/100 km obtained in the last section.

In fact, as Fig. 3.13 shows, the engine is operated with many different load/speed combinations. Such a variability offers many opportunities for energy optimization, particularly if more than one mechanical power source and energy storage device are available. Accordingly, for hybrid vehicles the

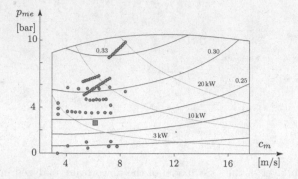

Fig. 3.13 Engine operating points of the example of this section for the MVEG–95 cycle. Also shown is the average operating point used in the previous section.

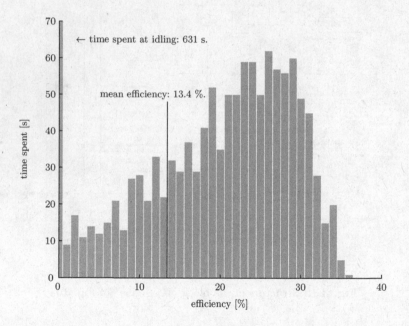

Fig. 3.14 A histogram showing the time spent at engine setpoints with a given efficiency during the WLTP test-cycle for a full-size vehicle

average-point method delivers only limited results. In these cases, reliable fuel consumption estimations are achieved only when using quasistatic simulations that include the supervisory control loops. Figure 3.13 must be analyzed with some care. In fact, the distribution of the load/speed points must be complemented by information on the frequency of these points. A representation similar to the one shown in Fig. 3.14 helps to understand which engine operation points are relevant.

3.3.4 Measures to Improve the Fuel Economy of IC-Engine Powertrains

Typically, in IC engine based vehicles, the engine is sized according to some relatively demanding peak power requirement, e.g., acceleration performance. As shown in Table 2.1, this results in a large discrepancy between peak and average power in regular driving conditions. Since IC engines have a very poor efficiency at low load (low torque), the average efficiency of conventional vehicles is very low. In this section we will present some solutions to this *part-load problem.*

- The first concept is *downsizing* the combustion engine, i.e., reducing its size. This helps to increase the relative load during driving such that the operating points will naturally come to lie in an operating region with a higher efficiency.
- The second concept is *recuperation*. Conventional vehicles have no means of recuperating energy and therefore waste a lot of energy. It has been proposed to equip such vehicles with a recuperation device, so as to *hybridize* them.
- Hybridization not only has the benefit that braking energy can be recuperated, but it also offers the possibility of *operating point shifting* – a third and powerful concept that helps to relax the part-load problem and further reduces the fuel consumption. Note that the *start-stop operation* of the prime mover, which helps to reduce idling losses during standstill, will here be considered to be part of the operating point shifting mode – simply by considering "off" as an additional operating point.

Obviously, decreasing the weight of vehicles and reducing aerodynamic drag and rolling friction will help to further cut fuel consumption values. These measures have already been discussed in Chap. 2. Furthermore, note that even though often mentioned in the context of fuel economy of hybrids, an *electric-only* driving mode is not a method to reduce fuel consumption. In fact, the temporary shut-down of the combustion engine during driving might come at the expense of increased fuel consumption[11].

With the tools derived in the last section the energy saving potential ΔE of any of the above mentioned methods with respect to a reference vehicle can be estimated using (2.43)

---

[11] The energy that was drawn from the battery has to be replenished after the zero-emission zone – likely at a higher current than would have been incurred by driving in hybrid mode. Therefore, more ohmic losses are incurred and overall more energy is needed to finish the cycle.

$$\Delta E = \frac{\bar{E}_f^{ref} - \bar{E}_f}{\bar{E}_f^{ref}}$$

$$= 1 - \frac{\bar{E}_f(\eta_{ft}, \eta_{rec})}{\bar{E}_f(\eta_{ft}^{ref}, \eta_{rec} = 0)}$$

$$= 1 - \frac{\eta_{ft}^{ref}}{\eta_{ft}} \cdot \frac{1 + (1 - \eta_{rec})\frac{\bar{E}_{circ}}{\bar{E}_{diss}}}{1 + \frac{\bar{E}_{circ}}{\bar{E}_{diss}}}, \tag{3.33}$$

where η_{ft}^{ref} is the average fuel-to-traction efficiency of the reference case used for comparison. Here we will use a conventional combustion engine based powertrain without a recuperation device as a reference case to benchmark the improvements that can be achieved by the methods mentioned above.

Downsizing

Assuming a reference vehicle without a recuperation device and with an average fuel-to-traction efficiency of 20%, (3.33) yields

$$\Delta E = 1 - \frac{\eta_{ft}^{ref}}{\eta_{ft}} = 1 - \frac{\eta_{ft}^{ref}}{\eta_{ft}^{ref} + \Delta\eta_{ft}}. \tag{3.34}$$

For an improvement of the fuel-to-traction efficiency of $\Delta\eta_{ft} = 1\%$ this equation predicts a fuel consumption reduction of almost 5%. This illustrates the strength of the concept of downsizing which, due to its simplicity, is often underestimated. Note that consumer acceptance is the main limiting factor of a market-wide adoption of drastic downsizing, even though it is among the most cost-effective methods to reduce fuel consumption.

Recuperation

As pointed out in Sect. 2.2.2, the values of $\bar{E}_{diss}$ and $\bar{E}_{circ}$ have been calculated by simulation for a full-size vehicle on the MVEG–95 and other driving cycles [240]. For mixed driving conditions the fraction of circulated energy and dissipation energy is $\frac{\bar{E}_{circ}}{\bar{E}_{diss}} \approx 15\%$. For urban driving this fraction increased to about 30%.

With a realistic recuperation efficiency of 60%, however, without operating point shifting, i.e., $\eta_{ft} = \eta_{ft}^{ref}$, (3.33) predicts a fuel consumption reduction of about 14% in urban driving and about 8% overall. These rather low numbers illustrate the fact that a recuperation device, which is usually quite expensive, should always be used for both recuperation and operating point shifting. Otherwise, the resulting fuel economy improvement is rather limited with respect to the investment required.

Operating Point Shifting

When operating point shifting is exploited together with recuperation, a much higher reduction of fuel consumption can be achieved. When assuming $\eta_{ft}^{ref} = \{0.2, 0.25\}$ (Diesel engine vehicle in urban driving and overall, respectively), together with $\eta_{ft}^{hyb} = 35\%$, (3.33) predicts a fuel consumption reduction potential of about 52% in urban driving and about 35% overall. These numbers impressively demonstrate the strength of the concept of operating point shifting. Note that operating point shifting is only useful if the prime mover can be shut off. Otherwise there will be many phases during which the combustion engine is in idling mode. Therefore the above results already include the intermittent operation of the combustion engine.

3.4 Problems

Gear-Box Models

Problem 3.1. Improve (3.9) in order to take into account the engine inertia and the transmission efficiency. *(Solution: A quadratic equation in γ_1 is obtained).*

Problem 3.2. Dimension the first gear of an ICE-based powertrain not based on a given drivability as in (3.9) but in order to obtain a given acceleration at vehicle take-off. Do the calculations according to Problem 3.1, using the following data: $m_v = 1100\,\text{kg}$, payload $m_p = 100\,\text{kg}$, equivalent mass of the wheels $m_{r,w} = 1/30$ of m_v, $c_r = 0.009$, transmission efficiency $\eta_t = 0.9$, $T_e = 142\,\text{Nm}$ at $\omega_{takeoff}$, desired acceleration $a = 4\,\text{m/s}^2$, engine inertia $\Theta_e = 0.128\,\text{kgm}^2$, $r_w = 30\,\text{cm}$. Compare with the approximate solution of (3.9). *(Solution: $\gamma_1 = 14.4$. Approximate solution: $\gamma_1 = 11.6$).*

Problem 3.3. Dimension the fifth gear in a six-gear transmission for maximum power using (3.10) for a vehicle with the following characteristics: curb $m_v = 1100\,\text{kg}$, performance mass $= 100\,\text{kg}$, $c_r = 0.009$, $A_f \cdot c_d = 0.65\,\text{m}^2$, $r_w = 30\,\text{cm}$, transmission efficiency $= 0.9$, $P_{e,max} = 80.4\,\text{kW}$ at 6032 rpm. *(Solution: $\gamma_5 = 3.4$).*

Problem 3.4. Consider again the system of Problems 3.2–3.3. Calculate the vehicle speed values, v_j at which the engine is at its maximum speed, for all the gears j. *(Solution: $v_1 = 47.5\,km/h$, $v_2 = 68\,km/h$ (geometric law) or $85.6\,km/h$ (arithmetic law), $v_3 = 97.2\,km/h$ (geometric) or $123.6\,km/h$ (arithmetic), $v_4 = 139.7\,km/h$ (geometric) or 161.3 (arithmetic), $v_4 = 199.4\,km/h$).*

Problem 3.5. Calculate the approximate efficiency of a clutch during a vehicle takeoff maneuver. *(Solution: efficiency $= 0.5$).*

Fuel Consumption of IC Engine Powertrains

Problem 3.6. Find the CO_2 emission factor (g/km) as a function of the fuel consumption rate (l/100 km) for gasoline and diesel fuels. Use these average fuels (gasoline, diesel) data: density $\rho = \{0.745, 0.832\}$ kg/l, carbon dioxide to fuel mass fraction $m = \{3.17, 3.16\}$. *(Solution: $\frac{g\,CO_2/km}{l/100\,km} = 26.3$ (gasoline), 23.6 (diesel)).*

Problem 3.7. Calculate the fuel consumption and the CO_2 emission rate for the MVEG–95 cycle for a vehicle having the following characteristics: $m_v = 1100$ kg, payload $= 100$ kg, $c_d \cdot A_f = 0.7$, $c_r = 0.013$, $e_{gb} = 0.98$, $P_{0,gb} = 3\%$, $P_{aux} = 250$ W, $v_{launch} = 3$ m/s, $e = 0.4$, $P_{e,0} = 1.26$ kW, $P_{e,max} = 66$ kW, diesel fuel ($H_f = 43.1$ MJ/kg, $\rho_f = 832$ g/l), idle consumption $\overset{*}{V}_{f,idle} = 150$ g/h. The declared CO_2 emission rate for this car is 99 g/km. *(Solution: $\overset{*}{V}_f = 3.6\,l/100\,km$, $\overset{*}{m}_{CO_2} = 95\,g/km$).*

Problem 3.8. Evaluate in a first approximation the contribution of stop-and-start, regenerative braking ($m_{rec} = 20\% \cdot m_v$, $\eta_{rec} = 0.5$) and optimization of power flows in reducing the fuel consumption when the system of Problem 3.7 is hybridized. Assume an 80% charge-discharge efficiency for the reversible storage system. *(Solution: Stop-and-start $\approx 3\%$, regenerative braking $\approx 3\%$, ideal optimization of power flows $\approx 47\%$).*

4

Electric and Hybrid-Electric Propulsion Systems

While in conventional ICE-based vehicles the energy carrier is a fossil fuel, electric and hybrid-electric propulsion systems are characterized by the presence of an electrochemical or electrostatic energy storage system. Moreover, at least one electric motor is responsible — totally or partially — for the vehicle propulsion.

In this chapter, first purely electric vehicles will be discussed briefly. Then various types of hybrid-electric vehicles will be introduced. The subsequent sections describe the quasi-stationary and the dynamic models of typical electric components of such vehicles, including electric motors/generators, electrochemical batteries, and supercapacitors. The modeling representations of an electric power bus, a torque coupler, and a planetary gear set are added as separate sections due to the importance of the mentioned components in most hybrid-electric powertrains. A mean-value analysis of the energy consumption of various powertrain configurations concludes the chapter.

4.1 Electric Propulsion Systems

Purely electric propulsion systems (*electric vehicles*, EVs, or *battery-electric vehicles*, BEVs) are characterized by an electric energy conversion chain upstream of the drive train, roughly consisting of a battery (or another electricity storage system) and an electric motor with its controller. The resulting vehicle is not autonomous (see the definition in the Introduction), since the energy density of batteries does not permit sufficient driving autonomy. Moreover, the time required for recharging is usually not negligible, and surely it is larger than the typical refueling time of ICE-based vehicles. For these reasons, only some brief considerations on the energy performance of such vehicles will be presented here.

4.1.1 Concepts Realized

After early attempts in the last decades of the 20th century, an interest in battery-electric vehicles was renewed in the early years of the 21st century. Key factors for such a revival are partially based on market pull, e.g., the increase of oil prices, and partially based on by technology push, e.g., the development of new technologies of lithium-ion batteries. Early battery-electric vehicles were in fact characterized by asynchronous AC or DC electric machines, with batteries usually of the nickel–metal hydride or lead–acid type exhibiting an energy density between 30 and 60 Wh/kg [147]. Recently introduced passenger cars mostly use permanent-magnet synchronous machines as traction motors, with a peak power ranging from 50 to 80 kW. Lithium-ion battery energy typically ranges from 15 to 25 kWh. For combined driving cycles, the top speed and range of such cars now comfortably reach 135 km/h and 160 km (100 miles), respectively.

Electric energy consumption is typically measured in kWh/100 km or Wh/km. This figure of merit depends on the efficiency of the powertrain components, mainly of the electric machine, the battery, and the transmission. Moreover, energy consumption is affected by energy recuperation as introduced in Sect. 2.2.2, that is, the availability of kinetic energy that is not dissipated during braking but rather converted into electricity and stored in the battery. To be fully exploited, such a concept ("regenerative braking") requires the complete controllability of the braking circuit, similarly to the anti-lock braking systems used in conventional cars. Typical values of the energy consumption of today's electric cars are 10–15 kWh/100 km for combined driving cycles.

Among the concepts realized and available on the market, notable examples are given by Mitsubishi [221], Renault Z.E. platform [270], and Nissan [230]. However, current limitations in recharge infrastructures and the required changes in the habits of drivers strongly delay the diffusion of such concepts. On the other hand, the arguments listed in Sects. 1.3 and 1.4 seem to suggest that the most suitable application of the EV concept is in small cars for use in urban surroundings, especially within car-sharing organizations. In summary, whether the technology of battery-electric vehicles will have a future in the coming years or not, and under which conditions, is still a matter of debate. This debate will be strongly influenced by the progress of battery technology and availability of "green" electric energy.

4.1.2 Modeling of Electric Vehicles

Modeling of battery-electric vehicles can be performed with the methods and tools introduced in Sect. 2.3 that will be detailed in Sect. 4.2. While in ICE-based powertrains the power flows are mechanical and thermal, in EVs electric power also plays a fundamental role. The constituents ("power factors") of electric power are voltage and current. Since the former can only have positive values, it plays a role analogous to that of speed in mechanical power. In contrast, current can switch direction to represent both power flows *from* (positive current) or *to* the storage system (negative current), the latter operation occurring, e.g., during regenerative braking. Thus current is the electric analogue of torque or force. The causality of power factors of an EV is illustrated in Fig. 4.1 for both quasistatic and dynamic modeling approaches.

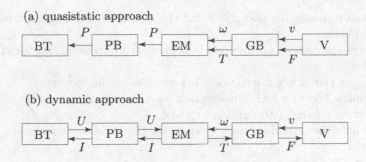

Fig. 4.1 Flow of power factors for a battery-electric powertrain, with the quasistatic approach (a) and the dynamic approach (b). Same block nomenclature as in Figs. 2.20 – 2.21 (PB: power link), F: force, I: current, P: power, T: torque, v: speed, U: voltage, ω: rotational speed.

4.2 Hybrid-Electric Propulsion Systems

4.2.1 Introduction

In contrast to ICE-based and battery-electric vehicles, hybrid vehicles are characterized by two or more prime movers and power sources. Usually, the term "hybrid vehicle" is used for a vehicle combining an engine and an electric motor. More appropriately, such a combination should be called a hybrid-electric vehicle (HEV), since other, different "hybrid" configurations have been proposed (see Chap. 5 for mechanical, pneumatic, and hydraulic hybrids and Chap. 6 for fuel cell vehicles).

In general, an HEV includes an engine (see Chap. 3) as a fuel converter or irreversible prime mover (fuel cells are treated in Chap. 6; other solutions are not considered here[1]). As electric prime movers, various types of motors are used, such as standard DC, induction AC, brushless DC, etc. In some configurations, a second electric machine is required, which acts primarily as a generator. The electric energy storage system is usually an electrochemical battery,

---

[1] Gas turbines and Stirling engines have been proposed in series hybrid configurations, where they are operated in their preferred steady-state conditions, see Sect. 4.4.

though supercapacitors may be used in some prototypes. Sections 4.3 – 4.6 describe motor, battery, and supercapacitor models, respectively.

One of the main motivations for developing HEVs is the possibility to combine the advantages of the purely electric vehicles, in particular zero local emissions, with the advantages of the ICE-based vehicles, namely high energy and power density. HEVs can profit from various possibilities for improving the fuel economy with respect to ICE-based vehicles. In principle, it is possible to:

1. downsize the engine and still fulfill the maximum power requirements of the vehicle;
2. recover some energy during deceleration instead of dissipating it in friction braking;
3. optimize the power distribution between the prime movers;
4. eliminate the idle fuel consumption by turning off the engine when no power is required (stop-and-start); and
5. eliminate the clutching losses by engaging the engine only when the speeds match.

These possible improvements are partially counteracted by the fact that HEVs are about 10–30% heavier than ICE-based vehicles.

Generally, not all the possibilities (1) – (5) are used simultaneously. The following section describes the different types of hybrid-electric vehicles developed or proposed and their modes of operation.

4.2.2 System Configurations

Hybrid-electric vehicles are classified into three main types:

- **Parallel hybrid:** Both prime movers operate on the same drive shaft, thus they can power the vehicle individually or simultaneously.
- **Series hybrid:** The electric motor alone drives the vehicle. The electricity can be supplied either by a battery or by an engine-driven generator.
- **Series-parallel, or combined hybrid:** This configuration has both a mechanical and an electrical link.

Additionally, certain new concepts have been introduced that cannot be adequately classified into any of the three basic types. For these concepts, the term "complex" hybrid is sometimes used [55, 144].

Series HEVs

Series hybrid propulsion systems utilize the internal combustion engine as an auxiliary power unit (APU) to extend the driving range of a purely electric vehicle. Using a generator, the engine output is converted into electricity that can either directly feed the motor or charge the battery (Fig. 4.2). Regenerative braking is possible using the traction motor as a generator and storing the electricity in the battery (point 2 in Sect. 4.2). The engine operation is not related to the power requirements of the vehicle (4), thus the engine can be operated at a point with optimal efficiency and emissions (3). An added advantage may be the fact that the transmission does not require a clutch, i.e., the engine is never disengaged since it is mechanically decoupled from the drive axle (5). However, a series hybrid configuration needs three machines: one engine, one electric generator, and one electric traction motor. At least the traction motor[2] has to be sized for the maximum power requirements of the vehicle. Thus a series hybrid in principle offers the possibilities of reducing fuel consumption following the approaches (2) to (5) listed above. The overall tank-to-wheel efficiency for series hybrid vehicles is on a par with the values of vehicles powered by modern, fuel-efficient IC engines.[3] The additional weight due to car body reinforcement, electric machines, battery, etc. may push the fuel consumption above the value of good ICE-based vehicles, however.

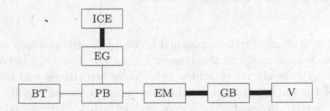

Fig. 4.2 Basic series hybrid configuration. Same block nomenclature as in Figs. 2.20 – 2.21 (EG: electric generator, PB: power link). Bold lines: mechanical link, solid lines: electrical link.

Parallel HEVs

While series hybrid vehicles may be considered as purely electric vehicles with an additional ICE-based energy path, parallel hybrid vehicles are rather ICE-based vehicles with an additional electrical path (Fig. 4.3). In parallel HEVs both the engine and the electric motor can supply the traction power either alone or in combination. This leaves an additional degree of freedom in fulfilling the power requirements of the vehicle, which can be used to optimize the power distribution between the two parallel paths (point 3 in the previous section). Typically, the engine can be turned off at idle (4) and the

---

[2] Possibly, the generator also in configurations with larger APUs.

[3] Although not a topic of this text, it is worth mentioning that series HEVs have extremely low pollutant emissions.

electric motor can be used to assist accelerations and, in general, high power demands. Both machines can therefore be sized for a fraction of the maximum power (1). Together with the fact that only two machines are needed, this is an advantage with respect to series hybrid vehicles. A disadvantage is the need for a clutch, since the engine is mechanically linked to the drive train (5). The electric motor can be utilized as a generator to charge the battery, being fed by regenerative braking (2) or by the engine. Even though the additional weight still plays an important role, all the possibilities (1) to (4) listed above in principle increase the system efficiency as compared to an ICE-based vehicle.

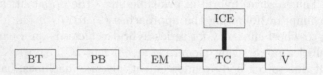

Fig. 4.3 Full parallel hybrid configuration. Same block nomenclature as in Figs. 2.20 – 2.21 (PB: power link, TC: torque coupler, here including GB). Bold lines: mechanical link, solid lines: electrical link.

Parallel hybrids are further classified according to the position of the electric machine with respect to the conventional drive train. In micro hybrids (see Sect. 4.2.4), the electric machine is typically belt-driven and mounted on the front of the engine (Fig. 4.4), thus its speed is always rigidly linked to that of the engine. In *pre-transmission* parallel hybrids the electric machine is mounted between the engine and the gearbox. Again, the two speed levels are linked, thus this configuration is also called *single-shaft*. Depending on the number and the position of the clutches, various functionalities can be achieved or not (see Problem 4.4). In *post-transmission* or *double-shaft* parallel hybrids the electric machine is mounted downstream of the gearbox, thus the two speed levels are decoupled. Finally, in *through-the-road* or *double-drive* parallel hybrids, the engine and the electric machine are mounted on two separate axles, thus the mechanical link between them is only through the road.

Fig. 4.4 Micro parallel hybrid configuration. Same block nomenclature as in Figs. 2.20 – 2.21 (PB: power link). Bold lines: mechanical link, solid lines: electrical link.

Combined HEVs

Somehow intermediate between series and parallel hybrids is the combined hybrid configuration. This is mostly a parallel hybrid, but it contains some features of a series hybrid. Actually, both a mechanical and an electric link are present, together with two distinct electric machines. As in a parallel hybrid configuration, one is used as a prime mover or for regenerative braking. The other machine acts like a generator in a series hybrid system. It is used to charge the battery via the engine or for the stop-and-start operation. Two different realizations of combined hybrids have been introduced. The first, see Fig. 4.5, has a power split device (PSD, Sect. 4.9). The second, see Fig. 4.6, combines two electric machines, both mechanically connected to the engine, coupled at the DC-link level.

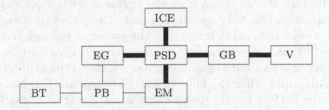

Fig. 4.5 Configuration of a combined hybrid with a planetary gear set. Same block nomenclature as in Figs. 2.20 – 2.21 (PB: power link, EG: electric generator, PSD: power split device). Bold lines: mechanical link, solid lines: electrical link.

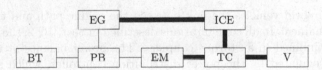

Fig. 4.6 Configuration of a combined hybrid without a planetary gear set. Same block nomenclature as in Figs. 2.20 – 2.21 (PB: power link, EG: electric generator, TC: torque coupler, here including GB). Bold lines: mechanical link, solid lines: electrical link.

4.2.3 Power Flow

Each HEV configuration allows for several operating modes. This section illustrates the power flows among the various components of a hybrid vehicle [55, 144].

Series HEVs

In the standard series hybrid configuration, the link between the engine path and the battery path is electrical, i.e., at the power link level, with the battery output voltage feeding the motor and the generator. The motor and generator currents balance the battery terminal current. The power balance at the power link is regulated by the power split controller, which selects the operating mode and the ratio u between the power from/to the battery and the total power at the link. Section 4.7 describes the modeling of electric power links, whereas control strategies for the energy management are described in detail in Chap. 7.

As illustrated in Fig. 4.7, series hybrid vehicles basically have four modes of operation. In urban driving, when the battery is sufficiently charged, the purely electric, zero-emission (ZEV) driving mode ($u = 1$) is usually selected. When the battery charge is too low, the engine is turned on and typically set to its maximum efficiency operating point. The power resulting from the difference between the engine power and the power at the link recharges the battery ($u < 0$) via the generator. Such a combination of battery discharge and charge represents a *duty-cycle operation*, which is typical of series hybrid vehicles. In principle, when the fuel-optimal engine power is below the power at the link, the battery could provide the missing power ($0 < u < 1$), though this mode of operation is rarely used in practice. Of course, during braking or deceleration some energy is recuperated in the battery by using the motor as a generator ($u = 1$).

Parallel HEVs

In parallel hybrid vehicles the link between the engine path and the electric path is mechanical. In all configurations described in Sect. 4.2.2, the two power flows are combined in a "torque coupler." The power balance at the torque coupler is regulated by the power distribution controller, which selects the operating mode and the ratio u between the power from/to the motor and the total power at the coupler. The modeling of torque couplers is described in Sect. 4.8.

Depending on the value of u, several operating modes are possible. During startup or acceleration, the engine provides only a fraction of the total power at the coupler, the rest is delivered by the motor ($0 < u < 1$). This operating mode is often referred to as *power assist* mode. During braking or deceleration, the motor acts in generator mode and recuperates energy into the battery ($u = 1$). It is also possible to shift the operating point of the engine towards higher efficiency. At light load, the engine would then provide more power than strictly demanded and the extra power would be used to charge the battery via the electric machine ($u < 0$). Both the pure engine operation ($u = 0$) and the purely electric operation ($u = 1$) are also possible in principle. Figure 4.8 illustrates these scenarios for a full parallel hybrid configuration.

(a) battery drive, $u = 1$

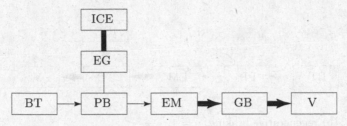

(b) battery recharging, $u < 0$

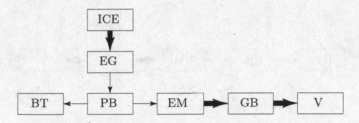

(c) hybrid drive, $0 < u < 1$

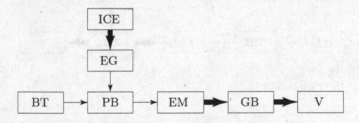

(d) regenerative braking, $u = 1$

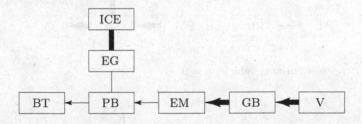

Fig. 4.7 Power flow for the modes of operation of series hybrid vehicles. Same nomenclature as in Fig. 4.2.

(a) power assist, $0 < u < 1$

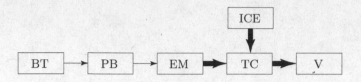

(b) regenerative braking, $u = 1$

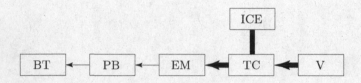

(c) battery recharging, $u < 0$

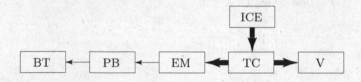

(d) ZEV, $u = 1$

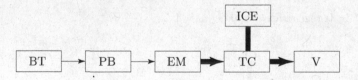

(e) conventional vehicle, $u = 0$

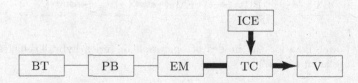

Fig. 4.8 Power flow for the modes of operation of parallel hybrid vehicles. Same nomenclature as in Fig. 4.3.

Combined HEVs

Using a power split device (Fig. 4.5), combined hybrids have the ability to operate as series or as parallel hybrids. Thus the possible modes of operation result from the combination of the modes already discussed in the previous sections. However, the use of a power split device adds some constraints to the possible energy flows. As will become clear in Sect. 4.9, the ICE-only mode is typically associated with a power flow through the generator and the motor. The other modes, comprising ZEV, regenerative braking, battery recharging, and power assist, are of course possible, as illustrated in Fig. 4.9.

4.2.4 Functional Classification

Regardless of their physical configuration, hybrids can be further classified in terms of degree of hybridization, see Fig. 4.10. The simplest (parallel) hybridization is the so-called *micro hybrid* concept, which is essentially an ICE-based powertrain with a small electric motor. Micro hybrids do not require a high battery capacity or complex power electronics, since their main purpose is the automatic engine stop-and-start. The motor may also act as an alternator for the electrical loads, thus it is sometimes referred to as an integrated starter-generator (ISG).

On the other hand, *full hybrids* allow all the modes of operation, including power assist, energy recuperation, and purely electric operation. Therefore, they need higher electric power levels leading to the use of high-voltage systems and complex power electronics. The consequent relevant size and weight of the electric components often poses several integration constraints.

Somehow intermediate are *mild hybrids*, where the size of the electric powertrain is such to allow engine boosting and energy recuperation to some extent, but not purely electric driving, at least not for any moderate or high vehicle speed.[4] Of course, there are no well-defined frontiers between these types of HEVs. A more continuous classification is obtained by defining the degree of hybridization [29] as the ratio between the electric power and the total power

$$DoH = \frac{P_{b,\max}}{P_{b,\max} + P_{e,max}}. \tag{4.1}$$

Generally speaking, the benefits, i.e., reduction of energy consumption, but also the additional cost associated with the hybridization both increase with the degree of hybridization.

Two types of HEVs are characterized by a relatively large energy storage system [219]. These HEVs have a larger capability of ZEV operation, with

---

[4] Historically, the term "mild hybrid" was also used for what are more often called "micro hybrids".

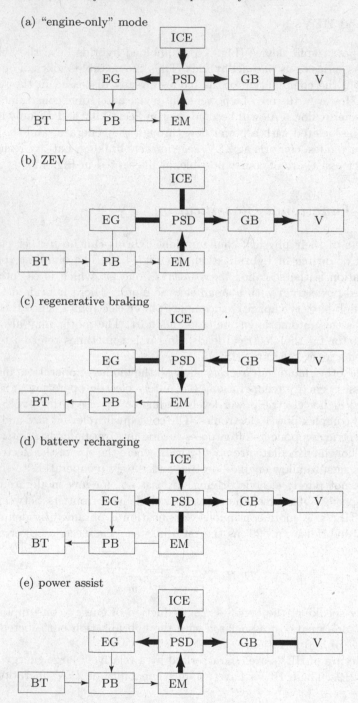

Fig. 4.9 Power flow for the operating modes of combined hybrid vehicles with a planetary gear set. Same nomenclature as in Fig. 4.5.

a purely electric range that is similar to that of battery-electric vehicles. In *plug-in hybrids* the battery can be recharged from the grid, like in BEVs. Therefore, they are not limited to charge-sustaining operation as are non-rechargeable hybrids. During driving, the battery can be discharged until a lower limit for battery charge is attained. The range covered in this so-called charge-depleting operation is the all-electric range (AER). Often these vehicles are designated by PHEV-[*number*], where the number indicates the AER in miles or km. The typical operation of plug-in HEVs comprise a pure charge-depleting phase where the system basically works as a BEV, followed by a charge-sustaining phase. Of course, a continuous blend of these phases is possible, and it has been shown that this blended-mode operation helps reducing the energy consumption [330, 308]

Extended-range electric vehicles can be recharged from the grid like electric vehicles but also, for longer missions, internally by an Auxiliary Power Unit (APU, see Sect. 4.4). In fact they are series hybrids that can be operated in charge-depleting mode until a lower limit for the battery charge is reached. After that, charge-sustaining operation is resumed, typically using the duty-cycle concept of series hybrids.

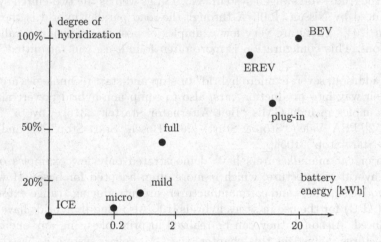

Fig. 4.10 Functional classification of HEVs in terms of degree of hybridization and battery capacity (typical values)

4.2.5 Concepts Realized

In the early years of the 21st century several passenger cars have been demonstrated that use one of the hybrid configurations discussed in this chapter. The models that already have entered mass production or that are considered reasonably ready for the market are parallel or series-parallel hybrids. Usually gasoline engines, permanent-magnet synchronous AC (brushless DC) motor/generators, gearboxes or PSD-based transmissions, and nickel–metal hydride or lithium-ion batteries are used in these hybrid powertrains, though there are a few systems equipped with different components, like a Diesel engine, a lithium-polymer battery, a continuously variable transmission, etc. The degree of hybridization ranges from 15% to 50% for full hybrids or from 5% to 15% for mild hybrids. The battery typically features capacities of between 400 Wh and 1.8 kWh.

The wider family of combined hybrid concepts is probably the Toyota "Hybrid Synergy Drive," which includes a first generation with a simple PSD [283], a slightly modified second generation [324], a third generation with a compound PSD, possibly with all-wheel drive [326], and a two-mode compounded PSD [327]. Other combined hybrid architectures include the Two-mode system from Global Hybrid Cooperation [232] jointly developed and adopted by GM [116], Daimler [73], BMW [35], and formerly by Chrysler [60], as well as the Ford hybrid system [103].

Notable examples of mild, pre-transmission parallel hybrids are Honda's "Integrated Motor Assist" system [140], the single-clutch system shared by Daimler [74] and BMW [36], the similar Hyundai "Blue Hybrid" technology [145], the Audi–Volkswagen system [343, 18], as well as the two-clutch system introduced by Nissan [150]. A through-the-road parallel hybrid is the PSA system [264]. There are very few examples of post-transmission parallel realizations. This configuration is more often found, e.g., in retrofitted buses [96, 75].

In addition, several "micro hybrid" or stop-and-start technologies are finding their way into production cars, also to equip non-hybrid powertrains. A few examples include GM's "Belt Alternator Starter" [316], Toyota "THS-M" [152], PSA–Valeo "Stop & Start" [263], Bosch "Start-Stop" [38] and Ford "Auto Start-Stop" [105].

Major car manufacturers have demonstrated only few examples of the series hybrid architecture, which is more often adopted for buses. However, automotive suppliers and car manufacturers are introducing "range-extender" units (APU) for the use in series hybrids [19]. Also plug-in hybrids have been introduced. Although they can be realized in principle using any one of the architectures defined in this chapter, the GM Voltec system is based on a combined hybrid configuration [117]. In both extended-range and plug-in hybrids the battery capacity is much larger than in charge-sustaining hybrids, depending on the AER target (5–20 kWh).

4.2.6 Modeling of Hybrid Vehicles

The model of a complete HEV can be split into several submodels, each one representing a component of the system. A good and useful modeling practice consists of creating "autonomous" submodels with clearly defined interfaces representing actual power flows. With such a modular approach, it is straightforward to combine submodels to represent even complex configurations. Another advantage is that various arrangements of the HEV components can be modeled using the same basic submodels. In other words, the same "library" of submodels can be used to represent series, parallel, and combined hybrids.

The difference between quasistatic and dynamic modeling approaches was discussed in Chap. 2. The flow of power factors is illustrated in Figs. 4.11 – 4.13 for series, full parallel, and PSD-based combined hybrid configurations, as well as for both modeling approaches. Each block represents a submodel, and each arrow represents the information that is exchanged between the submodels via the interfaces. The power flow modeling of mild hybrids is the same as that of full parallel hybrids, and that of non-PSD combined hybrids is the same as that of PSD-based ones. Of course, what changes is the physical realization of the respective models, in particular the nature and the position of the "torque coupler."

(a) quasistatic approach

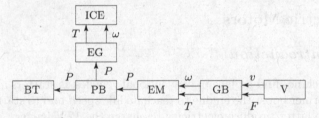

(b) dynamic approach

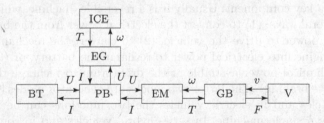

Fig. 4.11 Flow of power factors for a series hybrid configuration, with the quasistatic approach (a) and the dynamic approach (b). Same block nomenclature as in Fig. 4.2, F: force, I: current, P: power, T: torque, v: speed, U: voltage, ω: rotational speed.

(a) quasistatic approach

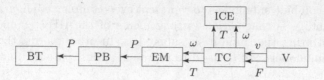

(b) dynamic approach

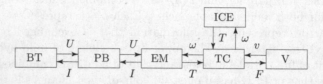

Fig. 4.12 Flow of power factors for a parallel hybrid configuration, with the quasistatic approach (a) and the dynamic approach (b). Same block nomenclature as in Fig. 4.3, F: force, I: current, P: power, T: torque, v: speed, U: voltage, ω: rotational speed.

4.3   Electric Motors

4.3.1   Introduction

Electric machines find a place in conventional vehicles as starters and alternators. The former boost the engine to reach its idle speed and to start delivering torque. The latter produce electricity to charge the 12 V battery and to feed the electric auxiliary loads. In electric and hybrid-electric vehicles, the electric machine is a key component. Usually it is a reversible machine, which can operate in several ways: (1) to convert the electrical power from the battery into mechanical power to drive the vehicle, (2) to convert the mechanical power from the engine into electrical power to recharge the battery, or (3) to recuperate mechanical power available at the drive train to recharge the battery (regenerative braking). The latter two modes are generator modes. In parallel hybrid vehicles and in electric vehicles the two functions can be fulfilled in principle by a single machine. In series hybrid vehicles and in combined hybrid vehicles two different machines are needed, the generator usually being a second reversible machine, smaller than the traction motor.

Characteristics of a good HEV motor generally include high efficiency, low cost, high specific power, good controllability, fault tolerance, low noise, and uniformity of operation (low torque fluctuations). Some performance and cost targets for typical hybrid applications are listed in Table 4.1.

(a) quasistatic approach

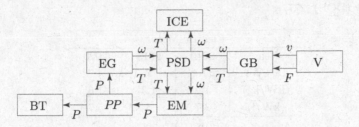

(b) dynamic approach

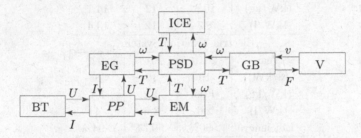

Fig. 4.13 Flow of power factors for a combined hybrid configuration, with the quasistatic approach (a) and the dynamic approach (b). Same block nomenclature as in Fig. 4.5, F: force, I: current, P: power, T: torque, v: speed, U: voltage, ω: rotational speed.

Types of Electric Motors

Electric motors basically can be organized in two main categories: direct-current (DC) motors and alternating-current (AC) motors (see Figs. 4.14 – 4.15). All types of motors have a stator and a rotor. The latter is connected to the output shaft on which the motor torque is acting. The electricity provided by the DC supply through the motor controller is applied at the motor terminals. Electromechanical energy conversion takes place as a consequence of Faraday's law and of Lorentz' law. The former describes the induction of an electromotive force (emf) in conductors being in relative motion with respect to a magnetic field. The latter describes the force generated on a current-carrying conductor inside a magnetic field.

In DC motors, the rotor surface hosts a number of conductors (rotor windings), which terminate with a collector. As a consequence of the application of DC voltage to the rotor windings by means of carbon brushes, which are in contact with the collector, a magnetic field is generated whose polarity is continuously changed by mechanical contact commutation. At the same time, a stationary magnetic field is generated in the stator using permanent magnets

Table 4.1 Technology targets for electric motors, motor controllers (power electronics) and their assembly (electric traction systems), as defined by the US Department of Energy (DOE) [277]

Electric Motor			
Year	2010	2015	2020
($/kW)	11.1	7	4.7
(kW/kg)	1.2	1.3	1.6
(kW/l)	3.7	5	5.7
Motor controller			
Year	2010	2015	2020
($/kW)	7.9	5	3.3
(kW/kg)	10.8	12	14.1
(kW/l)	8.7	12	13.4
Electric Traction System			
Year	2010	2015	2020
($/kW)	19	12	8
(kW/kg)	1.06	1.2	1.4
(kW/l)	2.6	3.6	4
Efficiency	>90%	>93%	>94%

or field windings. The interaction of the two magnetic fields causes a rotation of the rotor.

In AC motors, a rotating magnetic field is generated in the stator by loops of wires (stator windings). Three-phase motors have one or more sets of three windings on their stator. The number of these sets is called the number of poles of the motor. When three-phase AC voltage is applied to the stator, a magnetic field is generated, which changes its orientation according to the sign of the current flowing in the windings. Since this is continuously varying, the orientation of the magnetic field keeps varying, resulting in a rotating magnetic field. The speed of the rotating magnetic field is called the synchronous speed. It equals the pulsation of the three-phase AC voltage divided by the number of poles.

In synchronous AC motors the rotor operates at the same speed as the rotating magnetic field. This synchronization is achieved often using a permanent-magnet (PM) rotor.[5] The permanent magnets generate their own magnetic field which interacts with the rotating magnetic field generated by the stator windings. The position of the magnets may vary (on the surface of the rotor, in peripheral slots in the interior, radially in the interior, in the interior with a V-shape, etc.), resulting in various rotor constructions. A particular type of PM synchronous motor is often referred to as brushless

---

[5] The other possibility is to use a wound rotor. Machines that combine both excitations also exist.

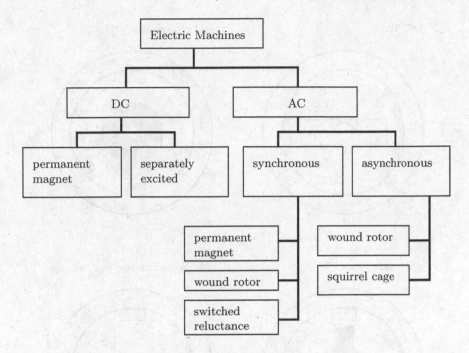

Fig. 4.14 Family tree of electric machines

DC motor. More details on this class of motors will be given below. Another particular embodiment of synchronous AC machine is the claw-pole alternator that is usually found in conventional powertrains.

Switched reluctance (SR) motors are a completely different type of synchronous AC motor, operating according to a different principle. In SR motors, both stator and rotor are designed with "notches" or "teeth" referred to as salient poles. Each stator pole carries an excitation coil. Opposite coils are connected to form one "phase," while the rotor has no windings. When DC voltage is supplied to a phase, the rotor rotates in order to minimize the reluctance of the magnetic path. Among various topologies adopted, the most popular is the one with six stator poles (i.e., three-phase) and four rotor poles.

In asynchronous AC motors, also called induction motors, the rotor usually hosts a set of conductors with end rings, an arrangement known as "squirrel cage."[6] Electromotive force and thus current is induced in the rotor windings by the interaction of the conductors with the rotating magnetic field generated by the stator. The rotor becomes an electromagnet with alternating poles, attracted by those of the rotating magnetic field of the stator. In order for torque to be produced, the speed of the rotor must be different from the speed of the rotating magnetic field.

---

[6] Wound rotor constructions are found as well.

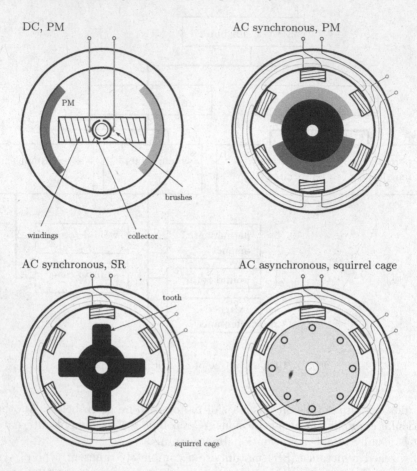

Fig. 4.15 Schematics of four types of motors: separately-excited DC, asynchronous AC (induction), PM synchronous AC, switched reluctance AC

The current motor technology for HEV applications is basically permanent-magnet synchronous AC, although induction motors are still sometimes used [284], and switched reluctance motors are being regarded as a very promising opportunity for the near future [351, 202]. Instead DC motors, although commonly adopted as conventional engine starters and also employed in the earlier generations of electrified vehicles, currently seem to have little future as traction machines. DC motors are known to be simpler and less expensive, since they need relatively uncomplicated control electronics to be fed using the DC supply already present on a vehicle. Their main disadvantage is the lower efficiency and high maintenance requirement, since brushes must be changed periodically.

AC motors are in general less expensive, but they require more sophisticated control electronics (inverters), which cause the overall cost to be higher than that of DC motors. However, they have higher power density and higher efficiency than DC motors. Among them, PM synchronous (and brushless DC) motors are characterized by higher efficiency and specific power. Interior-mounted PM rotors are particularly used for the enhanced controllability and wider speed range than surface-mounted PM rotors. Rotors with bar-shaped and V-shaped permanent magnets are common design choices [1, 324]. Nevertheless, the adoption of induction motors or wound-rotor synchronous motors could be justified since their use reduces the cost risk associated with magnets that are made of rare-earth elements whose prices have increased in recent years because of their scarcity and high demand.

Recently, SR motors have been gaining interest for their simple and rugged, cost-effective construction and hazard-free operation [269]. Moreover, SR motors have a high efficiency and a wider operating speed range. Complex power electronics, noise, and non-uniformity of operation due to torque ripples (depending on the number of phases) are their main disadvantages.

4.3.2 Quasistatic Modeling of Electric Motors

The causality representation of a motor/generator in quasistatic simulations is sketched in Fig. 4.16. The input variables are the torque $T_m(t)$ and the speed $\omega_m(t)$ required at the shaft. The output variable is the electric power at the DC link, $P_m(t) = I_m(t) \cdot U_m(t)$. A positive value of $P_m(t)$ is absorbed by the machine operating as a motor, a negative value of $P_m(t)$ is delivered by the machine operating as a generator.

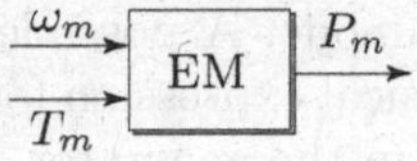

Fig. 4.16 Motor/generators: causality representation for quasistatic modeling

Motor Efficiency

The relationship between $P_m(t)$ and the mechanical power $T_m(t) \cdot \omega_m(t)$ can be calculated without a detailed model of the system when a stationary map of the machine efficiency η_m as a function of the input variables $T_m(t)$ and $\omega_m(t)$ is available. In such a case, the input power required is evaluated as

$$P_m(t) = \frac{T_m(t) \cdot \omega_m(t)}{\eta_m\left(\omega_m(t), T_m(t)\right)}, \quad T_m(t) \geq 0, \tag{4.2}$$

$$P_m(t) = T_m(t) \cdot \omega_m(t) \cdot \eta_m\left(\omega_m(t), T_m(t)\right), \quad T_m(t) < 0. \tag{4.3}$$

The efficiency map $\eta_m(\omega_m, T_m)$ is often measured only for the first quadrant (motor mode, positive speed and torque). Two examples of such maps are shown in Fig. 4.17 for a 32 kW permanent-magnet synchronous motor and a 30 kW induction motor, respectively [355]. Two ways are conceptually possible in order to extend the data available to the second quadrant (generator mode, positive speed and negative torque). The first method consists of mirroring the efficiency, assuming that

$$\eta_m\left(\omega_m, -|T_m|\right) = \eta_m\left(\omega_m, |T_m|\right). \tag{4.4}$$

The second method consists of mirroring the power losses. These are evaluated from an energy balance in the well-defined motor range,

$$P_l(\omega_m, |T_m|) = \frac{1 - \eta_m(\omega_m, |T_m|)}{\eta_m(\omega_m, |T_m|)} \cdot \omega_m \cdot |T_m|, \tag{4.5}$$

and then mirrored to the second quadrant, such that $P_l(\omega_m, -|T_m|) = P_l(\omega_m, |T_m|)$. Using the definition (4.3), the efficiency in the generator range is given by

$$\eta_m(\omega_m, -|T_m|) = 2 - \frac{1}{\eta_m(\omega_m, |T_m|)}. \tag{4.6}$$

Clearly, the two methods yield different results. In general, the result of a mirroring operation does not coincide with the data that can be obtained by measuring the motor efficiency also in the generator range, and the difference is typically more important for induction motors [16]. For example, the measured efficiency map shown in Fig. 4.18 illustrates a case in which neither the efficiency nor the power losses are mirrored when passing from the motor mode to the generator mode. A possible way to manage this general case consists of deriving a physical expression for the efficiency and the power losses, both in the motor and in the generator ranges, using phenomenological models of the electric machine. Since these models are better described in dynamic causality, they will be presented in Sect. 4.3.3.

Figure 4.18 also shows the typical limits to the input variables $T_m(t)$ and $\omega_m(t)$ under normal conditions.[7] The maximum torque curve (in motor mode) as a function of speed is constant up to a certain speed called "base speed", then it decreases hyperbolically with the speed. A similar behavior is observed in generator mode. These features are common to almost all types of motors used. The constant-torque part reflects a current limitation that protects the machine from overheating. The hyperbolically decreasing part reflects the limits imposed by the DC supply and the power converter and thus a limit to

---

[7] Electric motors can often be operated for a short period of time at higher-than-rated power. This peak power is determined to limit the thermal stress, see Problem 4.17.

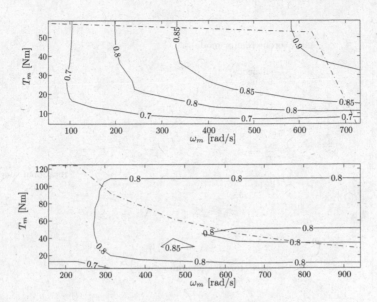

Fig. 4.17 Efficiency maps for a 32 kW PM motor (top) and a 30 kW AC motor (bottom), with curves of maximum torque (dashdot)

power. As a consequence, the shape of the maximum torque curve is totally different from that of engines, which typically is quadratic. Moreover, electric motors are also characterized by a maximum operating speed.

Willans Approach

As already discussed in Chap. 3, the use of efficiency maps to evaluate the energy required at the input stage of a power converter presents some disadvantages. At low loads in particular, the efficiency is not well defined, being the ratio of two quantities approaching zero, which makes it hard to measure or estimate the efficiency correctly. Like for other energy converters, an alternative quasistatic description, known as the Willans approach [272], can be used. For an electric motor, this approach takes the form

$$T_m(t) \cdot \omega_m(t) = e \cdot P_m(t) - P_0, \quad P_m(t) \geq 0 \tag{4.7}$$

$$T_m(t) \cdot \omega_m(t) = \frac{P_m(t)}{e} - P_0, \quad P_m(t) < 0 \tag{4.8}$$

where P_0 represents the power losses occurring after the energy conversion (friction, heat losses, etc.) and e is the "indicated" efficiency, i.e., the maximum efficiency that can be obtained when P_0 is zero. Thus e represents the efficiency of the energy conversion process only (electrical to mechanical energy and vice versa), while η_m takes into account the "idle" losses P_0 caused by friction effects.

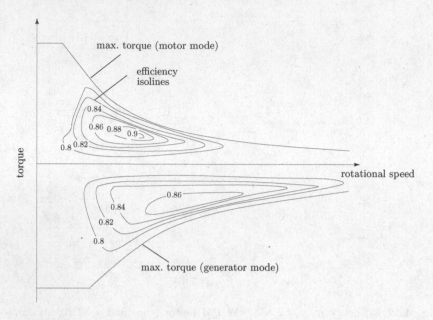

Fig. 4.18 Two-quadrant measured efficiency map for a typical traction motor

The model (4.7)–(4.8) implicitly states that neither the efficiency nor the power losses are mirrored from the motor to the generator range. This can be seen considering that in the motor mode

$$\eta_m(\omega_m, |T_m|) = \frac{e \cdot \omega_m \cdot |T_m|}{\omega_m \cdot |T_m| + P_0}, \tag{4.9}$$

$$P_l(\omega_m, |T_m|) = \left(\frac{1}{e} - 1\right) \cdot \omega_m \cdot |T_m| + \frac{P_0}{e}, \tag{4.10}$$

and showing that neither (4.4) nor (4.6) are verified.

Moreover, the description (4.7)–(4.8) correctly describes the effect of the "idle" losses P_0 for small values of power. In this circumstance, it may occur that $P_m > 0$ while being $\omega_m \cdot T_m < 0$, that is, even if some mechanical power is available from the downstream powertrain, still the energy source must supply a certain amount of electric power. Such a situation is not correctly represented using efficiency maps. Thus it is recommendable to either use a direct mapping of P_m as a function of speed and torque, or a Willans model.

The values of e and P_0 typically depend on the motor speed ω_m. Otherwise, the efficiency would be a function of the output power only, and the constant efficiency lines would be hyperbolae in the efficiency map of the motor. Despite this difficulty, it is possible to find with good approximation some average values for e and P_0 which are valid over the entire range of speeds for a given motor. This can be done by considering a large number of torque T_m

and speed ω_m operating points, evaluating the needed electric power P_m for each of them, and then fitting the dependency $P_m = f(\omega_m \cdot T_m)$ in an affine approximation (approach A). A second approach (approach B) consists of considering a certain number of driving profiles. For each of them the average values of $P_m(t)$ and $\omega_m(t) \cdot T_m(t)$ are evaluated and then fitted using an affine relationship.

If the approach A is used to identify e and P_0 for the same motors as shown in Fig. 4.17, the resulting affine relationships are the ones shown in Fig. 4.19a–b. The inner efficiency e is 0.78 for the induction motor and 0.91 for the permanent-magnet synchronous motor. For both motors, the friction losses are less than 1 kW. Similar numerical results can be obtained for e with motors of different size. With the approach B, for a permanent-magnet motor the values $e = 0.96$, $P_0 = 1.4$ kW were obtained by considering average energy values over several driving profiles.

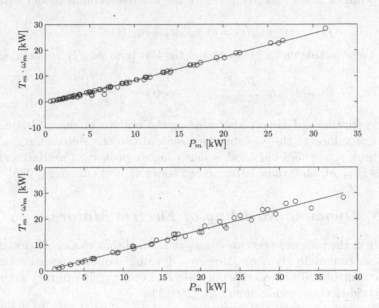

Fig. 4.19 Evaluation of the Willans parameters for a 32 kW permanent-magnet synchronous motor (top) and a 30 kW induction motor (bottom) of the ADVISOR database [355]

The Willans modeling approach can be made *scalable* by normalizing all the torque and speed quantities involved. Torque is substituted by electric mean effective pressure, in analogy to what is usually done for internal combustion engines. The mean speed $c_m(t)$, equivalent to the mean piston speed of engines, is defined as the tangential speed at the rotor radius r,

$$c_m(t) = \omega_m(t) \cdot r. \tag{4.11}$$

Thus, rated rotational speed is inversely proportional to rotor radius.[8] The rotor dimensions r and l and the mean effective pressure are related through the energy balance

$$p_{me}(t) \cdot (2\pi \cdot r \cdot l) \cdot c_m(t) = T_m(t) \cdot \omega_m(t), \tag{4.12}$$

where $2\pi \cdot r \cdot l$ is the rotor external surface. Based on this balance,

$$p_{me}(t) = \frac{T_m(t)}{2 \cdot V_r}, \tag{4.13}$$

where $V_r = \pi \cdot r^2 \cdot l$ is the rotor volume, i.e., the equivalent to the swept volume of engines. Thus, rated torque is proportional to rotor volume.

In terms of mean effective pressure the scalable Willans model is written as

$$p_{me}(t) = e \cdot p_{ma}(t) - p_{mr}(t), \tag{4.14}$$

where the available mean pressure and the loss term $p_{mr}(t)$ are defined as

$$p_{ma}(t) = \frac{P_m(t)}{2 \cdot V_r \cdot \omega_m(t)}, \quad p_{mr}(t) = \frac{P_0}{2 \cdot V_r \cdot \omega_m(t)}. \tag{4.15}$$

This description is formally the same as the one for engines, although the numerical values of the variables are quite different. Motors have a much lower peak p_{me} (0.5–1 bar) and a much higher peak c_m (50–100 m/s) than engines (p_{me} of more than 10 bar, mean speed of 10–15 m/s) [81, 356].

4.3.3 Dynamic Modeling of Electric Motors

Already in the previous sections it was mentioned that models of electric machines are essentially dynamic. However, dynamic models are used mainly for specific control and diagnostics purposes and only rarely are they embedded in a hybrid-electric vehicle simulator [259, 23].

In dynamic models, the correct physical causality should be used. The physical causality representation of a dynamic model of a motor/generator is sketched in Fig. 4.20. The input variables are the DC link voltage $U_m(t)$ and the motor shaft rotational speed $\omega_m(t)$. The output variables are the torque at the motor shaft $T_m(t)$ and the current exchanged at the DC link $I_m(t)$. A positive value of $I_m(t)$ is absorbed by the machine operating as a motor, while a negative value of $I_m(t)$ is delivered by the machine operating as a generator.

It is convenient to distinguish the following derivation of dynamic motor models according to the type of machine. Switched reluctance motors will

---

[8] This is not the only design rule used by motor manufacturers. Sometimes similarly designed machines have the same maximum speed, regardless of the diameter, see Problem 4.18.

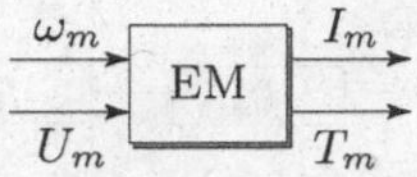

Fig. 4.20 Motor/generators: physical causality for dynamic modeling

be excluded from this analysis since, unlike DC, induction, and synchronous motors, their mathematical characterization is quite difficult to determine. Flux and torque in SR motors are in fact complex nonlinear functions of motor current and position. Consequently, these motors are conveniently described by means of tabular data [56].

DC Motors

Brush-type DC motors are classified according to the stator excitation. In permanent-magnet DC motors a static magnetic field excitation is generated on the stator using permanent magnets. In separately excited DC motors, the excitation is obtained using field windings that have a separate supply from the rotor windings. In both types, a back emf is induced in the rotor windings (armature).

The equivalent circuit of a separately excited DC motor is shown in Fig. 4.21 [111]. The Kirchhoff voltage equation for the armature circuit is written as

$$U_a(t) = L_a \cdot \frac{d}{dt} I_a(t) + R_a \cdot I_a(t) + U_i(t), \qquad (4.16)$$

where $I_a(t)$ is the armature current, $U_i(t)$ is the induced voltage or back emf, R_a is the armature resistance, L_a is the armature inductance, and $U_a(t)$ is the armature voltage. The armature resistance and inductance can be directly measured and are usually provided by the manufacturer. The armature voltage is the DC voltage applied to the rotor windings. It depends on the input DC voltage $U_m(t)$ and is regulated to fulfill the output requirements, usually with chopper converters (see below).

For the field circuit, the voltage equation is similarly written as

$$U_f(t) = L_f \cdot \frac{d}{dt} I_f(t) + R_f \cdot I_f(t), \qquad (4.17)$$

with obvious meanings of the variables. The field voltage $U_f(t)$ is also regulated with a chopper converter as a function of the DC supply.

Newton's second law applied at the motor shaft yields

$$\frac{d}{dt} \omega_m(t) = \frac{T_a(t) - T_m(t)}{\Theta_m}, \qquad (4.18)$$

where $T_a(t)$ is the armature torque and Θ_m is the moment of inertia of the motor.

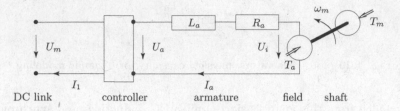

Fig. 4.21 Equivalent circuit of a separately excited DC motor

The induced voltage (back emf) is proportional to the field current and the rotor speed, $U_i(t) = L_m \cdot I_f(t) \cdot \omega_m(t)$, with L_m being the field–armature mutual inductance. The armature torque is proportional to the field current and the armature current, $T_a(t) = L_m \cdot I_f(t) \cdot I_a(t)$. In separately excited motors the field voltage is usually kept constant, at least for speeds below the rated or base speed.[9] For a constant field voltage and at steady state, as (4.17) indicates, the field current is constant as well. Thus back emf and torque are commonly expressed as

$$U_i(t) = \kappa_i \cdot \omega_m(t), \quad T_a(t) = \kappa_a \cdot I_a(t). \tag{4.19}$$

The two factors in (4.19), often called speed constant and torque constant, are equal in principle, $\kappa_i = \kappa_a = L_m \cdot I_f$. In practice, often κ_a is multiplied by a coefficient less than unity that accounts for friction and other losses. Equation (4.19) is valid also during transients for permanent-magnet DC motors, in which, through a constant magnetic flux, the back emf is proportional to the speed and the torque to the armature current.

The dependency between the armature quantities $U_a(t)$, $I_a(t)$ and the input quantities $U_m(t)$, $I_m(t)$ is determined by the motor controller. In DC motors, mostly DC–DC chopper converters are used [203]. To regulate $U_a(t)$ to values up to the supply voltage $U_m(t)$, while allowing single-quadrant operation (only motor mode), a step-down chopper is used as depicted in Fig. 4.22. It consists of a fast semiconductor switch and a free-wheeling diode. For low power levels (12–42 V systems), metal–oxide–semiconductor field-effect transistors (MOSFET) are common, for higher power levels, insulated gate bipolar transistors (IGBT) are typically used as switches.[10] The switch controls the armature voltage by "chopping" the supply DC voltage into segments. When the switch is on, the armature is directly fed by the supply voltage. When the switch is off, the armature current flows through the free-wheeling diode and the armature voltage is zero. The average value of the voltage is then regulated by the ratio of the time periods during which the switch is on or off

---

[9] Speeds above the base speed can be reached by applying "flux weakening" control, which implies a variable field voltage, see below and Problem 4.9.

[10] Gate turn-off thyristors (GTO) are reserved to higher voltage and power levels that are not usually found in cars.

("duty cycle"). Practical chopper converters are more complicated than the simple scheme of Fig. 4.22. For instance, they often include a low-pass filter between the switch network (switch plus diode) and the motor, to smooth the high-frequency switching harmonics.

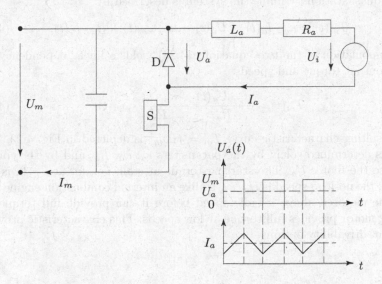

Fig. 4.22 Basic scheme of a DC chopper. D: freewheeling diode, S: semiconductor switch

If $\alpha(t)$ is the chopper duty cycle, for the single-quadrant or step-down chopper the armature voltage is

$$U_a(t) = \alpha(t) \cdot U_m(t). \tag{4.20}$$

For the two-quadrant operation, both motoring and regenerating, half-bridge chopper converters with two switches are used. The motor operation is the same as in step-down choppers, while in the generator range the average armature voltage is

$$U_a(t) = (1 - \alpha(t)) \cdot U_m(t). \tag{4.21}$$

If all the power losses, other than ohmic losses described by the resistance R_a, are lumped in the term $P_{l,c}(t)$, the power balance at the two sides of the converter can be written as

$$P_m(t) = U_m(t) \cdot I_m(t) = I_a(t) \cdot U_a(t) + P_{l,c}(t). \tag{4.22}$$

In summary, to solve the DC motor model in dynamic causality, the duty cycle $\alpha(t)$ must be determined by the control strategy to be a function of the DC voltage $U_m(t)$ and the desired motor performance, usually the achievement of a desired torque, see Problem 4.9. Since motor speed and acceleration are typically given from the vehicle or engine dynamics, (4.16)–(4.18) can be integrated to yield the armature current $I_a(t)$ and the output torque $T_m(t)$. A power balance across the motor controller, as in (4.22), yields $I_m(t)$.

Combining (4.16)–(4.22) also yields an expression for the motor efficiency. In the quasi-stationary limit, the system is described by

$$T_m(t) = \kappa_a \cdot I_a(t), \quad U_a(t) = \kappa_i \cdot \omega_m(t) + R_a \cdot I_a(t). \tag{4.23}$$

The combination of the two equations (4.23) yields a linear dependency between output torque and speed,

$$T_m(t) = \frac{\kappa_a \cdot U_a(t)}{R_a} - \frac{\kappa_a \cdot \kappa_i}{R_a} \cdot \omega_m(t). \tag{4.24}$$

The resulting characteristic curve $T_m = f(\omega_m)$ is depicted in Fig. 4.23. This curve is determined solely by the parameters κ_i, κ_a, R_a, and by the control input to the motor U_a. The starting torque at stall, i.e., for $\omega_m = 0$, is $\kappa_a \cdot U_a/R_a$, the no-load speed is U_a/κ_i. Unlike an internal combustion engine that must be motored above its idle speed before it can provide full torque, an electric motor provides full torque at low speeds. This characteristic provides excellent drivability options.

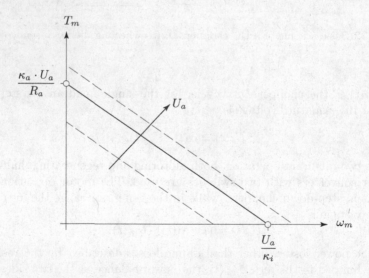

Fig. 4.23 Characteristic curve of a separately excited DC motor

The input power as a function of $\omega_m(t)$, $T_m(t)$ is

$$P_m(t) = \frac{T_m(t)}{\kappa_a} \cdot \left(\omega_m(t) \cdot \kappa_i + \frac{R_a \cdot T_m(t)}{\kappa_a} \right) + P_{l,c}(t), \tag{4.25}$$

and thus the efficiency η_m (in the motor mode) is

$$\eta_m(\omega_m(t), T_m(t)) = \cfrac{1}{\cfrac{\kappa_i}{\kappa_a} + \cfrac{R_a \cdot T_m(t)}{\kappa_a^2 \cdot \omega_m(t)} + \cfrac{P_{l,c}(t)}{\omega_m(t) \cdot T_m(t)}}. \qquad (4.26)$$

The power losses $P_l(t) = P_m(t) - \omega_m(t) \cdot T_m(t)$ are calculated as

$$P_l(t) = \omega_m(t) \cdot T_m(t) \left(\frac{\kappa_i}{\kappa_a} - 1 \right) + \frac{R_a}{\kappa_a^2} \cdot T_m^2(t) + P_{l,c}(t). \qquad (4.27)$$

The expression above shows that the overall power losses are due to ohmic resistance, losses in the controller, and if $\kappa_i \neq \kappa_a$ losses associated to the energy conversion process. If $\kappa_i = \kappa_a$ and $P_{l,c}(t) = 0$, then the power losses depend only on the torque squared. They can thus be mirrored from the motor to the generator quadrant, as in (4.5).

The simple model presented above also allows the calculation of the torque and speed limits of the motor that are, e.g., visible in Fig. 4.18. The armature current and voltage are limited by the values I_{max} and $U_{max} = U_m$, which are imposed by thermal considerations and by the operation of the step-down chopper, respectively. Thus the maximum torque that the motor can deliver is $T_{max} = I_{max}/\kappa_a$. Moreover, the operating region is limited by the characteristic curve (4.24) with $U_a = U_m$. The intersection between these two curves occurs at a speed

$$\omega_b(U_m) = \frac{U_m - R_a \cdot I_{max}}{\kappa_i} \qquad (4.28)$$

that is called *base speed*.

Operation above the base speed is possible by modifying the field current in separately-excited DC motors. In such a way, κ_i and κ_a can be adapted to obtain a desired torque without violating current and voltage limits. Such an operation mode is called *flux weakening* (or field weakening for DC motors). The flux weakening region is limited by the maximum power curve, that is $\omega_m \cdot T_m = I_{max} \cdot (U_m - R_a \cdot I_{max})$.[11]

Induction AC Motors

In induction AC motors the stator carries p sets of three-phase windings fed by external AC voltage, usually supplied by a DC source through a DC–AC converter ("inverter"). The rotor has three-phase shorted windings ("squirrel cage") and no external connections.

Instead of modeling the single phases, it is convenient to model the three-phase systems using a two-phase reference. In each reference frame, each electrical quantity can be described by its direct (d) and quadrature (q) component. The most convenient reference frame is the synchronous reference frame, which rotates at the same frequency as the stator magnetic field. Generally the transformation into a new reference set simplifies the mathematical

---

[11] See Problem 4.9 for a numerical example of DC motor control in these two regions.

manipulation, although the new quantities are not directly measurable. Moreover, the components in the synchronous reference frame can be treated as DC quantities [179, 205, 56, 197, 78, 233, 25, 345, 149].

The Kirchhoff voltage laws for the stator and rotor d-q axes are written as follows [179]. The stator is described by

$$
\begin{aligned}
U_q(t) = {} & \sigma \cdot L_s \cdot \frac{d}{dt} I_q(t) + \left(R_s + \frac{L_m^2 \cdot R_r}{L_r^2} \right) \cdot I_q(t) \\
& + \omega(t) \cdot \sigma \cdot L_s \cdot I_d(t) - \frac{L_m \cdot R_r}{L_r^2} \cdot \varphi_q(t) \\
& + \frac{L_m}{L_r} \cdot p \cdot \omega_m(t) \cdot \varphi_d(t),
\end{aligned}
\tag{4.29}
$$

$$
\begin{aligned}
U_d(t) = {} & \sigma \cdot L_s \cdot \frac{d}{dt} I_d(t) + \left(R_s + \frac{L_m^2 \cdot R_r}{L_r^2} \right) \cdot I_d(t) \\
& - \omega(t) \cdot \sigma \cdot L_s \cdot I_q(t) - \frac{L_m \cdot R_r}{L_r^2} \cdot \varphi_d(t) \\
& - \frac{L_m}{L_r} \cdot p \cdot \omega_m(t) \cdot \varphi_q(t),
\end{aligned}
\tag{4.30}
$$

The rotor is described by

$$
\begin{aligned}
0 = {} & \frac{d}{dt}\varphi_q(t) - \frac{L_m \cdot R_r}{L_r} \cdot I_q(t) + \frac{R_r}{L_r} \cdot \varphi_q(t) \\
& + (\omega(t) - p \cdot \omega_m(t)) \cdot \varphi_d(t),
\end{aligned}
\tag{4.31}
$$

$$
\begin{aligned}
0 = {} & \frac{d}{dt}\varphi_d(t) - \frac{L_m \cdot R_r}{L_r} \cdot I_d(t) + \frac{R_r}{L_r} \cdot \varphi_d(t) \\
& - (\omega(t) - p \cdot \omega_m(t)) \cdot \varphi_q(t).
\end{aligned}
\tag{4.32}
$$

In these equations, $I_d(t)$, $I_q(t)$ are the d-q axis stator currents, $U_d(t)$, $U_q(t)$ are the d-q axis stator voltages, $\varphi_d(t)$, $\varphi_q(t)$ are the d-q axis stator resolved rotor fluxes, L_r, L_s are the stator resolved rotor and stator inductance, R_r, R_s are the stator resolved rotor and stator resistance, L_m is the magnetizing (mutual) inductance, $\sigma = 1 - L_m^2/(L_s \cdot L_r)$, p is the number of pole pairs, $\omega(t)$ is the frequency of the stator voltage (i.e., the speed of the d-q frame), and $p \cdot \omega_m(t)$ is the frequency of the magnetic field induced in the rotor.

The torque generated at the rotor shaft is found using an energy balance

$$
T_r(t) = \frac{3}{2} \cdot p \cdot \frac{L_m}{L_r} \left(\varphi_d(t) \cdot I_q(t) - \varphi_q(t) \cdot I_d(t) \right).
\tag{4.33}
$$

Newton's second law applied at the motor shaft yields

$$
\frac{d}{dt}\omega_m(t) = \frac{T_r(t) - T_m(t)}{\Theta_m},
\tag{4.34}
$$

where Θ_m is the moment of inertia of the motor.

The dependency between $U_d(t)$, $U_q(t)$ and the DC supply voltage $U_m(t)$, as well as between $I_d(t)$, $I_q(t)$ and $I_m(t)$, is determined by the inverter. The most common inverter scheme comprises six semiconductor switches of the same type as that depicted in Fig. 4.22. Such as in chopper drives, the power level determines which type of switches is to be used (MOSFET, IGBT). The generation and the waveform of the three stator phase voltages depend on the particular sequence at which the switches are fired [203]. Common classes of switching schemes are the sinusoidal pulse-width modulation (PWM) and the space-vector modulation (SVM), whose detailed description is beyond the scope of this book [179, 49, 98]. However, a general expression for the stator voltages as a function of the DC voltage can be set as

$$U_d(t) = \alpha_d(t) \cdot m \cdot U_m, \quad U_q(t) = \alpha_q(t) \cdot m \cdot U_m(t), \qquad (4.35)$$

where $\alpha_d(t)$ and $\alpha_q(t)$ are d-q duty cycles and m is a numerical factor that depends on the type of inverter modulation. For sinusoidal PWM, $m = 1/2$, for SVM $m = 1/\sqrt{3}$.

A balance of power at the two sides of the inverter generally yields

$$P_m(t) = \frac{3}{2} \cdot (U_q(t) \cdot I_q(t) + U_d(t) \cdot I_d(t)) + P_{l,c}(t), \qquad (4.36)$$

where $P_{l,c}(t)$ is the term accounting for the losses in the inverter (and possibly losses other than the ohmic losses described by R_s and R_r), and the factor $3/2$ reflects the three-phase nature of the machine and its two-phase description in the d-q reference frame.

In summary, to solve the induction motor model in dynamic causality, the duty cycles and the stator frequency $\omega(t)$ must be determined by the control strategy to match a desired motor performance, such as the realization of a desired torque. In practice, intermediate reference quantities are selected and then realized using the control strategy. A common control strategy ("V/f" strategy) consists of keeping the ratio between the reference stator voltage $\sqrt{U_d(t)^2 + U_q(t)^2}$ and $\omega(t)$ constant and further choosing $\omega(t)$ to match the desired torque. In a more advanced strategy ("field-oriented control", FOC strategy) the reference rotor flux $\varphi_q(t)$ is set to zero, while the reference $\varphi_d(t)$ is set to an appropriate function of speed. The reference current $I_q(t)$ is further set to match the torque. Once the appropriate input quantities are known, equations (4.35) and (4.29)–(4.32) can be integrated to yield the currents $I_d(t)$, $I_q(t)$ and the fluxes $\varphi_d(t)$, $\varphi_q(t)$. Equations (4.33)–(4.34) yield the output torque $T_m(t)$. The power balance across the motor controller (4.36) finally yields $I_m(t)$.

A physical expression for the efficiency of an induction motor is now derived from (4.29)–(4.36). Within the quasi-stationary limit, it is possible to solve the set of equations above by expressing $I_q(t)$ and $I_d(t)$ as a function of $U_d(t)$, $U_q(t)$, $\omega(t)$ and $\omega_m(t)$. This procedure requires the inversion of a steady-state matrix [179], leading to

$$T_m(t) = \frac{3}{2} \cdot p \cdot \frac{\begin{array}{c} R_r \cdot L_m^2 \cdot (\omega(t) - p \cdot \omega_m(t)) \cdot \\ \cdot (U_d^2(t) + U_q^2(t)) \end{array}}{\begin{array}{c} [R_r \cdot L_s \cdot \omega(t) + R_s \cdot L_r \cdot (\omega(t) - p \cdot \omega_m(t))]^2 + \\ + [R_r \cdot R_s - \omega(t) \cdot \sigma \cdot L_s \cdot L_r \cdot (\omega(t) - p \cdot \omega_m(t))]^2 \end{array}}. \tag{4.37}$$

The characteristic curve $T_m = f(\omega_m)$ of an induction motor is shown in Fig. 4.24. At the synchronous frequency, $\omega_m = \omega/p$, the motor does not generate any torque. The starting or breakaway torque is obtained by setting $\omega_m = 0$ in (4.37). As the rotor speed increases, the torque reaches its maximum value, then it decreases to zero when the rotor rotates at the synchronous speed ($\omega = p \cdot \omega_m$). For higher speeds, the machine operates as a generator ($T_m < 0$). Notice the instability region from zero speed to the maximum-torque speed. Since the slope of the torque curve is positive in this region the rotor speed is prone to increase as long as a constant load torque is applied.

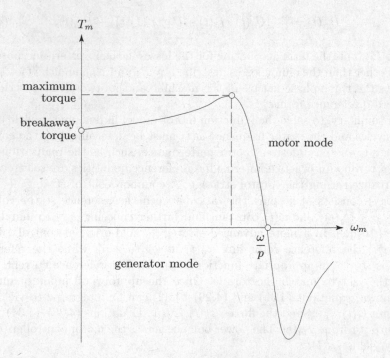

Fig. 4.24 Characteristic curve of an induction AC motor

The efficiency in the motor mode depends explicitly on the speed and only slightly (via the power losses in the controller) on the torque,

$$\eta_m(\omega_m(t), T_m(t)) = \left(\frac{\omega(t)}{p \cdot \omega_m(t)} + \frac{R_r \cdot R_s}{L_m^2 \cdot (\omega(t) - p \cdot \omega_m(t)) \cdot p \cdot \omega_m(t)} \right.$$
$$\left. + \frac{R_s}{R_r} \cdot \frac{L_r^2}{L_m^2} \cdot \frac{\omega(t) - p \cdot \omega_m(t)}{p \cdot \omega_m(t)} + \frac{P_{l,c}(t)}{\omega_m(t) \cdot T_m(t)} \right)^{-1}. \tag{4.38}$$

The corresponding expression of the power losses is

$$P_l(t) = \omega_m(t) \cdot T_m(t) \cdot \left(R_s \cdot \frac{R_r^2 + L_r^2 \cdot (\omega(t) - p \cdot \omega_m(t))^2}{R_r \cdot L_m^2 \cdot p \cdot \omega_m(t) \cdot (\omega(t) - p \cdot \omega_m(t))} \right.$$
$$\left. + \frac{\omega(t) - p \cdot \omega_m(t)}{p \cdot \omega_m(t)} \right) + P_{l,c}(t). \tag{4.39}$$

Thus $P_l(t)$ is the sum of three contributions due to ohmic resistance, slip (the difference between $\omega(t)$ and $p \cdot \omega_m(t)$), and controller efficiency, respectively. The latter equations would imply that, even neglecting $P_{l,c}(t)$, neither the efficiency nor the power losses could be mirrored from the motor to the generator quadrant. The experimental evidence is in agreement with the analysis carried out using the simple models discussed in this section. However, typical experimental efficiency maps for induction machines, like the one in Fig. 4.17a, clearly show that the efficiency depends also on the torque, especially at low loads.

Permanent-Magnet Synchronous Motors and Brushless DC Motors

Permanent-magnet synchronous motors (PMSM) are often confused in the literature with brushless DC motors (BLDC). Basically, these two machines are identical as the torque generation principle is concerned. Both types are synchronous machines that realize the excitation on the rotor with permanent magnets. The external three-phase AC voltage is applied to the armature windings on the stator, which is similar to the stator of an induction machine. The term brushless DC reflects the fact that this motor approximates the behavior of a brush-type DC motor with the power electronics taking the place of the brushes. On the other hand, the term permanent-magnet synchronous motor emphasizes the fact that the machine uses permanent magnets to create the field excitation. The main difference is the waveform of the stator currents, which are rectangular in BLDC motors and sinusoidal in PMSMs. Consequently, the back emf is trapezoidal in BLDC motors and sinusoidal in PMSMs. Historically, the first type is described similarly to DC motors, using a torque constant and a back emf constant. The second type is more often described in terms of synchronous reactance and d-q axis voltages.

For convenience, in the following, both types of motors will be described in the rotor natural d-q reference frame [179, 253, 212, 56, 169, 186], assuming that the back emf is sinusoidal. The equivalent circuit is depicted in Fig. 4.25. Kirchhoff's voltage law for the stator is written as

$$U_q(t) = R_s \cdot I_q(t) + L_s \cdot \frac{d}{dt} I_q(t) + p \cdot \omega_m(t) \cdot \varphi_m(t) + L_s \cdot p \cdot \omega_m(t) \cdot I_d(t), \quad (4.40)$$

$$U_d(t) = R_s \cdot I_d(t) + L_s \cdot \frac{d}{dt} I_d(t) - L_s \cdot p \cdot \omega_m(t) \cdot I_q(t), \quad (4.41)$$

where R_s is the stator resistance, L_s is the stator inductance, $\varphi_m(t)$ is the mutual flux linkage, $U_q(t)$, $U_d(t)$ are the d-q axis stator voltage components, and $I_q(t)$, $I_d(t)$ are the d-q axis stator current components.

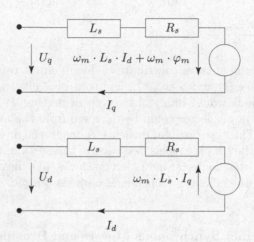

Fig. 4.25 Equivalent circuit of a brushless DC motor

The torque generated at the rotor shaft is

$$T_m(t) = \frac{3}{2} \cdot p \cdot \varphi_m(t) \cdot I_q(t), \quad (4.42)$$

where p is the number of pole pairs.

Newton's second law applied at the motor shaft yields

$$\frac{d}{dt} \omega_m(t) = \frac{T_m(t) - T_m(t)}{\Theta_m}, \quad (4.43)$$

where $T_m(t)$ is the load torque and Θ_m is the moment of inertia of the motor.

The dependency between $U_d(t)$, $U_q(t)$ and the DC supply voltage $U_m(t)$, as well as between $I_d(t)$, $I_q(t)$ and $I_m(t)$, is determined by the inverter model. Equations (4.35) and (4.36) can be applied for PMSMs as well.

To solve the PMSM model following its dynamic causality, the same steps as for induction machines should be followed. The duty cycles are determined by the control strategy to match a desired motor performance, such as the realization of a desired torque. A common control strategy consists of setting reference values of $I_d(t) = 0$ and $I_q(t)$ to match the torque, see Problems 4.11–4.14. It is easy to verify that with such a choice the stator current is minimized for a given torque, whence the name "maximum torque per Ampère" (MTPA) is given to this control strategy. Once the duty cycles are known, (4.35) and (4.40)–(4.41) can be integrated to evaluate the d-q components of voltage and current. Equations (4.42)–(4.43) yield the torque. Finally, (4.36) yields the DC current $I_m(t)$.

An expression for the efficiency can be derived solving (4.40)–(4.43) at the steady-state limit [179]. First of all, the characteristic curve $T_m = f(\omega_m)$ can be derived as

$$T_m(t) = \frac{3}{2} \cdot p \cdot \varphi_m \cdot \frac{R_s \cdot U_q(t) - p \cdot \omega_m(t) \cdot R_s \cdot \varphi_m - p \cdot \omega_m(t) \cdot L_s \cdot U_d(t)}{R_s^2 + p^2 \cdot \omega_2^2(t) \cdot L_s^2}.$$

(4.44)

If the stator inductance can be neglected, the expression above becomes an affine function of the rotor speed, similarly to that expressed by (4.24) for separately excited DC motors. A typical characteristic curve of a brushless DC motor is illustrated in Fig. 4.26.

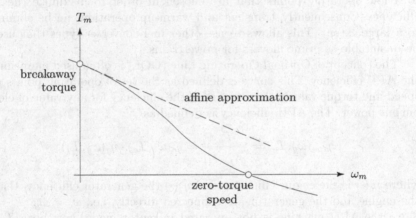

Fig. 4.26 Characteristic curve of a brushless DC motor

The efficiency (in the motor mode) is calculated with the additional assumption $I_d(t) = 0$ as

$$\eta_m(\omega_m(t), T_m(t)) = \left(1 + \frac{R_s}{\frac{3}{2} \cdot p^2 \cdot \varphi_m^2} \cdot \frac{T_m(t)}{\omega_m(t)} + \frac{P_{l,c}(t)}{\omega_m(t) \cdot T_m(t)}\right)^{-1}. \quad (4.45)$$

For a given speed, the power loss term $P_l(t) = P_m(t) - \omega_m(t) \cdot T_m(t)$ is a quadratic function of the torque,

$$P_l(t) = \frac{2 \cdot T_m^2(t)}{3 \cdot \varphi_m^2} \cdot \frac{R_s}{p^2} + P_{l,c}(t). \quad (4.46)$$

As a consequence of (4.46), the power losses can be mirrored from the motor to the generator quadrant (if $P_{l,c}$ is an even function of $T_m(t)$).

The simple model presented above also allows the calculation of the torque and speed limits of PMSM motors. However, these limits depend on the particular control strategy adopted as well and their derivation is straightforward but tedious. An example calculation is proposed in Problem 4.12.

4.4   Range Extenders

Range extenders or Auxiliary Power Units (APU) consist of an engine and an electric machine operated as a generator, which are mechanically linked. Such an assembly is a key component of series hybrids, as well as of those powertrain architectures classified as EREVs (see Sect. 4.2.4). Typically, the engine is a reciprocating internal combustion engine, often with two or three cylinders. However, rotary engines [19], gas turbines, fuel cells, and Stirling engines have been proposed as well. In fact, since the APU rotational speed is not linked to the vehicle speed, the operation of the engine can be restricted to a few operating points that are chosen in order to maximize the APU efficiency. Consequently, transient and warm-up operation can be eliminated to a large extent. This allows to use other fuel converter types that used to be unsuitable as prime movers for powertrains.

The concept of Optimal Operating Line (OOL) is often used in optimizing the APU efficiency. This curve is defined as the set of operating points (e.g., speed and torque values) that maximize the efficiency for any value of electric output power. The APU efficiency is defined as

$$\eta_{apu}(\omega_g, T_g) = \frac{P_g}{\overset{*}{m}_f \cdot H_l} = \eta_e(\omega_e, T_e) \cdot \eta_g(\omega_g, T_g), \qquad (4.47)$$

where $\eta_e(\cdot)$ is the engine efficiency and $\eta_g(\cdot)$ the generator efficiency. Usually, the engine and the generator are connected directly, i.e., $\omega_e = \omega_g$, $T_e = T_g$. Since the APU machine is not operated in motor mode, now both T_g and P_g are positive in generator mode, contrarily to the sign conventions defined in Sect. 4.3. Consequently, the relationship between the motor efficiency in generating mode and the generator efficiency is $\eta_g(\omega_g, T_g) = \eta_m(\omega_g, -T_g)$.

With the definitions above, the OOL is given by

$$\hat{\omega}(P_g) = \arg \max_{\omega_g} \eta_{apu}(\omega_g, T_g), \quad \text{s.t.} \quad T_g = \frac{P_g}{\omega_g \eta_g}. \qquad (4.48)$$

To evaluate the overall efficiency, each APU component can be modeled separately, using the approaches presented in Sect. 3 and 4.3, respectively. However, the causality representation of the generator must be inverted with respect to that of the traction motor. This becomes clear observing the flow of power factors in Fig. 4.11.[12]

---

[12] Similar considerations also apply to the second electric machine that is used in combined hybrids, primarily acting as a generator, see Fig. 4.13.

In quasistatic simulations, the input variable is the electric power $P_g(t)$, while the output variables are the rotational speed $\omega_g(t)$ and the torque $T_g(t)$. One relationship between these variables is given by the efficiency map or the Willans model of the generator. The missing degree of freedom can be provided by the further requirement that the rotational speed follows the OOL (4.48). Therefore, a speed command $\hat{\omega}(t)$ is supplied to the generator model, such that $\omega_g(t) = \hat{\omega}(t)$, and from the second expression of (4.48) the torque $T_g(t)$ also can be calculated [292].

In dynamic models, the input variables are the voltage $U_g(t)$ and the torque $T_g(t)$, while the output variables are the speed $\omega_g(t)$ and the current $I_g(t)$. In this case the motor equations introduced in Sect. 4.3 still can be used, provided that the generator sign convention is adopted.

Alternatively, the same causality of Figs. 4.16 – 4.20 can be used, provided that a model of the supervisory controller is explicitly inserted between the engine and the generator models (Fig. 4.11). An example of such a causality representation is illustrated in the case study 7 and particularly in Fig. A.28.[13]

4.5 Batteries

4.5.1 Introduction

Electrochemical batteries are a key component of both electric vehicles (EV) and hybrid-electric vehicles (HEV). Batteries are devices that transform chemical energy into electrical energy and vice versa. They represent a reversible electrical energy storage system.

Traction batteries are primarily characterized in terms of power, which has to match the power of the electric path, and nominal capacity, which has to match the desired driving range specification. The latter, usually expressed in Ah, is the integral of the current that could be delivered by a full battery when completely discharged under certain reference conditions. A dimensionless parameter is the state of charge (SoC), which describes the capacity remaining in the battery, expressed as a percentage of its nominal capacity.

Desirable attributes of traction batteries for EV and HEV applications are high specific energy, high specific power, long calendar and cycle life, low initial and replacement costs, high reliability, wide range of operating temperatures, and high robustness. The specific energy is the amount of useful energy that can be stored in the battery per unit mass, typically expressed in Wh/kg. This parameter takes into account the fact that not the whole capacity can be actually used. In fact, the battery operation is typically defined by a certain

---

[13] In that case, the causality representation of the electric power link for quasistatic modeling changes with respect to that described in Sect. 4.7. However, the modifications to the model equations are straightforward and are not discussed here.

SoC window, whose limits are the minimum SoC that can be attained during discharging and the maximum SoC that can be reached during charging. The specific energy of a battery is important for the range of a purely electric vehicle. For HEVs, possibly more important is the specific power, typically expressed in W/kg, which determines the acceleration and top speed that the vehicle can achieve.

Types of Batteries

Batteries are composed of a number of individual cells in which three main components are present: two electrodes, where half-reactions take place resulting in the circulation of electrons through an external load, and a medium that provides the ion transport mechanism between the positive and negative electrodes. The cathode is the electrode where reduction (gain of electrons) takes place. It is the positive electrode during discharging, the negative electrode during charging. Contrarily, the anode is the electrode where oxidation (loss of electrons) takes place. It is the negative electrode during discharging, the positive electrode during charging (see Fig. 4.27).

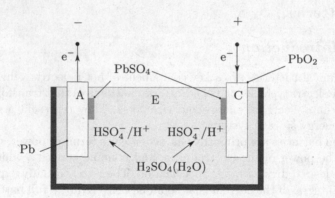

Fig. 4.27 Schematics of a lead–acid battery cell under discharge. A: anode, C: cathode, E: electrolyte. Reduction semi-reaction: $PbO_2 + HSO_4^- + 3H^+ + 2e^- \rightarrow PbSO_4 + 2H_2O$. Oxidation semi-reaction: $Pb + HSO_4^- \rightarrow PbSO_4 + H^+ + 2e^-$.

Batteries used in automotive applications are all rechargeable (secondary batteries). They can be divided in two categories, according to the type of the electrolyte. Ambient-temperature operating batteries have either aqueous (flooded) or non-aqueous electrolytes. High-temperature operating batteries have molten or solid electrolytes. Presently, more than ten different technologies have been proposed. The most commonly used are: (i) lead–acid, (ii) nickel–metal hydride, (iii) lithium-based, among which lithium-ion and lithium-metal polymer, (iv) molten salt, among which sodium–nickel chloride,

Table 4.2 Electrochemical features of various traction battery technologies

battery	anode	cathode	electrolyte	cell voltage
lead–acid	Pb	PbO_2	H_2SO_4	2 V
nickel–metal hydride	metal hydride	$Ni(OH)_2$	KOH	1.2 V
lithium-ion	carbon	Li oxide	lithiated solution	3.6 V
lithium-metal-polymer	Li	plastic composite	solid polymer	3.7 V
sodium–nickel chloride	Na	$NiCl_2$	Al_2O_3	2.58 V
lithium–air	Li	O_2	organic solution	3.4 V

(v) metal–air, among which lithium–air and zinc–air. Table 4.2 summarizes the electrochemical aspects of these technologies.

Lead–acid batteries have a sponge metallic lead anode, a lead dioxide (PbO_2) cathode, and an aqueous sulfuric acid (H_2SO_4) electrolyte. The cell reactions imply the formation of water. Above a certain voltage, called gassing voltage, this water dissociates into hydrogen and oxygen. This is a heavy limitation in the lead–acid battery's operation, since water has to be continuously replaced. This problem was solved with sealed batteries[14], where hydrogen and oxygen are converted back into water, which no longer needs to be replaced. Lead–acid batteries are widely used within the 12 V network of conventional vehicles, but also in some earlier HEV and EV applications, mainly due to their low cost, robustness, and reliability. Their main disadvantages are the low cycle life and the low energy density, which limit their use presently. In contrast, power densities of up to 600 W/kg are obtained with some advanced high-power lead–acid batteries that are being developed for HEV applications.

Nickel–metal hydride (NiMH) batteries work with an anode made of alloys with metals that can store hydrogen atoms. A typical rare earth–nickel alloy used is $LaNi_5$. The cathode is nickel oxyhydroxide, and the electrolyte is potassium hydroxide. Compared to lead–acid batteries, nickel–metal hydride batteries have a higher cost, but they also have a higher energy density and power density, and a higher life. The undesirable "memory effect" (a temporary loss of cell capacity that occurs when a cell is recharged without being

---

[14] Because they cannot be sealed completely, they are more appropriately called valve-regulated lead–acid (VRLA) batteries.

fully discharged, which can cause the battery life to be shortened) is present only under certain conditions. For these reasons, NiMH batteries have been used successfully in production EVs and recently in HEVs.

While already dominating laptop and cell-phone markets, lithium-ion batteries are also rapidly becoming the standard choice in HEV and EV application because of their high specific energy and specific power. However, this technology has not yet reached full maturity and further development is expected, including improvements in calendar and cycle life and reduction of costs.

Today's Li-based cells usually have a carbon-based anode, typically made of graphite, in which lithium ions are intercalated in the interstitial spaces of the crystal. For stability and safety reasons lithium titanate ("LTO") may be preferred, while substantial improvements in specific energy are expected with novel anode materials, which are based on silicon and nanostructures [201]. The specific energy of the battery cell is mainly limited by that of the cathode. The traditional choice for the cathode is lithium cobalt oxide ($LiCoO_2$), particularly for its relatively high energy density. However, this technology has several disadvantages, including a relatively low specific power, high cost of raw material, and a fast ageing that shortens the cycle life, caused by the increase of internal resistance. Battery research therefore has been focusing on the cathode material, and several alternatives to cobalt cathodes have been introduced [34]. Lithium manganese oxide ($LiMn_2O_4$) cathodes in the form of a *spinel* structure yield higher specific power [190, 192] and more improvements are expected with high-voltage ("5 V") spinels [208]. On the other hand, higher specific energy is obtained with nickel-based ($LiNiO_2$) or multi-metal cathodes based on lithium nickel cobalt oxide, also with aluminum ("NCA") [7] or manganese ("NCM") doping [138, 95]. Specific energy is expected to increase to a larger extent with vanadium oxides [198]. However, the choice of the cathode material for current HEV and EV applications is more often dictated by other criteria, such as material cost, life span, environmental friendliness, thermal stability, and safety. In this respect, lithium iron phosphate ($LiFePO_4$) cathodes may be an additional choice [107]. Another material that has potential to improve the cell performance is the electrolyte, which is typically made of a lithiated liquid solution, e.g., $LiPF_6$ dissolved in an organic solvent. A promising alternative consists of solid or gel polymer electrolytes [145].

A different type of lithium-based battery is the lithium-metal-polymer technology, developed by a few manufacturers for its interesting specific energy and stability.

The sodium–nickel chloride battery, also known as the "zebra" battery, is based on another advanced technology. The anode consists of molten sodium, while the cathode is made of nickel-chloride. The electrolyte is a solution of $NaAlCl_4$, with a solid separator (beta-alumina). This cell has been studied extensively for EVs because of its inexpensive materials, high cycle life, and high specific energy and power. However, the cell must operate at high temperatures of 270–350 °C to keep the reactants in liquid form. Advantages over other molten-salt batteries are that the zebra battery has fewer problems concerning cycle life, safety, and hazards to the environment.

Metal–air systems are of current research interest. The lithium–air technology uses a catalytic air cathode, where the active material is oxygen from the air, in combination with an electrolyte and a lithium anode. It is a semi-open system, since the cathode air is fed as a continuous flow, similarly to fuel cells. To make it a reversible system, thus allowing recharge, both nonaqueous and aqueous cells have been developed. Based on their high theoretical specific energy, lithium–air batteries are a very attractive technology. However, their impact has so far fallen short of its potential due to several challenges, mainly concerning the recharging ability and the cyclability. For these reasons, it is expected that Li–air batteries will primarily remain a research topic for the next several years [114, 307, 59]. The Zn–air technology is a more established alternative, which has the advantage of using less expensive and more available materials than lithium, and an intrinsically enhanced safety. On the other hand, specific energy limits of Zn–air batteries are much lower than those of Li–air batteries and they are comparable with nowadays lithium-ion technologies.

Several of these battery types can be designed to maximize either specific power or specific energy, typically acting on the electrode thickness.[15] A comparison of the features of various types of batteries is shown in Table 4.3. The data shown are intended to be average values of present-day technology. The specific energy values shown in Table 4.3 refer to single battery *cells*. Battery systems made of several cells and including cell-monitoring circuits, cooling and heating devices, safety systems, etc., have substantially lower energy densities (see below).

Table 4.3 Comparison of battery cells for electrical and hybrid propulsion (various sources). [a]: depending on the use. [b]: with respect to lead–acid. n.a. = not available.

battery	Wh/kg	Wh/l	W/kg	life (cycles)[a]	cost ratio[b]
Pb–acid	< 40	100	> 250	600	1
Ni–MH (high energy)	65	150	200	1500	3–4
Ni–MH (high power)	55	110	1200	1500	3–4
Li-ion (high energy)	> 150	350	500–750	1200	> 5
Li-ion (high power)	100	250	1800	1200	> 5
Li-MP	120	140	320	n.a.	n.a.
zebra	110	160	170	1000	3
Li–air	> 1000	n.a.	n.a.	n.a.	n.a.

---

[15] This concept is illustrated by "Ragone" plots that are introduced in Chap. 5

Specific Energy of Batteries

The theoretical quantity of charge that an electrode could deliver until completely discharged only depends on the nature of the semi-reaction taking place. For a unit mass of electrode, the theoretical specific capacity is calculated as

$$q_{th,\{p,n\}} = \frac{F}{M_{\{p,n\}}}, \tag{4.49}$$

where $F = 96.48 \cdot 10^3 \, C/mol$ is the Faraday constant and M is the molar mass of the electrode (p: cathode, n: anode) substance. Specific capacity is typically expressed in Ah/kg units.

The reversible or practical capacity $q_{rev,\{p,n\}}$ takes into account the additional losses that appear during charge/discharge cycles and thus it is much lower than $q_{th,\{p,n\}}$. Table 4.4 lists values of the reversible specific capacity for some common lithium-ion electrode materials. Assuming a perfect stoichiometric coupling of the electrodes, the ideal specific capacity of the cell is

$$q_{rev} = \frac{q_{rev,n} \cdot q_{rev,p}}{q_{rev,n} + q_{rev,p}}. \tag{4.50}$$

This value is clearly limited by the reversible capacity of the cathode (see Problem 4.27).

However, the cell does not only contain active materials in the electrodes but also additional electrode components, separator, electrolyte, and connections. The practical specific capacity of the cell is therefore

$$q_{cell} = q_{rev} \cdot \frac{\text{active mass}}{\text{cell mass}} = q_{rev} \cdot \eta_m, \tag{4.51}$$

where η_m is the ratio of the active mass over the cell mass.

The ideal or open-circuit cell voltage is calculated as

$$U_{cell} = U_p - U_n, \tag{4.52}$$

where $U_{\{p,n\}}$ is the electrode potential measured versus a standard reference electrode (usually the Li/Li$^+$ electrode). These values depend on the concentrations in the electrodes, i.e., the charge level of the cell. Nominal potentials are usually given for a 50% charged cell. The specific energy (Wh/kg) of the cell is thus

$$\left(\frac{E}{m}\right)_{cell} = U_{cell} \cdot q_{cell}. \tag{4.53}$$

For typical automotive applications, the voltage output of a single cell is too low. Thus cells are arranged in "packs" that are usually made of intermediate units called modules (e.g., an assembly of six cells). When N identical cells are connected in series, the pack voltage is simply N times the cell voltage. The overall capacity coincides with the cell practical capacity; therefore the

Table 4.4 Electrochemical characteristics of various lithium-ion materials (various sources). [a]: reversible specific capacity; [b]: versus Li/Li^+.

Cathode	Capacity[a] (Ah/kg)	Potential[b] (V)
$LiCoO_2$ (LCO)	140	3.85
$LiMn_2O_4$ (LMO)	115	4
$Li\{Ni,Co,Al\}O_2$ (NCA)	190	3.85
$LiNi_{1/3}Mn_{1/3}Co_{1/3}O_4$ (NMC)	190	4
$LiFePO_4$ (LFP)	150	3.4
$LiNi_{0.4}Mn_{1.6}O_4$ (high-voltage spinel)	145	4.7

Anode	Capacity[a] (Ah/kg)	Potential[b] (V)
Graphite	290	0.25
$Li_4Ti_5O_{12}$ (LTO)	150	1.45
Si/C	1950	0.5

specific energy should equal the practical specific energy of each cell. In reality, additional weight of connections, cooling, housing, as well as management electronics ("Battery Management System", BMS) must be taken into consideration. Moreover, only useful energy corresponding to the used SoC window must be accounted for. These effects lead to a specific energy for the whole battery system,

$$\left(\frac{E}{m}\right)_{battery} = \left(\frac{E}{m}\right)_{cell} \cdot \frac{N \cdot \text{cell mass}}{\text{system mass}} \cdot \eta_u, \qquad (4.54)$$

where η_u is ratio of the useful energy to the total energy, i.e., the SoC window width. Typical values of η_u are about 20% for HEV batteries, while they can reach 70% and more for lithium-ion batteries powering plug-in HEVs, EREVs, and BEVs.

4.5.2 Quasistatic Modeling of Batteries

The causality representation of a battery in quasistatic simulations is sketched in Fig. 4.28. The input variable is the terminal power $P_b(t)$. The output variable is the battery charge $Q(t)$.

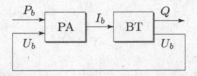

Fig. 4.28 Batteries: causality representation for quasistatic modeling

The charge variation can be calculated directly from the terminal power P_b when semi-empirical data are available from the manufacturer [14]. More generally, charge variation is related to terminal current, which in turn is calculated from the terminal power. Besides the trivial equality

$$I_b(t) = \frac{P_b(t)}{U_b(t)}, \qquad (4.55)$$

a relationship between the terminal voltage U_b and the terminal current I_b must be used.

Capacity and C-Rate

The battery (practical) capacity defined in the previous section is usually expressed in Ah rather than in coulomb. It is actually a function of the current applied and is determined with constant-current discharge/charge tests.

In a typical constant-current discharge test, the battery is initially fully charged and the voltage equals the open-circuit voltage for fully charged cells. A constant discharge current I_b is then applied. After a certain time t_f (called discharge time), the voltage reaches a value (e.g., the 80% of the open-circuit voltage) called *cut voltage* at which the battery is considered as discharged. The discharge time is thus a function of the discharge current. A widely used dependency is that of the Peukert equation

$$t_f = \text{const} \cdot I_b^{-n}, \qquad (4.56)$$

where n is the so-called Peukert exponent, which varies between 1 and 1.5 (1.35 for typical lead–acid batteries). The dependency expressed by (4.56) means that the battery capacity depends on the discharge current. If the *nominal* capacity Q_0^* for a given discharge current I_b^* is known, then the capacity at a different discharge current is given by

$$\frac{Q_0}{Q_0^*} = \left(\frac{I_b}{I_b^*}\right)^{1-n}. \qquad (4.57)$$

Other, more sophisticated models for battery discharge have been developed [302, 163], including neural network-based estimations [48]. A modified Peukert equation for low currents is [46]

$$\frac{Q_0}{Q_0^*} = \frac{K_c}{1 + (K_c - 1) \cdot \left(\frac{I_b}{I_b^*}\right)^{n-1}}, \qquad (4.58)$$

where K_c is a constant.

Often, a non-dimensional value called C-rate is used instead of the discharge current. It is defined as

$$c(t) = \frac{I_b(t)}{I_0}, \quad I_0 = \frac{Q_0}{1\text{h}}, \tag{4.59}$$

where I_0 is the current that discharges the battery in one hour, which has the same numerical value as the battery capacity Q_0. The C-rate is often written as C/x, where x is the number of hours needed to discharge the battery with a C-rate $c = 1/x$, i.e., a current x times lower than I_0. For example, C/3 corresponds to a current that discharges the battery in three hours, C/5 in five hours.

The capacity of a battery does not only vary with the C-rate applied but also as a consequence of charge-discharge cycling and temperature excursions. Several *ageing* mechanisms have been identified that deteriorate the properties of the materials (structural changes, decomposition/dissolution, corrosion, formation of films, etc.). Their net effect is an irreversible decrease of capacity with the number of cycles and calendar time.

State of Charge

The state of charge ξ is the ratio of the electric charge Q that can be delivered by the battery to the nominal battery capacity Q_0,

$$\xi(t) = \frac{Q(t)}{Q_0}. \tag{4.60}$$

A parameter which is often used in alternative to ξ is the depth of discharge (DoD) $1 - \xi(t)$.

Direct measurement of Q is usually not possible with automotive battery systems. However, the variation of the battery charge can be approximately related to the discharge current I_b by charge balance,

$$\dot{Q}(t) = -I_b(t). \tag{4.61}$$

In case of charge, the evaluation of the state of charge must take into account the fact that a fraction of the current I_b is not transformed into charge. This fraction is due to irreversible, parasitic reactions taking place in the battery. Often [81, 189, 159] such an effect is modeled by a charging or Coulombic efficiency η_c,

$$\dot{Q}(t) = -\eta_c \cdot I_b(t). \tag{4.62}$$

The combination of (4.60)–(4.62) yields a method to determine the state of charge by measuring the terminal current. This method, known as "Coulomb counting" has the advantage of being easy and reliable as long as the current measurement is accurate. Practically, however, this method requires frequent recalibration points, to compensate the effects neglected by (4.60)–(4.62) [254]. Modern BMS attempt in some cases to determine the SoC by estimating some of these effects, namely, losses of current during discharge, which is due to reaction kinetics and diffusion processes, battery self-discharge (loss of charge even when the applied current is zero), as well as capacity loss as the battery ages [256].

More advanced methods of SoC determination include adaptive methods based on physical models of batteries [255, 254, 256, 80]. If the state of charge is a state of the model, it can be estimated by comparing the measurement data available (terminal voltage, current), with the model outputs, using well-known techniques such as Kalman filtering.

Equivalent Circuit

A basic physical model of a battery can be derived by considering an equivalent circuit of the system such as the one shown in Fig. 4.29. In this circuit, the battery is represented by an ideal open-circuit voltage source in series with an internal resistance. Kirchhoff's voltage law for the equivalent circuit yields the equation

$$U_{oc}(t) - R_i(t) \cdot I_b(t) = U_b(t). \tag{4.63}$$

The steady-state battery equivalent circuit has been originally applied for various lead–acid batteries [81, 97, 159, 189], but later also for nickel–metal hydride and lithium-ion batteries [159, 76].

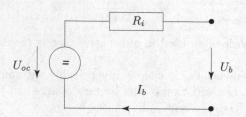

Fig. 4.29 Equivalent circuit of a battery

The open-circuit voltage $U_{oc} = N \cdot U_{cell}$ represents the equilibrium potential of the battery. It is a function of the battery charge, and a possible parameterization of this function is the affine relationship [81] written as

$$U_{oc}(t) = \kappa_2 \cdot \xi(t) + \kappa_1. \tag{4.64}$$

The coefficients κ_2 and κ_1 depend only on the battery design and the number of cells, but not on operative variables, thus they can be considered as constant with time. As can be easily proven with physical considerations, they must be the same both for charging and discharging. More complex parameterizations may be derived from the nonlinear Nernst equation [189, 358]. Alternatively, U_{oc} can be tabulated as a function of the state of charge [159, 97].

The linear dependency of (4.64) is equivalent to a constant voltage source in series with a capacitor with constant capacitance. In fact, in a capacitor the voltage is proportional to the charge stored. A bulk capacitance C_b is thus an alternative way to represent the battery potential when it varies linearly with the state of charge [81].

As for the internal resistance R_i, it takes into account several phenomena. In principle, it is the combination of three contributions. The first is the ohmic resistance R_o, i.e., the series of the ohmic resistance in the electrolyte, in the electrodes, and in the interconnections and battery terminals. The second contribution is the charge-transfer resistance R_{ct}, associated with the "charge-transfer" (i.e., involving electrons) reactions taking place at the electrodes. The third contribution is the diffusion or concentration resistance R_d, associated with the diffusion of ions in the electrolyte due to concentration gradients. Thus, in principle, the internal resistance of a battery is calculated as

$$R_i = R_d + R_{ct} + R_o. \tag{4.65}$$

The internal resistance can be evaluated as a function of the state of charge (and possibly temperature) only [159, 189], conveniently distinguishing between charge and discharge. The fact that the resistance does not depend on the battery current is a serious limitation of these models, since the processes described by (4.65) are in fact highly nonlinear. Nonlinear models are derived in literature by fitting experimental data with semi-physical equations such as the Tafel equation [81, 97].

Often, instead of modeling the various electrochemical processes of a battery, experimental data from a constant-current discharge test are used to derive a black-box model. Simple curve fitting techniques or more sophisticated neural networks are used to derive these input/output representations of the battery behavior. A typical voltage profile as a function of time for a constant discharge current test is shown in Fig. 4.30. Since for a constant current the state of charge varies linearly with time, often the voltage profile is given as a function of ξ instead of time, with the C-rate or the current as a parameter. The voltage variation in Fig. 4.30 shows an initial voltage drop, occurring when the current is applied, which can be considered as instantaneous in a first approximation. Subsequently, the voltage varies linearly with the state of charge. The initial voltage drop clearly increases with the discharge current.

The behavior shown by the battery voltage during the discharge tests is in agreement with the steady-state model of (4.63), with U_{oc} expressed by (4.64), and with a similar affine relationship for the internal resistance,

$$R_i(t) = \kappa_4 \cdot \xi(t) + \kappa_3. \tag{4.66}$$

In fact, combining (4.63), (4.64) and (4.66), the following equation is obtained for the battery voltage

$$U_b(t) = (\kappa_1 - \kappa_3 \cdot I_b(t)) + (\kappa_2 - \kappa_4 \cdot I_b(t)) \cdot \xi(t). \tag{4.67}$$

This equation expresses U_b as the sum of a fully charged battery voltage $\kappa_1 + \kappa_2$, a voltage drop occurring when the current is applied at $\xi = 1$,

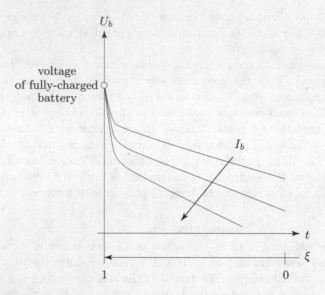

Fig. 4.30 Voltage profile of a battery for a constant discharge current

$(\kappa_3+\kappa_4)\cdot I_b$, and a voltage drop which increases as the state of charge decreases, $(\kappa_2 - \kappa_4 \cdot I_b) \cdot (1 - \xi)$. This description is in agreement with the trend observed in Fig. 4.30. The parameters κ_3 and κ_4 may vary from charging to discharging, as in some models for NiMH batteries described in literature [356, 81].

Now, substituting (4.55) into (4.67), a quadratic equation is obtained for the battery voltage

$$U_b^2(t) - (\kappa_2 \cdot \xi(t) + \kappa_1) \cdot U_b(t) + P_b(t) \cdot (\kappa_4 \cdot \xi(t) + \kappa_3) = 0. \qquad (4.68)$$

The solution of this equation yields

$$U_b(t) = \frac{\kappa_1 + \kappa_2 \cdot \xi(t)}{2} \pm \sqrt{\frac{(\kappa_1 + \kappa_2 \cdot \xi(t))^2}{4} - P_b(t) \cdot (\kappa_4 \cdot \xi(t) + \kappa_3)}, \quad (4.69)$$

For consistency reasons, it is clear that the minus sign is to be used in (4.69). Once U_b is calculated from (4.69), I_b can be calculated from (4.55) and consequently ξ from (4.60)–(4.62). A quantity that is relevant for control purposes is the *electrochemical power* $P_{ech} = I_b \cdot U_{oc}$, which describes the potential power output of the battery not considering internal losses.

Operating Limits

Keeping the dependency on SoC implicit, (4.68) is rewritten as

$$U_b^2(t) - U_{oc}(t) \cdot U_b(t) + P_b(t) \cdot R_i(t) = 0, \qquad (4.70)$$

and (4.69) as

$$U_b(t) = \frac{U_{oc}(t)}{2} + \sqrt{\frac{U_{oc}^2(t)}{4} - P_b(t) \cdot R_i(t)}. \tag{4.71}$$

The terminal variables are subjected to power and current limitations which, for the discharge case, are $P_b > 0$ and $U_b < U_{oc}$. The power as a function of the voltage is calculated as

$$P_b(t) = \frac{-U_b^2(t) + U_b(t) \cdot U_{oc}(t)}{R_i(t)}. \tag{4.72}$$

The power is zero both for $U_b = 0$ and for $U_b = U_{oc}$ (see Fig. 4.31). For values of voltage between these two limits, the power is positive and thus the curve $P_b(U_b)$ has a maximum in that range. The condition for maximum power is obtained by setting to zero the derivative of P_b with respect to U_b,

$$\frac{dP_b}{dU_b} = \frac{U_{oc} - 2 \cdot U_b}{R_i} = 0. \tag{4.73}$$

This equation leads to a maximum power available from the battery

$$P_{b,max}(t) = \frac{U_{oc}^2(t)}{4 \cdot R_i(t)}, \tag{4.74}$$

to which the following values of voltage and current correspond

$$U_{b,P}(t) = \frac{U_{oc}(t)}{2}, \quad I_{b,P}(t) = \frac{U_{oc}(t)}{2 \cdot R_i(t)}. \tag{4.75}$$

In practice, the battery voltage is limited to a relatively narrow band around U_{oc}, $U_b \in (U_{b,min}, U_{b,max})$. Typical values of $U_{b,min}$ are higher than $U_{b,P}$. Therefore, (4.74) is substituted by

$$P_{b,max}(t) = \frac{U_{oc}(t) \cdot U_{b,min} - U_{b,min}^2}{R_i(t)}. \tag{4.76}$$

The corresponding limit for the discharge current is

$$I_{b,max}(t) = \frac{U_{oc}(t) - U_{b,min}}{R_i(t)}. \tag{4.77}$$

In the case of charging, the terminal power would increase (in absolute value) indefinitely with the terminal voltage $U_b > U_{oc}$. In this case, the power is limited solely by the maximum allowed battery voltage $U_{b,max}$,

$$P_{b,min}(t) = -\frac{U_{b,max}^2 - U_{oc}(t) \cdot U_{b,max}}{R_i(t)}. \tag{4.78}$$

The corresponding limit for the charge current (negative) is

$$I_{b,min}(t) = -\frac{U_{b,max} - U_{oc}(t)}{R_i(t)}. \tag{4.79}$$

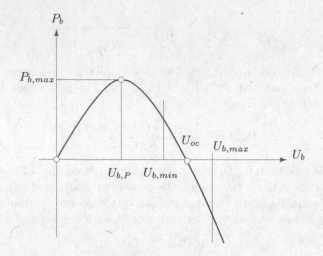

Fig. 4.31 Dependency between battery power and voltage

Battery Efficiency

The efficiency of a battery is not an obvious concept, since the battery is not an energy converter, but rather an energy storage system. Nevertheless, at least two definitions can be formulated for practical purposes. The global efficiency is defined on the basis of a full charge/discharge cycle as the ratio of total energy delivered by the battery to the total energy that is necessary to recharge it. Such a definition is dependent on the details of the charge/discharge cycle, i.e., whether the battery is charged/discharged at constant current (Peukert test) or at constant power (Ragone test). Assuming that the battery has a Coulombic efficiency of 1, that it can be represented by the steady-state equivalent circuit model of (4.63), with a constant open-circuit voltage source and an internal resistance that varies with the sign of the current (charge or discharge), the global efficiency can be evaluated in the "Peukert" case. At constant-current discharge, the battery is depleted in a time $t_f = Q_0/I_b$. The discharge energy is therefore

$$E_d = \int_0^{t_f} P_b(t)\, dt = t_f \cdot (U_{oc} - R_i \cdot I_b) \cdot I_b. \qquad (4.80)$$

Charging the battery with a current of the same intensity, $I_b = -|I_b|$, requires an energy that is evaluated as

$$|E_c| = \int_0^{t_f} |P_b|(t)\, dt = t_f \cdot (U_{oc} + R_i \cdot |I_b|) \cdot |I_b|. \qquad (4.81)$$

The ratio of E_d to E_c is by definition the global efficiency, which is a function of I_b,

$$\eta_b = \frac{E_d}{|E_c|} = \frac{U_{oc} - R_i \cdot |I_b|}{U_{oc} + R_i \cdot |I_b|}. \tag{4.82}$$

The "Ragone" charge/discharge cycle with constant power P_b can be treated in a similar way. During discharge, the current constant value is found from (4.63), (4.71) as

$$I_{b,d} = \frac{U_{oc}}{2R_i} - \sqrt{\left(\frac{U_{oc}}{2R_i}\right)^2 - P_b}, \tag{4.83}$$

the discharge time is $t_{f,d} = Q_0/I_{b,d}$ and the energy discharged is $E_d = t_{f,d} \cdot P_b$. During charge, the power is $-|P_b|$, the constant current is

$$I_{b,c} = \frac{U_{oc}}{2R_i} - \sqrt{\left(\frac{U_{oc}}{2R_i}\right)^2 + P_b}, \tag{4.84}$$

the charge time is $t_{f,c} = Q_0/|I_{b,c}|$ and the charged energy is $|E_c| = t_{f,c} \cdot P_b$. Therefore, the global efficiency is

$$\eta_b = \frac{E_d}{|E_c|} = \frac{t_{f,d}}{t_{f,c}} = \frac{|I_{b,c}|}{I_{b,d}} = \frac{U_{oc} - \sqrt{U_{oc}^2 + 4R_i \cdot |P_b|}}{U_{oc} - \sqrt{U_{oc}^2 - 4R_i \cdot |P_b|}}. \tag{4.85}$$

If the assumption of constant U_{oc} and R_i is removed, then the calculation of η_b made by integrating the power is no longer valid, since now P_b varies with time alongside with ξ. If the affine parameterization of (4.64)–(4.66) is used to express U_b as a function of the state of charge ξ, it can be proven[16] that (4.82) still holds, but with the values of U_{oc} and R_i calculated at the middle point $\xi = 0.5$, i.e., for a battery half full. However, for nonlinear parameterizations or for the "Ragone" case of constant power, (4.82) cannot be used in this way anymore. An example of numerical integration of E_d, E_c is shown in Fig. 4.32 for two batteries, a 25 Ah lead–acid battery and a 45 Ah NiMH battery [355]. The global efficiency for the first battery is $\eta_b = 0.89$, for the second $\eta_b = 0.92$.

In alternative to global efficiency, which always depends on the cyclic pattern used, a *local efficiency* can be also defined, as a power ratio instead of an energy ratio. This yields

$$\eta_b(t) = \frac{P_{b,d}(t)}{|P_{b,c}|(t)} = \frac{U_{oc}(t) - R_i(t) \cdot |I_b|(t)}{U_{oc}(t) + R_i(t) \cdot |I_b|(t)}, \tag{4.86}$$

which is formally the same relationship as that of (4.82). The difference is that in (4.86) the open-circuit voltage and internal resistance may depend on the state of charge and on the current I_b as well. In other terms, the local efficiency is an instantaneous function of both current and charge, $\eta_b(I_b, \xi)$. This concept can be easily represented in terms of "efficiency maps" similar to the maps used to describe other energy converters. Figure 4.33 shows the local efficiency maps of the same two batteries as those of Fig. 4.32. The figures clearly show that the local efficiency is strongly dependent on the C-rate and only very weakly on the state of charge.

---

[16] See Problem 4.29.

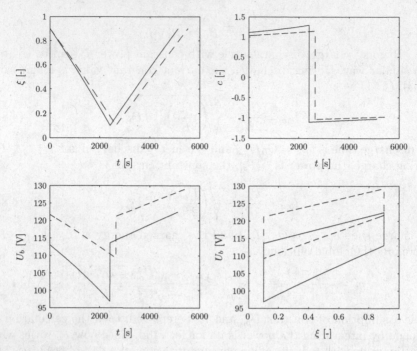

Fig. 4.32 Simulated state of charge ξ, C-rate c, terminal voltage U_b for a discharge/charge cycle at constant power. Solid lines: 25 Ah lead–acid battery ($\kappa_1 = 11.7$, $\kappa_2 = 1.18$, $\kappa_3^{(d)} = 0.0369$, $\kappa_4^{(d)} = -0.0314$, $\kappa_3^{(c)} = 0.0272$, $\kappa_4^{(c)} = 0.0047$, $|P_b| = 3.14\,\text{kW}$). Dashed lines: 45 Ah NiMH battery ($\kappa_1 = 12.7$, $\kappa_2 = 1.40$, $\kappa_3 = 0.0140$, $\kappa_4 = -0.0051$, $|P_b| = 5.65\,\text{kW}$).

The definition (4.86) is based on the comparison of a discharge power to a charge power, both occurring at the same current level. An alternative definition of local efficiency, which has the advantage of being entirely based on the present operating conditions, considers the electrochemical power P_{ech} defined above as the battery "input" during discharge, whereas the battery output is the electric power P_b. During charge, the opposite is true. Consequently, the efficiency is defined as

$$\eta_b(t) = \frac{P_b(t)}{P_{ech}(t)} = \frac{U_b(t)}{U_{oc}(t)}, \quad P_b(t) \geq 0 \tag{4.87}$$

$$\eta_b(t) = \frac{P_{ech}(t)}{P_b(t)} = \frac{U_{oc}(t)}{U_b(t)}, \quad P_b(t) \leq 0 \tag{4.88}$$

Such a definition is preferred for use in supervisory controllers, as described in Chapter 7.

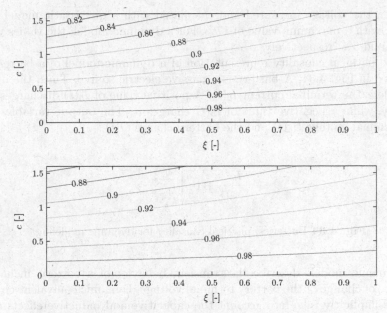

Fig. 4.33 Local efficiency map for a 25 Ah lead–acid battery (top) and a 45 Ah NiMH battery (bottom)

Battery Scalable Models

The quasistatic battery model presented above is inherently scalable by considering the number of cells (or modules) and their arrangement in the battery pack. In this respect, the open-circuit voltage and the internal resistance for the single cell, U_{cell} and R_{cell}, play the role of normalized variables. The corresponding values for the whole pack are

$$U_{oc} = N \cdot U_{cell}, \quad R_i = R_{cell} \cdot \frac{N}{N_p}, \tag{4.89}$$

where N_p is the number of parallel branches each of which has N series-connected cells. Similarly, the capacity and the current limits for the battery pack are $Q_0 = Q_{cell} \cdot N_p$ and $I_{b,\{max,min\}} = I_{cell,\{max,min\}} \cdot N_p$, with an obvious meaning of the normalized variables.

4.5.3 Dynamic Modeling of Batteries

The observed limitations of steady-state models are that the voltage response to load changes is too prompt and that several key phenomena are not adequately taken into account. For more accurate calculations, dynamic models are required. For instance, it was observed [261] that for a typical variable-current test (mean value 7.5 A) of a single Li-ion 2.3 Ah cell, the mean error between the voltage measured and calculated with a quasistatic model was 1.35%, with a maximum value of 17%. For a dynamic model, the values were 0.42% and 5%, respectively.

The physical causality representation of a dynamic model of a battery is sketched in Fig. 4.34. A battery is a passive electric source. Thus, the input variable is the terminal current $I_b(t)$. A positive value of $I_b(t)$ discharges the battery, while a negative value of $I_b(t)$ charges it. The output variables are the terminal voltage $U_b(t)$ and the battery charge $Q(t)$.

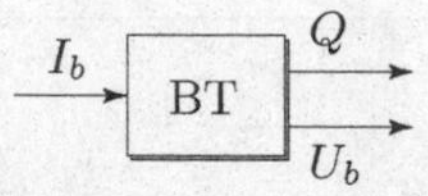

$$I_b \longrightarrow \boxed{\text{BT}} \begin{array}{l} Q \\ \\ U_b \end{array}$$

Fig. 4.34 Batteries: physical causality for dynamic modeling

Dynamic models describe the transient behavior of a battery, including the rate of change of the battery terminal voltage. Dynamic equivalent-circuit models implicitly take into account the capacitive and inductive effects that are neglected in the steady-state circuits. With a different approach, electrochemical models explicitly represent the variations in concentration of the relevant chemical species. Here only lumped-parameter electrochemical models are described. Both dynamic equivalent circuits and lumped-parameter electrochemical models are suitable for the model-based determination of the state of charge (SoC observers).

Dynamic Equivalent Circuit

The simplest of the dynamic models of practical use is the Randles or Thevenin model [189, 279]. This circuit represents a modification of the "steady-state battery" of (4.63), i.e., a voltage source in series with an internal resistance. In this equivalent circuit, depicted in Fig. 4.35, the ohmic voltage drop is distinguished from the non-ohmic overpotential or polarization.[17] The latter is the sum of *charge-transfer (or surface) overpotential* and *diffusion overpotential*. It drives two parallel branches in which the capacitive current and the charge-transfer current flow, respectively. The capacitive current flows across a *double-layer capacitance* C_{dl}, which describes the capacitive effects of the charge accumulation/separation that occurs at the interface between electrodes and electrolyte. The charge-transfer current, caused by chemical reactions, crosses the diffusion resistance and the charge-transfer resistance. The dynamic equations for this circuit are derived from Kirchhoff's voltage and current laws,

---

[17] The terms "overpotential" and "polarization" are both traditionally used in electrochemistry to denote a voltage drop caused by the passage of current. Here the former is used for batteries, the latter for fuel cells.

$$U_b(t) = U_{oc} - R_o \cdot I_b(t) - U_o(t)$$

$$R_o \cdot C_{dl} \cdot \frac{d}{dt} U_o(t) = U_{oc} - U_b(t) - U_o(t) \cdot \left(1 + \frac{R_o}{R_d + R_{ct}}\right) , \qquad (4.90)$$

where U_o is the non-ohmic overpotential. Clearly, these equations coincide at steady state with (4.63), i.e., with $R_i = R_o + R_{ct} + R_d$.

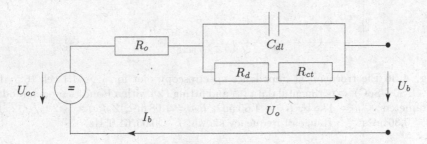

Fig. 4.35 Randles model for batteries

The parameters of the Randles model are experimentally tuned by comparing the model output against the impedance response. The latter is obtained with a technique called Electrochemical Impedance Spectroscopy (EIS), which consists of the application of a periodic current excitation to the system, the monitoring of the system's voltage response, and finally the evaluation of the complex impedance. The current frequency typically ranges from a few mHz to several kHz. The amplitude of the voltage response typically is in the order of a few mV. Often, data obtained by EIS are represented graphically in a Bode plot or a Nyquist plot. EIS, and consequently Randles modeling, can be applied separately to each one of the electrodes or to the whole cell or pack. An example comparison between modeled and experimentally determined impedance values is shown in Fig. 4.36.

Many authors have adopted refinements to the Randles model to take into account additional phenomena or to improve the dynamic description of the basic processes. One of these refinements is to describe the diffusion layer by a so-called *Warburg impedance* Z_d instead of a purely resistive element R_d [43, 315, 318, 180]. This impedance is a constant phase element with a phase angle of 45°, independent of the frequency of the excitation waveform. The magnitude is inversely proportional to the square root of the frequency. In some cases the Warburg impedance is moved for simplicity from the charge-transfer branch to the main circuit branch crossed by the whole battery current [43, 318, 180]. Vice versa, in some other cases all the resistances are placed in the charge-transfer branch [50]. Often a parallel current branch is included to simulate battery *self-discharge* or parasitic currents. Such effects are due to impurities and side reactions, and they occur while charging.

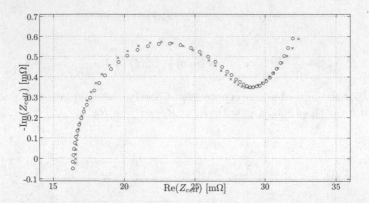

Fig. 4.36 Electrochemical impedance spectroscopy plot for a NiMH cell (6.5 Ah, 20°C, 50% SoC), experimental data (○) and fitting (×) with a Randles model. Model parameters: $C_{dl} = 43.8\,\text{F}$, $R_o = 1.65\,\text{m}\Omega$, $R_{ct} = 1.08\,\text{m}\Omega$, $Z_d(\omega) = \sqrt{2}\sigma_d/\sqrt{j2\pi f}$, $\sigma_d = 0.30\,\text{m}\Omega\text{s}^{-1/2}$. Range of frequency shown: $f \in [0.05, 61.0]\,\text{Hz}$.

Therefore, they can be taken into account by introducing the Coulombic efficiency (4.62) [43, 50, 46, 279]. Sometimes, a bulk capacity is used to represent an open-circuit voltage that varies with the state of charge, as observed while discussing (4.64) [50, 279, 315, 159]. More complex circuit schemes may have multiple RC parallel elements in series [43, 289, 315, 180], typically as an attempt to implement the Warburg impedance in the time domain. In many cases, some of the resistances introduced are actually functions of the current itself, representing the nonlinear nature of the processes involved [43, 50]. In some other cases [113] it is the battery potential that depends on the current.

Another approach consists of representing the battery transient behavior by means of "black-box" dynamic circuits, where the resistive and capacitive elements have no physical meaning but are identified from transient battery response. The resulting topology is usually different from that of the Randles circuit consisting of various parallel capacitor branches. Example topologies include two capacitors and three resistors [159], two capacitors and two resistors along with a voltage source [260], or two capacitors and four resistors [81]. State equations are derived from electric circuit analysis. For example, Johnson's model (see Fig. 4.37) reads

$$\frac{d}{dt}U'(t) = \frac{U''(t) - U'(t) - R'' \cdot I_b(t)}{C' \cdot (R' + R'')},$$

$$\frac{d}{dt}U''(t) = \frac{U'(t) - U''(t) - R' \cdot I_b(t)}{C'' \cdot (R' + R'')}, \tag{4.91}$$

$$U_b(t) = \frac{R''}{R' + R''} \cdot U'(t) + \frac{R'}{R' + R''} \cdot U''(t) - R''' \cdot I_b(t) - \frac{R' \cdot R''}{R' + R''} \cdot I_b(t).$$

All the resistances are tabulated functions of the state of charge of the battery, which is calculated as a weighted sum of the charge on both capacitors, $Q' = C' \cdot U'$, $Q'' = C'' \cdot U''$.

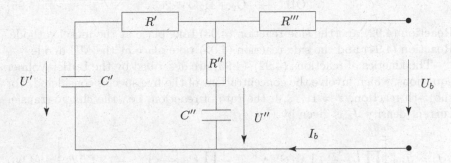

Fig. 4.37 Johnson's capacitive battery model

Dynamic models have been widely validated for lead–acid batteries [43, 50, 289, 46, 260, 279, 81], but also for nickel–metal hydride batteries [289, 159, 318, 180] and lithium-ion batteries [113, 159, 315].

Lumped Parameter Electrochemical Models

Batteries can also be modeled from a purely electrochemical point of view, representing the complex physical processes that take place in the various constituting regions (electrodes, electrolyte, gas reservoir, etc.). This type of modeling is inherently multi-species and multi-phase, since in a battery the solid, liquid, and gaseous phases may coexist. Several detailed electrochemical models can be found in the literature, see [246, 123] for NiMH batteries and [88, 118] for lithium batteries. Due to their complexity, these methods are neither suitable for vehicle system-level simulations, nor for the online determination of the SoC. However, simpler lumped-parameter counterparts of electrochemical models can be derived under certain assumptions. The advantages of this lumped parameter approach are that the resulting models derive from first-principle equations, they are simple to solve, and they can be easily integrated in system simulators or SoC observers. In contrast, their drawbacks are that model assumptions may not be valid under some situations, and the effect of several cell parameters cannot be evaluated [358].

Considering a discharging NiMH battery, the terminal electric quantities and the variations of the SoC mainly depend on the concentrations of five chemical species, namely nickel hydroxide $Ni(OH)_2$, metal hydride MH, oxygen O_2, OH^- ions and nickel oxyhydroxide NiOOH. The reactions involving these species can be summarized as

$$NiOOH + H_2O + e^- \rightarrow Ni(OH)_2 + OH^-, \tag{4.92}$$

$$\frac{1}{2}O_2 + H_2O + 2e^- \rightarrow 2OH^-, \tag{4.93}$$

$$MH + OH^- \rightarrow H_2O + e^- + M, \tag{4.94}$$

$$2OH^- \rightarrow \frac{1}{2}O_2 + H_2O + 2e^-. \tag{4.95}$$

Reaction (4.92) and the side reaction (4.93) take place at the nickel cathode. Reaction (4.94) and the side reaction (4.95) take place at the MH anode.

The kinetics of reactions (4.92)–(4.95) are described by the Butler-Volmer equations, which involve the concentrations of the five species considered. For the z-th reaction, $z = 1, \ldots, 4$, the rate of reaction, i.e., the charge-transfer current density J_z is given by [79]

$$J_z = J_{z,0} \left[\prod_i \left(\frac{c_i}{c_{i,ref}} \right)^{\kappa_i} \cdot e^{\alpha_{a,z} \cdot K \cdot \eta_z} - \prod_j \left(\frac{c_j}{c_{j,ref}} \right)^{\kappa_j} \cdot e^{-\alpha_{c,z} \cdot K \cdot \eta_z} \right], \tag{4.96}$$

where $J_{z,0}$ is the *exchange current density* at reference reactant concentrations (positive in the direction of oxidation), $\alpha_{a,z}$ and $\alpha_{c,z}$ are the anodic and cathodic *transfer coefficients*, η_z is the surface overpotential that drives the charge-transfer reaction, $K = F/(R \cdot \vartheta_b)$, F is Faraday's constant, ϑ_b is the cell temperature, and R is the gas constant. The subscripts i and j denote reducers and oxidizers of the reaction considered, respectively. The c's are the species concentrations, the κ's are their respective numbers of moles participating in the reaction (stoichiometric coefficients), and the subscript ref denotes reference concentrations.

Frequently used simplifying assumptions consist of neglecting the variations of the OH^- concentration in the electrolyte and relating the concentration of $NiOOH$ to the $Ni(OH)_2$ concentration. Moreover, due to the large electrochemical driving force, reaction (4.95) is considered to be limited only by the oxygen concentration. Under these assumptions, (4.96) is specified as [358, 123, 26]

$$J_1(t) = J_{1,0} \cdot \left\{ \left(\frac{c_n(t)}{c_{n,ref}} \right) \cdot \left(\frac{c_e}{c_{e,ref}} \right) \cdot e^{0.5 \cdot K \cdot \eta_1(t)} \right. \\ \left. - \left(\frac{c_{n,max} - c_n(t)}{c_{n,max} - c_{n,ref}} \right) \cdot e^{-0.5 \cdot K \cdot \eta_1(t)} \right\}, \tag{4.97}$$

$$J_2(t) = J_{2,0} \cdot \left\{ \left(\frac{c_e}{c_{e,ref}} \right)^2 \cdot e^{K \cdot \eta_2(t)} - \left(\frac{p_o(t)}{p_{o,ref}} \right)^{1/2} \cdot e^{-K \cdot \eta_2(t)} \right\}, \tag{4.98}$$

$$J_3(t) = J_{3,0} \cdot \left\{ \left(\frac{c_m(t)}{c_{m,ref}} \right)^{\mu} \cdot \left(\frac{c_e}{c_{e,ref}} \right) \cdot e^{0.5 \cdot K \cdot \eta_3(t)} - e^{-0.5 \cdot K \cdot \eta_3(t)} \right\}, \tag{4.99}$$

$$J_4(t) = -J_{4,0} \cdot \left(\frac{p_o(t)}{p_{o,ref}} \right), \tag{4.100}$$

where $c_n(t)$ is the nickel hydroxide concentration, c_e is the constant concentration of KOH electrolyte representing the concentration of OH^- ions, $c_m(t)$ is the concentration of hydrogen in metal hydride material and μ its stoichiometric coefficient, and $p_o(t)$ is the oxygen partial pressure.

The surface overpotentials are

$$\eta_1(t) = \Delta\Phi_{pos}(t) - \phi_1(t), \quad \eta_2(t) = \Delta\Phi_{pos}(t) - \phi_2(t), \tag{4.101}$$

$$\eta_3(t) = \Delta\Phi_{neg}(t) - \phi_3(t), \tag{4.102}$$

where $\Delta\Phi_{pos}(t)$ and $\Delta\Phi_{neg}(t)$ are the potential differences at the solid–liquid interface on the positive and negative electrodes, respectively, while $\phi_m(t), \ldots, \phi_3(t)$ are the equilibrium potentials at the reference state of the reactions (4.92)–(4.94). The latter terms are conveniently parameterized as a function of the species concentrations and the temperature. The typical hysteresis behavior of NiMH batteries can be simulated by distinguishing $\phi_1(t)$ during charging and discharging phases and relaxing the switching with an exponential term [358].

The charge balance at the electrodes imposes the two constraints

$$I_b(t) = S_{pos} \cdot (J_1(t) + J_2(t)), \tag{4.103}$$

$$I_b(t) = -S_{neg} \cdot (J_3(t) + J_4(t)), \tag{4.104}$$

where S_{pos}, S_{neg} are the equivalent surfaces of the positive and negative electrode, respectively. The combination of (4.97)–(4.104) yields a solution for the four current densities $J_1(t), \ldots, J_4(t)$, $\Delta\Phi_{pos}(t)$, and $\Delta\Phi_{neg}(t)$.

The terminal voltage results from the contribution of equilibrium potential, surface overpotential, ohmic losses, and concentration overpotential. The lumped-parameter approach does not consider the latter term, while the first two terms are combined in the quantities $\Delta\Phi_{pos}$ and $\Delta\Phi_{neg}$ through (4.101)–(4.102). Consequently, the terminal voltage can still be calculated with an equation of the type (4.63), however, $U_{oc}(t)$ is replaced by $\Delta\Phi_{pos}(t) - \Delta\Phi_{neg}(t)$ and R_i is represented only by the ohmic resistance R_o.

Once the current densities are known, the mass balance of the nickel active material under the lumped parameter assumption yields

$$\frac{d}{dt}c_n(t) = -\frac{J_1(t)}{l_{y,pos} \cdot F}, \tag{4.105}$$

where $l_{y,pos}$ is the effective thickness of the nickel active material. The mass balance of metal hydride material reads

$$\frac{d}{dt}c_m(t) = -\frac{J_3(t)}{l_{y,neg} \cdot F}, \tag{4.106}$$

where $l_{y,neg}$ is the effective thickness of the metal hydride material. The mass balance of oxygen reads

$$\frac{d}{dt}p_o(t) = -\frac{R \cdot \vartheta_b}{V_{gas}} \cdot \frac{S_{pos} \cdot J_2(t) + S_{neg} \cdot J_4(t)}{F}, \qquad (4.107)$$

where V_{gas} is the gas volume of the cell. Typical numerical values of the model parameters are listed in Table 4.5.

To take into account in a lumped-parameter way the diffusion in the solid phase, the concentrations calculated with (4.105)–(4.107) can be distinguished from the concentrations used in the Butler-Volmer equations (4.97)–(4.100), for instance using the diffusion length approach [123, 32]. Double-layer capacitive effects also can be introduced in a straightforward way by treating $\Delta\Phi_{pos}$, $\Delta\Phi_{neg}$ as additional state variables and transforming (4.103)–(4.104) into their state equations [32].

Finally, this model allows the direct calculation of the SoC

$$\xi(t) = 1 - \frac{c_n(t)}{c_{n,max}}. \qquad (4.108)$$

Table 4.5 Typical numerical values for the parameters of the NiMH model presented (various sources)

parameter	value	parameter	value
S_{pos}	$600\,\mathrm{cm}^3$	$l_{y,pos}$	$0.0245\,\mathrm{cm}$
S_{neg}	$600\,\mathrm{cm}^3$	$l_{y,neg}$	$0.02\,\mathrm{cm}$
c_e	$0.006\,\mathrm{mol/cm}^3$	μ	0.167
$c_{e,ref}$	$0.006\,\mathrm{mol/cm}^3$	R_o	$0.0016\,\Omega$
$c_{m,ref}$	$0.666\,\mathrm{mol/cm}^3$	ϕ_1	$0.465\,\mathrm{V}$ (charge)
$P_{o,ref}$	$1\,\mathrm{atm}$	ϕ_1	$0.363\,\mathrm{V}$ (discharge)
$c_{n,max}$	$0.25\,\mathrm{mol/cm}^3$	ϕ_2	$0.34\,\mathrm{V}$
$c_{n,ref}$	0	ϕ_3	$-0.93\,\mathrm{V}$
$J_{1,0}$	$2.1 \cdot 10^{-4}\,\mathrm{A/cm}^2$	R	$82.1\,\mathrm{atm \cdot cm}^3/\mathrm{mol/K}$
$J_{2,0}$	$8 \cdot 10^{-12}\,\mathrm{A/cm}^2$	h	$15 \cdot 10^{-4}\,\mathrm{W/cm}^2/\mathrm{K}$
$J_{3,0}$	$5.4 \cdot 10^{-4}\,\mathrm{A/cm}^2$	A	$80\,\mathrm{cm}^3$
$J_{4,0}$	$1.5 \cdot 10^{-15}\,\mathrm{A/cm}^2$	$C_{t,b}$	$330\,\mathrm{J/K}$

For lithium-ion batteries, similar concepts have been applied [80, 287]. The model complexity is reduced, since there are only two relevant state variables, namely, the concentrations of the active material at both electrodes. Consequently, only two reactions are considered, leading to two Butler-Volmer equations and two charge balance equations. In contrast, diffusion effects must be handled more carefully than in NiMH batteries. For the diffusion in the solid phase the diffusion length approach can be still used. However, the diffusion in the liquid phase is usually not negligible, which implies the introduction of additional voltage states.

Battery Thermal Models

Detailed dynamic simulations require a battery submodel that evaluates how the battery temperature ϑ_b varies during vehicle operation. Temperature variations in general affect many aspects of a battery's operation, including efficiency, cycle life, and capacity. From the point of view of energy use, temperature variations are important to evaluate the thermal power to be removed via the cooling system. In battery modeling, temperature is often used to correct ambient-temperature data for capacity, open-circuit voltage, and internal resistance.

Practical thermal models are of the lumped-capacitance type [248], i.e., the battery pack is treated as a single reservoir with a thermal capacitance $C_{t,b}$. The thermal balance for the battery is written as

$$\frac{d}{dt}\vartheta_b(t) = \frac{q_{in}(t) - q_{out}(t)}{C_{t,b}}. \tag{4.109}$$

Inflow thermal power is heat generation in the battery core that is due mainly to resistive heating. In the simplest battery equivalent circuit, this term is approximated by

$$q_{in}(t) = R_i \cdot I_b^2(t). \tag{4.110}$$

Another source of heat generation are the parasitic reactions modeled with Coulombic efficiency η_c, see (4.62). If electrochemical models are used, the generated heat flow can be evaluated with more precision, including irreversible losses due to overpotentials and reversible losses due to the variation of the equilibrium potentials with the temperature [261, 32, 80].[18] For the Ni-MH lumped-parameter model of the previous section, that yields

$$q_{in}(t) = \sum_{z=1}^{4} J_z(t) \cdot S_z \cdot \left(\phi_z(t) - \vartheta_b(t) \cdot \frac{\partial \phi_z}{\partial \vartheta_b}(t) \right) - U_b(t) \cdot I_b(t), \tag{4.111}$$

where S_z is either S_{neg} or S_{pos}, according to the z-th reaction ($\phi_4 = \Delta\Phi_{neg}$). It is straightforward to verify that (4.111) simplifies to (4.110) if all the overpotentials $\eta_z = 0$ and the irreversible losses are neglected. Similar considerations apply for lithium-ion batteries [80].

Outflow thermal power is due to thermal conduction through the battery case and convective heat transfer to the cooling air,

$$q_{out}(t) = \frac{\vartheta_b(t) - \vartheta_{air}(t)}{R_{th}}. \tag{4.112}$$

---

[18] The latter contribution can be positive (exothermic) or negative (endothermic), depending on SoC.

The effective air temperature often is taken as the average between the temperatures measured at the airflow inlet and outlet,

$$\vartheta_{air}(t) = \vartheta_a + \frac{1}{2} \cdot \frac{q_{out}(t)}{\overset{*}{m}_a \cdot c_{p,a}}. \tag{4.113}$$

where $\overset{*}{m}_a$ is the mass flow rate of cooling air, and $c_{p,a}$ and ϑ_a its specific heat and inlet temperature. The equivalent thermal resistance R_{th} is the sum of two contributions, a conductive term s/kA and a convective term $1/hA$, where A is the effective battery surface, and s, h and k are the case thickness, the convective heat transfer coefficient, and the thermal conductivity. The most difficult parameter to estimate is h, for which various empirical correlations are available, depending on the type of air flow (natural convection, forced convection), cooling air flow rate and temperature, geometrical features, etc. [257].

An example of the output of a lumped-capacitance model is shown in Fig. 4.38 [80].

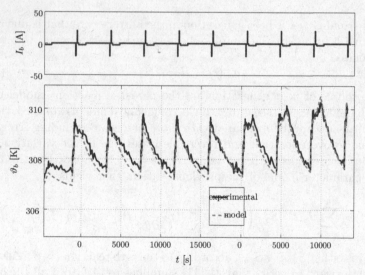

Fig. 4.38 Calculated temperature of a lithium-ion cell over experimental data (bottom) for a current charge-discharge test (top). Cell data: $C_{t,b} = 77 \, \text{J/K}$, $R_i \approx 8.7 \, \text{m}\Omega$, $R_{th} = 10.5 \, \text{K/W}$ [80].

4.6 Supercapacitors

4.6.1 Introduction

Supercapacitors (also termed electrochemical capacitors, electric double-layer capacitors, or ultracapacitors) store energy in the electric field of an electrochemical double layer. While their specific power is much higher than that of batteries, their specific energy is substantially lower. As principal energy storage systems, these devices are being developed for power assist during acceleration and hill climbing, as well as for the recovery of braking energy of conventional vehicles [121, 227]. Another emerging application is in "micro" hybrids. Together with an integrated starter-generator, they act as low-energy buffers that are also capable of high power recuperation [119, 66, 210]. Supercapacitors are also potentially useful as secondary energy storage systems in HEVs, providing load-leveling power to electrochemical batteries, which may be downsized [218, 37]. Another advantage in this case would be the additional degree of freedom they add to the vehicle energy management, which allows for an optimization of the operating conditions of the main energy storage system. Delivering short-term power to components such as power steering and air conditioning compressors is also a possible application of supercapacitors.

A supercapacitor differs from conventional capacitors both in the materials of which it is made and in the physical processes involved while storing energy. In a supercapacitor, the dielectric is represented by an ion-conducting electrolyte interposed between conducting electrodes, see Fig. 4.39. The energy is stored by the charge separation taking place in the layers that separate the electrolyte and the electrodes. Since the voltage that can be applied is limited to a few volts by the physical characteristics of the electrolyte, the storage capacity is increased by raising the capacitance, i.e., increasing the surface and decreasing the thickness of the layer. The surface is increased by using electrodes made of a porous material. Electrode materials with the required very high specific area ($10^3 \, \mathrm{m}^2/\mathrm{g}$) are active carbon and some metallic oxides (ruthenium, iridium). The porous carbon electrodes are connected to metallic plates that collect the charge. The electrodes are separated by an insulating, ion-conducting membrane, referred to as the separator. The separator also has the function of storing and immobilizing the liquid electrolyte. The electrolyte may be an aqueous acid solution or an organic liquid filling the porous electrodes.

Compared to electrochemical batteries, supercapacitors typically show a relatively high specific power of 500–10000 W/kg, but a very low specific energy of 0.2–5 Wh/kg, although much higher values have been obtained with advanced materials [199, 349]. However, similarly to batteries, any increase of specific energy happens at the expense of specific power ("Ragone plot", see Chap. 5), and the compromise depends on the materials used. With respect to high-power battery systems, the future use of supercapacitors seems indeed to be dependent on cost issues [119, 47].

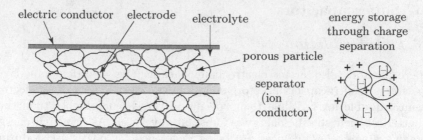

electric conductor   electrode   electrolyte   energy storage
                                                                    through charge
                                                                    separation

                                                      porous particle

                                            separator
                                            (ion
                                            conductor)

Fig. 4.39 Schematic of a supercapacitor

4.6.2 Quasistatic Modeling of Supercapacitors

The causality representation of a supercapacitor in quasistatic simulations is
sketched in Fig. 4.40. The input variable is the power $P_{sc}(t)$ at the terminals.
The output variable is the charge $Q(t)$.

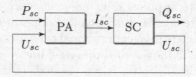

Fig. 4.40 Supercapacitors: causality representation for quasistatic modeling

Similarly to batteries, the state of charge is evaluated from the terminal
current I_{sc} and the nominal capacity Q_0. The former may be calculated from
the terminal power P_{sc} using the trivial equality

$$I_{sc}(t) = \frac{P_{sc}(t)}{U_{sc}(t)} \tag{4.114}$$

and a relationship between current and voltage U_{sc}.

Equivalent Circuit

A basic physical model of a supercapacitor can be derived from a description
of the system in terms of equivalent circuit. The simplest equivalent circuit
consists of a capacitor representing the double-layer capacitance and a resistor
in series representing ohmic losses in the electrodes and electrolyte [57, 159].
The basic equivalent circuit is depicted in Fig. 4.41. Kirchhoff's voltage law
yields

$$R_{sc} \cdot I_{sc}(t) - \frac{Q_{sc}(t)}{C_{sc}} + U_{sc}(t) = 0, \quad I_{sc}(t) = -\frac{d}{dt}Q_{sc}(t). \tag{4.115}$$

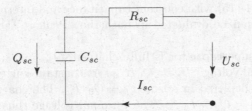

Fig. 4.41 Equivalent circuit of a supercapacitor

Substitution of (4.114) into (4.115) yields a quadratic equation for the supercapacitor voltage,

$$U_{sc}^2(t) - \frac{Q_{sc}(t)}{C_{sc}} \cdot U_{sc}(t) + P_{sc}(t) \cdot R_{sc} = 0. \qquad (4.116)$$

To let Q_{sc} vanish, both terms of (4.116) are differentiated and the second of the equations (4.115) is used. This leads to the following differential equation for U_{sc}

$$\left(1 - \frac{R_{sc} \cdot P_{sc}(t)}{U_{sc}^2(t)}\right) \cdot \frac{d}{dt} U_{sc}^2(t) = -\frac{2 \cdot P_{sc}(t)}{C_{sc}}. \qquad (4.117)$$

Based on (4.114) the current I_{sc} and consequently the charge Q_{sc} may be calculated.

An alternative approach is quite similar to the one discussed for electrochemical batteries. After having substituted the open-circuit voltage U_{oc} with the ratio Q_{sc}/C_{sc}, (4.71) can be used. The resulting equation for the voltage is

$$U_{sc}(t) = \frac{Q_{sc}(t)}{2 \cdot C_{sc}} + \sqrt{\frac{Q_{sc}^2(t)}{4 \cdot C_{sc}^2} - P_{sc}(t) \cdot R_{sc}}. \qquad (4.118)$$

Using numerical integration methods, this equation may be evaluated at any time using the value of Q_{sc} at the previous time step. The maximum power available may be found with the approach used to derive the same quantity for batteries, (4.73)–(4.77).

Supercapacitor Efficiency

The definition of the efficiency of supercapacitors is similar to that of batteries, both elements being energy storage systems rather than energy converters. On the basis of a full charge/discharge cycle, the global (or "round-trip") efficiency is defined as the ratio of total energy delivered to the energy that is necessary to charge the device [83]. Such a definition is dependent on the features of the charge/discharge cycle, i.e., whether the energy storage system is charged/discharged at constant current (Peukert test) or at constant power (Ragone test) [57, 58]. If the supercapacitor is represented by the equivalent circuit model of (4.115) with constant capacitance and internal resistance, the global efficiency can be evaluated in both the "Peukert" and the "Ragone" cases.

Initially, the supercapacitor is fully charged, $Q_{sc} = Q_0$, and the voltage equals the nominal voltage, $U_{sc} = Q_0/C_{sc}$. At constant-current discharge, the supercapacitor is depleted in a time $t_f = Q_0/I_{sc}$. The charge varies linearly with time, $Q_{sc}(t) = Q_0 - I_{sc} \cdot t$. The terminal voltage thus varies according to (4.115). The discharge energy is therefore

$$E_d = \int_0^{t_f} U_{sc}(t) \cdot I_{sc}\, dt = I_{sc} \cdot \left(\frac{Q_0^2}{2 \cdot C_{sc} \cdot I_{sc}} - R_{sc} \cdot Q_0 \right). \tag{4.119}$$

Charging the supercapacitor with a current of the same magnitude, i.e., $I_{sc} = -|I_{sc}|$, the charge varies as $Q_{sc} = |I_{sc}|t$. The charge energy is evaluated as

$$|E_c| = \int_0^{t_f} U_{sc}(t) \cdot |I_{sc}|\, dt = |I_{sc}| \cdot \left(\frac{Q_0^2}{2 \cdot C_{sc} \cdot |I_{sc}|} + R_{sc} \cdot Q_0 \right). \tag{4.120}$$

By definition the ratio of E_d to E_c is the global efficiency, which is a function of I_{sc}:

$$\eta_{sc} = \frac{E_d}{E_c} = \frac{Q_0 - 2 \cdot R_{sc} \cdot C_{sc} \cdot |I_{sc}|}{Q_0 + 2 \cdot R_{sc} \cdot C_{sc} \cdot |I_{sc}|}. \tag{4.121}$$

At constant-power discharge (Ragone test), the current varies with time, thus (4.119)–(4.120) cannot be used. Instead, (4.117) may be solved for constant power, yielding an implicit dependency of the terminal voltage on time:

$$t = \frac{C_{sc}}{2 \cdot P_{sc}} \cdot \left(R_{sc} \cdot P_{sc} \cdot \ln \left(\frac{U_{sc}}{U_0} \right)^2 + U_0^2 - U_{sc}^2(t) \right). \tag{4.122}$$

The initial voltage U_0 follows from (4.118) with $Q_{sc} = Q_0$. From (4.118), the discharge ends when $Q_{sc} = 2 \cdot C_{sc} \cdot \sqrt{P_{sc} \cdot R_{sc}}$ which, based on (4.115), corresponds to a terminal voltage $U_f = \sqrt{P_{sc} \cdot R_{sc}}$. Note that, in contrast to the Peukert discharge, the supercapacitor voltage is not zero at the time t_f. From (4.122), the final time is calculated as

$$t_f = \frac{C_{sc}}{2 \cdot P_{sc}} \cdot \left(R_{sc} \cdot P_{sc} \cdot \ln \left(\frac{R_{sc} \cdot P_{sc}}{U_0^2} \right) + U_0^2 - R_{sc} \cdot P_{sc} \right). \tag{4.123}$$

The discharge energy is given by

$$E_d = t_f \cdot P_{sc} = \frac{C_{sc}}{2} \cdot \left(-R_{sc} \cdot P_{sc} \cdot \ln \left(\frac{U_0^2}{R_{sc} \cdot P_{sc}} \right) + U_0^2 - R_{sc} \cdot P_{sc} \right). \tag{4.124}$$

For a charge with a constant power of the same intensity, $P_{sc} = -|P_{sc}|$, the initial voltage is $\sqrt{R_{sc} \cdot |P_{sc}|}$, while the final voltage equals the value of U_0 calculated previously. The charge energy is evaluated as

$$|E_c| = t_f \cdot |P_{sc}| = \frac{C_{sc}}{2} \cdot \left(R_{sc} \cdot |P_{sc}| \cdot \ln\left(\frac{U_0^2}{R_{sc} \cdot |P_{sc}|}\right) + U_0^2 - R_{sc} \cdot |P_{sc}| \right).$$
(4.125)

The efficiency is therefore calculated from

$$\eta_{sc} = \frac{E_d}{E_c} = \frac{U_0^2 - R_{sc} \cdot |P_{sc}| - R_{sc} \cdot |P_{sc}| \cdot \ln\left(\dfrac{U_0^2}{R_{sc} \cdot |P_{sc}|}\right)}{U_0^2 - R_{sc} \cdot |P_{sc}| + R_{sc} \cdot |P_{sc}| \ln\left(\dfrac{U_0^2}{R_{sc} \cdot |P_{sc}|}\right)}.$$
(4.126)

A third possibility is to charge and discharge the supercapacitor with maximum power. Equation (4.118) states that the discharge power is limited by the state of charge, $P_{sc}(t) < Q_{sc}^2(t)/4/C_{sc}^2/R_{sc}$. Examples of supercapacitor discharges at constant current, at constant power, and at maximum power are illustrated in Fig. 4.42.

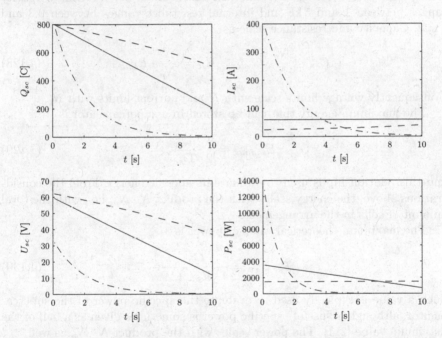

Fig. 4.42 Calculated discharge tests for a supercapacitor ($C_{sc} = 12.5\,\mathrm{F}$, $R_{sc} = 0.08\,\Omega$, $Q_0 = 800\,\mathrm{C}$). Solid lines: discharge with constant current $I_2 = 60\,\mathrm{A}$. Dashed lines: discharge with constant power $P_2 = 1500\,\mathrm{W}$. Dashdot lines: discharge at maximum power.

The local definition of supercapacitor efficiency is based on a power ratio rather than an energy ratio. If the discharge and charge powers are expressed in terms of charge and current, then the local efficiency is evaluated as

$$\eta_{sc}(I_{sc}) = \frac{P_{2,d}}{|P_{2,c}|} = \frac{Q_{sc} - R_{sc} \cdot C_{sc} \cdot |I_{sc}|}{Q_{sc} + R_{sc} \cdot C_{sc} \cdot |I_{sc}|}. \tag{4.127}$$

Note that, if in (4.127) an average charge $Q_{sc} = Q_0/2$ is used, the result is the efficiency of (4.121). The advantage of the local definition is that it does not require any assumption on the type of charge/discharge, while global expressions like (4.121) or (4.126) are strictly valid for Peukert and Ragone cycles, respectively. Similarly to batteries and energy converters, the local supercapacitor efficiency can be easily represented in terms of "efficiency maps" as well.

Supercapacitor Scalable Models

Considerations similar to those described for batteries apply. Supercapacitor "packs" are built as N_p parallel modules, each of which has N equal cells in series. The typical voltage of commercial cells is about 2.5–3 V, while capacity ranges between 1 and 5 kF and internal resistance ranges between 0.3 and 3 mΩ. Capacity and resistance scale as

$$C_{sc} = \frac{N_p}{N} \cdot C_{cell}, \quad R_{sc} = \frac{N}{N_p} \cdot R_{cell}. \tag{4.128}$$

Consequently voltage limits scale with N and current limits with N_p.

The maximum energy that can be stored in a supercapacitor is

$$E_{sc,max} = \frac{1}{2} \cdot \frac{Q_0^2}{C_{sc}} \tag{4.129}$$

and this relationship is used to evaluate its specific energy. From the considerations above, the energy scales with the product $N \cdot N_p$, i.e., with the total amount of cells in the arrangement.

The maximum theoretical value of power is

$$P_{sc,max} = \frac{Q_0^2}{4 \cdot R_{sc} \cdot C_{sc}}. \tag{4.130}$$

Such a value is typically used to evaluate the specific power of the supercapacitor, although a "useful" specific power is sometimes given as a half of the maximum value [211]. The power scales with the product $N \cdot N_p$ as well.

4.6.3 Dynamic Modeling of Supercapacitors

The physical causality representation of a supercapacitor is the same as that of a battery and is sketched in Fig. 4.43. The input variable is the terminal current $I_{sc}(t)$. A positive value of $I_{sc}(t)$ discharges the supercapacitor, while a negative value of $I_{sc}(t)$ charges it. The output variables are the terminal voltage $U_{sc}(t)$ and the charge $Q(t)$.

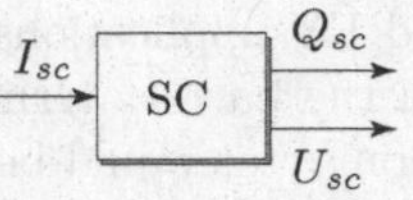

Fig. 4.43 Supercapacitors: physical causality for dynamic modeling

The basic equivalent circuit model represented by (4.115) can be used to calculate the terminal voltage as a function of the terminal current. Integrating the current yields the supercapacitor charge and thus the non-dimensional state of charge.

More complex dynamic equivalent circuits describe the distributed nature of the resistance and of the charge stored in a porous electrode with multiple transmission-line impedances [247, 183]. Self-discharge is also considered with a leakage resistance in parallel. A frequent simplification, which is valid for typical frequencies found in HEV applications, leads to Randles-type topologies, with a few additional impedances in series with the equivalent circuit of Fig. 4.41 [360]. These impedances basically consist of a capacitor and a resistance in parallel. Similarly to electrochemical batteries, parameters of these models are determined in the frequency domain from EIS.

4.7 Electric Power Links

4.7.1 Introduction

In a conventional vehicle architecture, all the electric loads are supplied by a 14 V DC link connected to a 12 V battery and to an engine-driven alternator. The most common electric loads are: ignition system, lighting system, and electric starter motor. The remaining accessories are usually belt-driven by the engine, although electric versions also exist. In common passenger cars today the peak electric power demand is around 1 kW. Luxury cars may require a maximum electrical load of 2 kW.

However, the trend in HEVs is to have more and more electric accessories that are supplied by the high-voltage electric link [18, 342]. Additional electric loads include power steering motors, ABS actuators, cooling fans and pumps, air conditioning compressors, active suspension actuators, catalyst heaters, throttle actuators, etc. Consequently, the electric accessory load may rise up to 2–5 kW.

To match the voltage level of the traction battery (hundreds of volts) with those of the electric loads (up to 20–40 V for some accessories), DC–DC "buck" or step-down converters are used that are similar in principle to the devices introduced in Sect. 4.3.3 as DC motor controllers.[19]

---

[19] Some electric loads, e.g., the air-conditioning compressor motor, could be powered by AC electricity. In that case, an inverter (DC–AC converter) is necessary.

In some parallel HEV and EV applications, the operating voltage of the traction motor is higher than the battery terminal voltage. In those cases, a "boost" or step-up DC–DC converter would be necessary, at least in traction mode. In their simpler topology, the transistor switch is in parallel to the load, while the free-wheeling diode is inserted in series (compare with Fig. 4.22). However, to allow for a bidirectional flow of electricity and recharge the battery, "buck-boost" topologies that combine both actions are used in practice. Similarly, in combined hybrids, a bidirectional buck-boost converter may be found between the battery and the DC supply of both motor inverters. Buck-boost converters are also used to interface the battery with the APU generator in series hybrids and let the machine act as the engine starter.

Different topologies of DC links may be found when multiple electric sources are present in a vehicle [182]. Examples are series hybrids with a generator and a battery, purely electric hybrids with a battery and a supercapacitor, or hybrid fuel-cell vehicles with a battery or a supercapacitor. In the latter two cases, the possibility of controlling the power distribution depends on the DC link topology. In the passive topology, the two sources are just connected in parallel with each other and the load (the powertrain). In this configuration the power distribution between the sources is uncontrolled; it is only determined by their internal resistances. To control the power distribution, semi-active topologies are required, where a DC–DC converter is inserted either on the load side or on the side of one of the sources. In active topologies, two DC–DC converters are employed, again with three possible configurations depending on the placement of the converters.

Notice that the band of voltage levels at which supercapacitors operate is much wider than the one for batteries. This effect is inherently connected to the physical principles of supercapacitors. Accordingly, the power electronic devices connecting the supercapacitor banks with the remaining system components must be more complex and, thus, costly.

4.7.2 Quasistatic Modeling of Electric Power Links

The causality representation of a power link (with m power sources and a single load) in quasistatic simulations is sketched in Fig. 4.44 for $m = 2$ (usually, a battery and a generator or a supercapacitor).[20] The input variable is the power $P_{m+1}(t)$ at the load port. The output variables are the power $P_j(t)$ at each source port, $j = 1, \ldots, m$.

In the general case (in most practical cases $m = 1, 2$), one equation is given by the power balance across the link,

$$P_{m+1}(t) = \sum_{j=1}^{m} P_j(t) - P_l(t), \qquad (4.131)$$

---

[20] The extension to the configurations with multiple loads, e.g., combined hybrids with two electric motors, is straightforward.

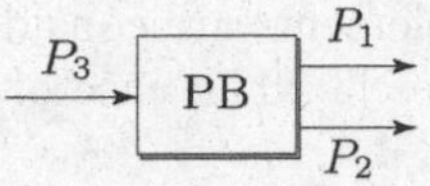

Fig. 4.44 Electric power links: causality representation for quasistatic modeling

where $P_l(t)$ is the power to the electric loads, which generally is a quantity that varies in time according to the vehicle operation (from a control point of view, it may be regarded as a disturbance). There are still $m - 1$ degrees of freedom available. These can be represented by $m - 1$ control variables u_j, which are conveniently defined as the power-split ratios

$$u_j(t) = \frac{P_j(t)}{P_{m+1}(t) + P_l(t)}. \tag{4.132}$$

The combination of (4.131) and (4.132) allows the evaluation of the power at the source ports,

$$P_j(t) = u_j(t) \cdot (P_{m+1}(t) + P_l(t)), \quad j = 1, \ldots, m - 1, \tag{4.133}$$

$$P_m(t) = \left(1 - \sum_{j=1}^{m-1} u_j(t)\right) \cdot (P_{m+1}(t) + P_l(t)). \tag{4.134}$$

Of course, in the common case of $m = 1$ (EVs and parallel HEVs), this model simply implies the equality of the power delivered by the battery with the power absorbed by the motor and the electric loads.

4.7.3 Dynamic Modeling of Electric Power Links

The physical causality representation of a power link is sketched in Fig. 4.45. The input variables are the voltage at the first main port (typically the battery), $U_1(t)$, and the current at the other ports, $I_j(t)$, $j = 2, \ldots, m + 1$. The output variables are the current at the main source port, $I_1(t)$, and the voltage at the other ports, $U_j(t)$, $j = 2, \ldots, m + 1$.

The causality of the secondary sources ($j = 2, \ldots, m$) is reversed with respect to that of the main source. In the case of a supercapacitor, its causality is thus the opposite of that illustrated in Sect. 4.6.3 [275]. The same causality

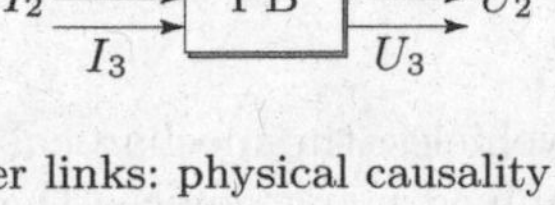

Fig. 4.45 Electric power links: physical causality for dynamic modeling.

can be recovered if dynamic elements are considered in the DC link topology, such as smoothing inductances [3]. The case of a generator in series hybrids was already discussed in Sect. 4.4.

A schematic active topology of a DC power link with $m = 2$ is shown in Fig. 4.46. Each DC–DC converter can be ideally described by means of its voltage ratio $R(t)$ that is controlled by the duty cycle $\alpha(t)$ of the switch. For example, buck converters are described by (4.20)–(4.21), while for step-up converters the voltage ratio is

$$R(t) = \frac{1}{1 - \alpha(t)} \qquad (4.135)$$

for the direct flow of electricity (e.g., in the traction mode), and

$$R(t) = \frac{1}{\alpha(t)} \qquad (4.136)$$

for the reverse electricity flow (e.g., in the generating mode). Note that (4.136) is the reciprocal of (4.20) and (4.135) is the reciprocal of (4.21).

Power links have capacitive dynamics (see Fig. 4.46) as well as inductive dynamics. However, typical capacitance ($\approx$ mF) and inductance ($\approx$ μF) values are so small that models aimed at the evaluation of vehicle energy consumption do not consider any dynamic effects. Consequently, m equations are given by the equalization of voltage

$$U_j(t) = \frac{U_1(t)}{R_j(t)}, \quad j = 2, \dots, m. \qquad (4.137)$$

$$U_{m+1}(t) = R_{m+1}(t) \cdot U_1(t). \qquad (4.138)$$

Another equation is given by the quasistatic balance of currents flowing across the link

$$I_1(t) = R_{m+1}(t) \cdot I_{m+1}(t) - \sum_{j=2}^{m} \frac{I_j(t)}{R_j(t)} + \frac{P_l(t)}{U_1(t)}. \qquad (4.139)$$

Since these are $m + 1$ equations in the $m + 1$ model variables, the variables u_j introduced in the previous section no longer are necessary inputs.

More refined models used for control purposes may include efficiencies of the single converters [182] or enhance the dynamic description of each converter unit to include parasitic resistance, current-smoothing inductance, leakage resistance, and voltage-smoothing capacitance [181, 3].

4.8 Torque Couplers

4.8.1 Introduction

In parallel hybrid-electric vehicles, the mechanical power outputs from different power sources are combined using devices that can be generically called

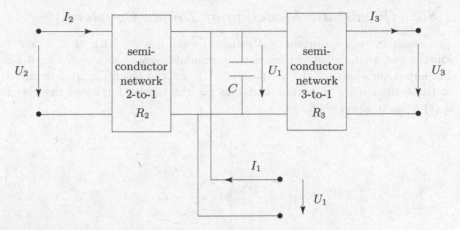

Fig. 4.46 Schematic of a DC power link with two sources of electricity (active topology with two DC–DC converters). Electrical loads are not represented.

"torque couplers." These include three-sprocket gears driven by belts or chains and direct coupling onto the crankshaft. The case of different prime movers that power different wheel axles ("through-the-road" hybrids) is a particular case of torque coupling.

The entry level to hybridization is the belt-driven architecture typically implemented by starter/generator "micro" hybrids. In this configuration, a reversible electric machine, capable of not only generating electricity but also acting as a motor to crank the engine, replaces the traditional alternator. Mechanical modifications are required to mount the larger electric machine: a reinforced belt, a reversible tensioner system, and possibly larger pulleys on belt-driven accessories [263, 316]. Electrical modifications are of course needed as well.

Another concept for torque coupling is implemented in full parallel hybrids. It consists of mounting the rotor of the electric machine on the same engine shaft, between the engine and the transmission. Depending on the position of the clutch, active dampening of the oscillations at the drive train, battery recharge at vehicle stop, and other features are feasible. The electric machine can also be linked to the primary shaft of the transmission through a reduction gear. In that case, the machine inertia transferred to the transmission shaft can substantially slow down the gear changing phases, unless an active compensation is performed.

Through-the-road coupling can be realized in a four-wheel-drive vehicle architecture, where one of the axles can be in traction or braking mode. Particular care must be taken when splitting the total power between the axles during battery recharge, since the constraints imposed by vehicle stability must not be violated.

4.8.2 Quasistatic Modeling of Torque Couplers

The causality representation of a generic torque coupler with $m = 3$ power sources and a single load in quasistatic simulations is sketched in Fig. 4.47. The input variables are the torque $T_{m+1}(t)$ and the speed $\omega_{m+1}(t)$ required at the load shaft. The output variables are the torque $T_j(t)$ and the speed $\omega_j(t)$ at each power shaft, $j = 1, \ldots, m$.

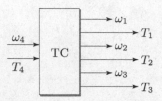

Fig. 4.47 Torque couplers: causality representation for quasistatic modeling

In the general case (in all practical cases $m = 2, 3$), m equations are given by the equalization of the rotational speed

$$\omega_j(t) = \gamma_j \cdot \omega_{m+1}(t), \quad j = 1, \ldots, m, \tag{4.140}$$

where the values of γ_j are the transmission ratios of the power sources to the output gear. Usually, the main power sources have a unitary transmission ratio to the output. In the case of direct coupling on the same crankshaft, all the transmission ratios are unitary. The power balance across the device yields another equation,

$$T_{m+1}(t) = \sum_{j=1}^{m} \gamma_j \cdot T_j(t) - T_l(t), \tag{4.141}$$

where $T_l(t)$ is a loss term that, when not neglected, is usually modeled as a constant [355]. There are still $m - 1$ degrees of freedom available. These can be represented by $m - 1$ control variables u_j which are conveniently defined as the torque-split ratios

$$u_j(t) = \frac{\gamma_j \cdot T_j(t)}{T_{m+1}(t) + T_l(t)}. \tag{4.142}$$

The combination of (4.141) and (4.142) permits the calculation of the torque at the power shafts,

$$T_j(t) = \frac{u_j(t)}{\gamma_j} \cdot (T_{m+1}(t) + T_l(t)), \quad j = 1, \ldots, m - 1. \tag{4.143}$$

$$T_m(t) = \frac{1 - \sum_{j=1}^{m-1} u_j(t)}{\gamma_m} \cdot (T_{m+1}(t) + T_l(t)). \tag{4.144}$$

4.8.3 Dynamic Modeling of Torque Couplers

The physical causality representation of a torque coupler is sketched in Fig. 4.48. The input variables are the torque at the power shafts, $T_j(t)$, $j = 1, \ldots, m$ and the rotational speed at the output shaft, $\omega_{m+1}(t)$. The output variables are the rotational speed at the input shafts, $\omega_j(t)$, $j = 1, \ldots, m$ and the torque at the output shaft, $T_{m+1}(t)$.

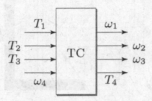

Fig. 4.48 Torque couplers: physical causality for dynamic modeling

Usually in modeling torque couplers, no dynamic effects are considered. The rotational inertia is conveniently attributed to the power and the load shafts, respectively. Moreover, any elasticity of the shafts inside the torque coupler is usually neglected. Consequently, the same quasistatic equations, (4.140)–(4.141) can be used. In particular, they constitute $m + 1$ equations in the $m + 1$ model variables. Therefore, the torque split ratios are completely defined by the torque inputs and there is no need to introduce explicitly the additional inputs u_j as in the previous section.

4.9 Power Split Devices

4.9.1 Introduction

Power split devices (PSDs) are widely used in automatic transmissions. A power split device is also found in many hybrid vehicles to combine mechanical power from various (usually, two) power sources to various (usually, two) mechanical loads. In the typical configuration of a combined hybrid vehicle, a PSD connects an engine, a motor, a generator, and the drive train.

The core of most PSDs is formed by a *planetary gear set*, often referred to also as *epicyclic gearing*. A basic planetary gear has three main rotating parts (Fig. 4.49). The inner part is the *sun*, the outer part is the *ring*. The intermediate part carries rotating elements (planets) and is the *carrier*. More complex configurations are possible but they are not considered here. Each of the three parts can be connected to the input shaft or the output shaft, or it can be held stationary. Choosing which piece plays which role determines the gear ratio for the gear set.

Common automatic transmissions use a compound planetary gear set in combination with an hydraulic torque converter. In contrast, several PSDs for HEVs use one or two planetary gear sets in combination with two electric machines. The latter case is analyzed in detail in the next section; however, the modeling of these devices can be applied in both cases.

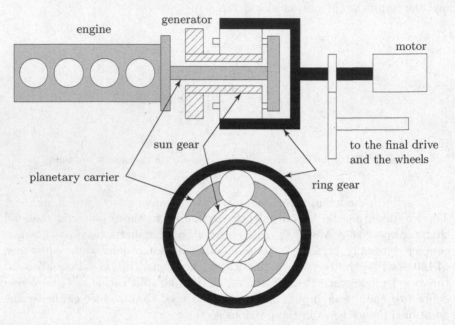

Fig. 4.49 Schematics of a planetary gear set arrangement (as in the Toyota Prius [324])

4.9.2 Quasistatic Modeling of Power Split Devices

The causality representation of a power split device in quasistatic simulations is sketched in Fig. 4.50. The input variables are the torque and speed at the load shafts, $\omega_f(t)$, $\omega_g(t)$, $T_f(t)$, $T_g(t)$, where the subscripts f and g stand for "final driveline" and "generator", respectively. The output variables are the torque and speed at the power source shafts, $\omega_e(t)$, $\omega_m(t)$, $T_e(t)$, $T_m(t)$, where the subscripts e and m stand for "engine" and "motor", respectively. Often in combined hybrids, the roles of the electric machines as generator and motor are interchangeable.

Simple Power Split Devices

The basic equation to consider in analyzing the quasistatic behavior of a planetary gear set is the relationship between the speeds of the three main

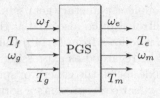

Fig. 4.50 Power split devices: causality representation for quasistatic modeling

parts, which can be derived according to the Willis formula. The planetary gear set can be considered as an ordinary gear set in a rotating frame that is attached to the carrier. Thus, the ratio of the relative speeds of the ring and of the sun can be written as

$$\frac{\omega_{ring}(t) - \omega_{carrier}(t)}{\omega_{sun}(t) - \omega_{carrier}(t)} = -z, \quad z = \frac{n_{sun}}{n_{ring}}, \tag{4.145}$$

where n is the number of teeth. A typical value of the epicyclic gear ratio z is 0.385 [283].

The connection of the machines to the various ports may differ from one system to another. In the Toyota Hybrid System (THS-II) [324], the sun gear is connected to the generator, the planetary carrier to the engine, and the ring gear to the motor shaft. On the output shaft, the power transmitted by the ring gear and the power from the motor are combined. In other hybrid-electric vehicles, the connections can be different [206, 347].[21]

Assuming $\omega_e(t) = \omega_{carrier}(t)$, $\omega_g(t) = \omega_{sun}(t)$, $\omega_f(t) = \omega_{ring}(t)$, then (4.145) is specified as [283]

$$\omega_m(t) = \omega_f(t), \tag{4.146}$$

$$\omega_e(t) = \frac{z \cdot \omega_g(t) + \omega_f(t)}{1 + z}. \tag{4.147}$$

The balance of power applied to the four ports,

$$T_f(t) \cdot \omega_f(t) + T_g(t) \cdot \omega_g(t) = T_e(t) \cdot \omega_e(t) + T_m(t) \cdot \omega_m(t), \tag{4.148}$$

combined with (4.146)–(4.147) yields

$$T_e(t) = \frac{1+z}{z} \cdot T_g(t), \tag{4.149}$$

$$T_m(t) = T_f(t) - \frac{T_g(t)}{z}, \tag{4.150}$$

which are the remaining two equations.

---

[21] Usually, in combined hybrids, the two electric machines are simply referred to as motor–generator 1 and 2, without distinguishing their prevalent functions.

In contrast to torque couplers, planetary gear sets do not have any available degrees of freedom to control. Notice also that if the generator shaft is not connected, i,e., $T_g(t) = 0$, it follows from (4.149) that also $T_e(t) = 0$.

Equation (4.147) may be regarded as expressing the relationship between engine speed and output axle speed, the generator speed being a parameter. In other words, the planetary gear set represents a sort of continuously variable transmission, where the generator controller regulates the transmission ratio. Four typical modes of operation are illustrated in Fig. 4.51, which clearly shows the bilinear relationship between the speed levels [324].

A simple PSD may also have an output stage gearing to combine the torque from the motor to the torque from the ring gear of the planetary gear set. An example of such a case is found in the Ford Hybrid System (FHS) [103], for which (4.146)–(4.150) are modified to

$$\omega_m(t) = \gamma_m \cdot \omega_f(t), \tag{4.151}$$

$$\omega_e(t) = \frac{z \cdot \omega_g(t) + \gamma_{ring} \cdot \omega_f(t)}{1 + z}, \tag{4.152}$$

$$T_e(t) = \frac{1 + z}{z} \cdot T_g(t), \tag{4.153}$$

$$T_m(t) = \frac{1}{\gamma_m} \cdot T_f(t) - \frac{\gamma_{ring}}{\gamma_m} \cdot \frac{T_g(t)}{z}. \tag{4.154}$$

where γ_{ring} and γ_m are the transmission ratios of the motor and ring speeds, respectively, to the final shaft. These ratios are easily calculated as a function of the number of teeth of the three torque-coupling gears.

Compound Power Split Devices

Compound power split devices are increasingly used, especially in large hybrid sport utility vehicle applications, since smaller electric machines can be used compared to simple PSDs.

An example of a compound PSD is found in GM's Two-mode Hybrid System (previously known as AHS-II) shown in Fig. 4.52a [232]. The gears of two planetary gearings are mutually connected and linked to the engine, to the two electric machines, and to the final driveline. Choosing the locking status of a pair of controlled clutches E1 and E2 permits the realization of two different operating modes.

In the first operating mode, used at low loads and relatively low vehicle speeds, E1 is disengaged while E2 is engaged, thus the mechanical connections can be schematized as in Fig. 4.52b. Applying (4.145) and (4.148) to both planetary gear sets, one obtains for the speeds

$$\omega_e(t) = -z_1 \cdot \omega_g(t) + \omega_f(t) \cdot (1 + z_1) \tag{4.155}$$

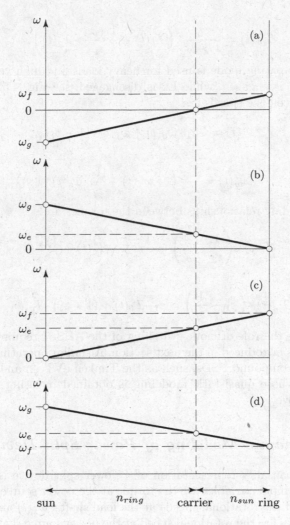

Fig. 4.51 Relationship between rotational speeds in four typical operation modes: (a) purely electric ($\omega_e = 0$), (b) engine/generator operation (battery recharge with vehicle stopped, $\omega_m = \omega_f = 0$), (c) electric assist (normal drive, $\omega_g = 0$), (d) acceleration

and

$$\omega_m(t) = \frac{1 + z_2}{z_2} \cdot \omega_f(t), \qquad (4.156)$$

where z_1 and z_2 are the epicyclic ratios of the first and of the second planetary gear set, respectively. The corresponding relationships between the torques are

$$T_e(t) = -\frac{1}{z_1} \cdot T_g(t) \qquad (4.157)$$

and

$$T_m(t) = \frac{z_2}{1 + z_2} \cdot \left(T_f(t) + \frac{1 + z_1}{z_1} \cdot T_g(t) \right). \qquad (4.158)$$

The second operating mode is used for heavy loads and high vehicle speeds. The clutch E1 is engaged, while E2 is disengaged, see Fig. 4.52c. The relationships between the speeds are

$$\omega_e(t) = -z_1 \cdot \omega_g(t) + \omega_f(t) \cdot (1 + z_1) \qquad (4.159)$$

and

$$\omega_m(t) = \frac{1 + z_2}{z_2} \cdot \omega_f(t) - \frac{1}{z_2} \cdot \omega_g(t). \qquad (4.160)$$

The corresponding relationships between the torques are

$$T_e(t) \cdot \left(\frac{1}{z_2} - z_1 \right) = \frac{1 + z_2}{z_2} \cdot T_g(t) + \frac{T_f(t)}{z_2} \qquad (4.161)$$

and

$$T_m(t) \cdot \left(z_1 - \frac{1}{z_2} \right) = z_1 \cdot T_f(t) + (1 + z_1) \cdot T_g(t). \qquad (4.162)$$

Understanding the role of mode-switching of the AHS-II requires a dynamic analysis that is introduced in the next section. A similar operating principle is used in other compound PSDs, such as the Timken eVT [2] and the Renault e-IVT [340], whose quasistatic modeling is obtained by using the methods illustrated above.

4.9.3 Dynamic Modeling of Power Split Devices

The physical causality representation of a power split device is sketched in Fig. 4.53. The input variables are the torque at the power source shafts, $T_e(t)$ and $T_m(t)$, and the rotational speeds at the load shafts, $\omega_f(t)$ and $\omega_g(t)$. The output variables are the rotational speed at the power source shafts, $\omega_e(t)$ and $\omega_m(t)$, and the torques at the load shafts, $T_f(t)$ and $T_g(t)$.

Dynamic models of PSDs include the inertia effects of the gears. Equations (4.146)–(4.147) are still valid and already in the required causality representation. Conversely, (4.149)–(4.150) are substituted by the more general equations

$$T_f(t) = T_m(t) + \frac{1}{1 + z} \cdot T_e(t) - \dot{\omega}_g(t) \cdot \frac{z \cdot \Theta_{carrier}}{(1 + z)^2} - \dot{\omega}_f(t) \cdot \left(\frac{\Theta_{carrier}}{(1 + z)^2} + \Theta_{ring} \right)$$
$$(4.163)$$

and

$$T_g(t) = \frac{z}{1 + z} \cdot T_e(t) - \dot{\omega}_g(t) \cdot \left(\frac{z^2 \cdot \Theta_{carrier}}{(1 + z)^2} + \Theta_{sun} \right) - \dot{\omega}_f(t) \cdot \frac{z \cdot \Theta_{carrier}}{(1 + z)^2},$$
$$(4.164)$$

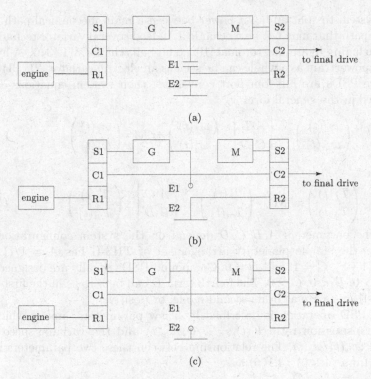

(a)

(b)

(c)

Fig. 4.52 Schematic of a Two-Mode Hybrid System (a) and of its two modes of operation (b) and (c). The figure shows the connections of the engine, the "speeder" machine or generator (G), the "torquer" machine or motor (M), via two planetary gear sets, for which S: sun gear, C: planetary carrier, R: ring gear. Two control clutches are labelled E1 and E2.

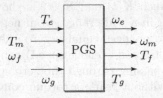

Fig. 4.53 Planetary gear sets: physical causality for dynamic modeling

where $\Theta_{...}$ are the moments of inertia. Despite the generality of (4.163)–(4.164), in planetary gear sets the dynamic effects normally are not considered, as in the case of torque couplers. The rotational inertia of the planetary gear set itself is indeed negligible when compared with those of the machines and shafts connected.

However, the physical causality representation is still useful to describe the role and the operation of a PSD within a generic "variator" architecture, i.e., a system to split an input power between a main mechanical path and a parallel path that may not be mechanical. Indeed several variator technologies exist, including hydraulic, toroidal [109], and belt/chain CVT [346]. A hybrid-electric powertrain as a whole can be seen as an electric variator [217, 340, 65].

Since PSDs are full four-port converters, their system equations can be expressed in the general form

$$\begin{pmatrix} \omega_e(t) \\ \omega_m(t) \end{pmatrix} = \begin{bmatrix} A & B \\ C & D \end{bmatrix} \cdot \begin{pmatrix} \omega_f(t) \\ \omega_g(t) \end{pmatrix} = \mathbf{M} \cdot \begin{pmatrix} \omega_f(t) \\ \omega_g(t) \end{pmatrix} \tag{4.165}$$

and

$$\begin{pmatrix} T_f(t) \\ T_g(t) \end{pmatrix} = \mathbf{M}^T \cdot \begin{pmatrix} T_e(t) \\ T_m(t) \end{pmatrix} = \begin{bmatrix} A & C \\ B & D \end{bmatrix} \cdot \begin{pmatrix} T_e(t) \\ T_m(t) \end{pmatrix}, \tag{4.166}$$

where the parameters A, B, C, D depend on the system configuration. For example, the simple planetary arrangement of THS-II has $A = 1/(1 + z)$, $B = z/(1+z)$, $C = 1$, and $D = 0$. Compound PSDs usually are designed such that $A > 0$, $B < 0$, $C > 0$. The fourth term D can be zero as in the first mode of AHS-II, negative as in its second mode, or positive.

The PSD operation is described by a few parameters, among which the speed transmission ratio $K(t) = \omega_f(t)/\omega_e(t)$ and the variator speed ratio $K_v(t) = \omega_m(t)/\omega_g(t)$. The relationship between these two parameters is calculated from (4.165)–(4.166) as

$$K_v(t) = D \cdot \frac{1 - \dfrac{K(t)}{K_1}}{1 - \dfrac{K(t)}{K_2}} \tag{4.167}$$

where $K_1 = (A - B \cdot C/D)^{-1}$ and $K_2 = 1/A$.

The relationship between $K(t)$ and $K_v(t)$ is shown in Fig. 4.54a. If the system allows K_v varying in a wide range, from negative to positive values, enabling a generator or motor speed equal to zero, then at a given input speed (engine speed), the output speed (vehicle speed) can vary continuously from reverse to forward speed passing through the zero speed. In other words, K may vary "infinitely"[22] without using a torque converter or any equivalent device.

The variator ratio K_v not only controls the speed transmission ratio, but also the split of power between the mechanical path and the electrical path. In variator theory, the power split ratio $r(t) = P_g(t)/P_e(t)$ is conveniently used to describe this effect. In the ideal case of unitary efficiency of the electrical path and no power supply from the battery, $P_m(t) = P_g(t)$.[23] After some

---

[22] Whence the name Infinitely Variable Transmission (IVT) often given to these systems.

[23] This assumption is obviously not realistic and is used here only for illustrating the general operation of PSDs. However, this point has to be handled carefully since it might leads to the conclusion that for $K \to 0$, $T_m \to \infty$.

manipulations of (4.165)–(4.166), the power split ratio thus can be expressed as a function of $K(t)$ as

$$r(t) = \frac{\left(1 - \dfrac{K(t)}{K_2}\right) \cdot \left(1 - \dfrac{K(t)}{K_1}\right)}{K(t) \cdot \left(\dfrac{1}{K_2} - \dfrac{1}{K_1}\right)}. \tag{4.168}$$

The dependency between $r(t)$ and $K(t)$ is illustrated in Fig. 4.54b for $K_1 < K_2$. The figure clearly shows that K_1 and K_2 are particular values of the speed transmission ratio, referred to as *nodes*, such that $r(t) = 0$, a condition that corresponds to a purely mechanical transmission. Between the nodes, the electric power exhibits a finite maximum. Outside of this range, however, the electric power would have nonphysically high levels [340]. Therefore, variators are designed to operate properly for values of K between nodes.

The power split ratio of simple PSDs is calculated as a special case of (4.168) with $D = 0$. This condition applies also to represent the first operating mode of a compound PSD such as the AHS-II, see (4.155)–(4.158). In both cases,

$$r(t)^{(\text{simple})} = 1 - \frac{K(t)}{K_2}, \tag{4.169}$$

that is to say, simple PSDs exhibit only one node K_2, while $K_1 \to 0$ as $D \to 0$. For $K = 0$ the system works as a series hybrid, with $r = 1$. In Fig. 4.54b the function $r(t) = r(K(t))$ of a simple PSD is plotted together with those of a compound PSD characterized by the same K_2. In the nominal operating range of K, i.e., between K_1 and K_2, the power split ratio of the simple PSD is always greater than that of the compound PSD. In other terms, the use of a compound PSD permits the reduction of the electric power for the same speed span. Consequently, the sizing of the electric machines can be reduced.

The analysis above also explains how dual-mode PSDs allow the range of K to be increased without increasing the power split ratio. Some systems, like for instance the e-IVT system, have two operating modes with two nodes each. The two modes share the same node K_2, while the other node, K_1, switches from $K_1^{(1)} < K_2$ to $K_1^{(2)} > K_2$. Consequently, the curve $r(t) = r(K(t))$ results from the superimposition of the two curves labeled A and B in Fig. 4.55a, one for each mode, with the switching occurring at K_2. In this way, large values of K are obtained without any excessive increase of r. The operation of a system such as the AHS-II is described as in Fig. 4.55a with $K_1^{(1)} \to 0$, thus with curve A given by (4.169).

The nodes of the PSD also play the role of vehicle speed thresholds at which the directions of energy flows between the linked machines are inverted and the system operation mode is changed [2, 340]. This behavior is observed in Fig. 4.55b, which shows ω_g and ω_m as a function of K for a dual-mode compound PSD. For $K < K_1^{(1)}$, the motor and the generator rotate in the same negative direction and they operate in their reverse power direction,

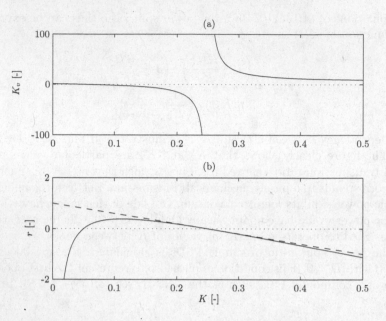

Fig. 4.54 Variator ratio (top) and power split ratio (bottom) as a function of the speed transmission ratio. Dashed curve is representative of a simple PSD with $D = 0$. Solid curves are representative of a compound PSD with $K_1 < K_2$. Numerical values: $K_1 = 0.05$, $K_2 = 0.25$.

with the "generator" working as a motor and the "motor" as a generator. As the vehicle speed increases, $K_1^{(1)} < K < K_2$, the machines rotate in opposite directions and in their positive power direction. Thus, the "generator" works as a generator and the "motor" as a motor. For $K_2 < K < K_1^{(2)}$ both machines rotate in the same positive direction; their power directions are inverted again. Finally, for $K > K_1^{(2)}$, the two machines counter-rotate again, while keeping their positive power direction.

4.10   Problems

Electric Propulsion Systems

Problem 4.1. Design an electric powertrain for a small city car having the following characteristics: curb mass $= 840$ kg, payload $= 2 \cdot 75$ kg, tires: 155/65/14T, $c_d \cdot A_f = 1.85$ m^2, rolling resistance coefficient $= 0.009$, to meet the following performance criteria: (i) max speed $= 65$ km/h, (ii) max grade $= 16\%$, (iii) 100 km range. Assume perfect recuperation, overall efficiency of 0.6, and 85% SoC window. Choose motor size in a class with a maximum speed of 6000 rpm and the number of battery modules having a capacity of

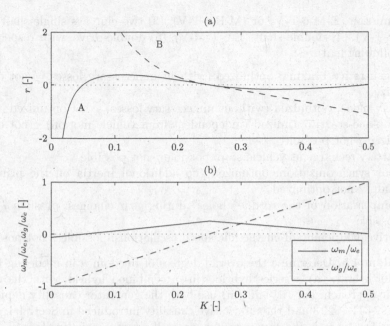

Fig. 4.55 Power split ratio (top), generator and motor speed (bottom) as a function of the speed transmission ratio. In (a) the solid curve is the resulting superimposition of curve A for the first mode and curve B for the second mode, with a switching occurring at K_2. Numerical values: $K_1^{(1)} = 0.05$, $K_2 = 0.25$, $K_1^{(2)} = 0.5$.

1.2 kWh each. *(Solution: motor rated power = 8 kW, motor maximum torque = 45 Nm, battery capacity = 16.8 kWh or 14 modules).*

Problem 4.2. Find an equation for the AER D_{ev} of a full electric vehicle as a function of its vehicle parameters, battery capacity and powertrain efficiency. Then evaluate the D_{ev} for a bus with the following characteristics: $\eta_{rec} = 100\%$, $\eta_{sys} = 0.45$ (including unused SoC), $c_r = 0.006$, $A_f \cdot c_d = 6.8 \cdot 0.62$, $Q_{bat} = 89$ Ah, $U_{bat} = 600$ V, $m_v = 14.6$ t, without payload and with a load of 60 passengers, respectively. Assume a MVEG-type speed profile. *(Solution: $D_{ev} = 48$ km without payload, $D_{ev} = 42$ km with payload).*

Problem 4.3. The 2011 Nissan Leaf electric vehicle has been rated by the EPA as achieving 99 mpg equivalent or 34 kWh/100 miles. Justify this rating. *(Hint: Use energy content of gasoline).*

Hybrid-Electric Propulsion Systems

Problem 4.4. Classify the five different parallel hybrid architectures, (1) single-shaft with single clutch between engine and electric machine (E-c-M-T-V), (2) single-shaft with single clutch between engine–electric machine and transmission (E-M-c-T-V) or (M-E-c-T-V), (3) two-clutches single-shaft (E-c-M-c-T-V), (4) double-shaft (E-c-T-M-V), (5) double-drive, with respect to the following features:

- regenerative braking: optimized (without unnecessary losses) / not optimized
- ZEV mode: optimized (without unnecessary losses) / not optimized
- stop-and-start: optimized (independent from vehicle motion) / not optimized / not possible
- battery recharge at vehicle stop: possible / not possible
- gear synchronization: optimized (no additional inertia on the primary shaft) / not optimized
- compensation of the torque "holes" during gear changes: possible / not possible
- active dampening of engine idle speed oscillations: possible / not possible

Problem 4.5. Determine the overall degrees of freedom u in modeling (i) a parallel hybrid, (ii) a series hybrid, (iii) a combined hybrid, with the quasistatic approach. For (ii) and (iii) use both the generator causality depicted in Figs. 4.11 – 4.13 and the alternative causality introduced in Sect. 4.4. *(Solution: For (i) u is the torque-split ratio at the torque coupler. For (ii) and (iii) u = \{u_1, u_2\} where u_1 is the power-split ratio at the power link and u_2 is the generator speed. In the alternative generator causality, u_1 is the generator torque, u_2 is the generator speed).*

Problem 4.6. Perform the same analysis as in Problem 4.5 with the dynamic approach. Calculate the number n_v of variables in the flowcharts of Figs. 4.11 – 4.13. Then calculate the number n_e of the equations available using the simple models presented in this chapter. Finally evaluate the manipulated variables that are necessary to realize the degrees of freedom (DOF) determined in Problem 4.5. *(Solution: For (i), n_e = 10, n_v = 10 and to realize the torque-split ratio DOF, u represents the engine control input. For (ii), n_e = 12, n_v = 12 and to realize the two DOF, u = \{u_1, u_2\}, where u_1 is the engine control input and u_2 is the control input of the DC–DC converter on either the APU or the battery side. For (iii), n_e = 16, n_v = 16 and then the same as (ii) holds).*

Problem 4.7. Perform the same analysis as in Problems 4.5 – 4.6 for an electric powertrain powered by a battery and a supercapacitor. *(Solution: For quasistatic simulation u is the power-split ratio at the power link. For dynamic simulation n_e = 10, n_v = 10, thus at least a DC–DC converter must be used to realize the DOF).*

Problem 4.8. For a plug-in hybrid, the fuel consumption according to UN/ECE regulation [91] is

$$C = \frac{D_e \cdot C_1 + D_{av} \cdot C_2}{D_e + D_{av}},$$

where C_1 is the fuel consumption in charge-depleting mode, C_2 is the consumption in charge-sustaining mode, D_e is the electric range, and D_{av} is 25 km, the assumed average distance between two battery recharges. Estimate the fuel consumption of the electric system of Problem 4.1 equipped with a range extender having a max power of 5 kW and an efficiency of 0.4. *(Solution: $C_2 = 3.4\,l/100\,km$, $C = 0.68\,l/100\,km$).*

Motor and Motor Controller

Problem 4.9. Consider a separately-excited DC motor having the following characteristics: $R_a = 0.05\,\Omega$, battery voltage = 50 V (neglect battery resistance), rated power = 4 kW, nominal torque constant $\kappa_a = \kappa_i = 0.25$ Wb, aimed at propelling a small city vehicle. Calculate the motor voltage and current limits, then the flux weakening region limit (maximum torque and base speed). Calculate the step-down chopper duty-cycle α for the following operating points: (i) $\omega_m = 100$ rad/s and $T_m = 15$ Nm; (ii) $\omega_m = 300$ rad/s and $T_m = 8$ Nm. *(Solution: $U_{max} = 50$ V, $I_{max} = 88$ A, $T_{max} = 22$ Nm, $\omega_b = 182$ rad/s, (i) $\alpha = 56\%$, (ii) $\alpha = 100\%$ and $\kappa_a = 0.17$ Wb).*

Problem 4.10. For the DC motor of Problem 4.9, evaluate the approximation of mirroring the efficiency from the first to the fourth quadrant, for the two operating points (i) $\omega_m = 50$ rad/s and $T_m = 22$ Nm; (ii) $\omega_m = 300$ rad/s, $T_m = 8$ Nm. Assume further that $P_{l,c} = 0$. *(Solution: (i) $\eta_m(50, 22) = 0.74$, $\eta_m(50, -22) = 0.65$, $P_l = 387$ W, (ii) $\eta_m(300, 8) = 0.98$, $\eta_m(300, -8) = 0.98$, $P_l = 51$ W. The approximation of mirroring the efficiency becomes better as the losses decrease).*

Problem 4.11. Using the PMSM model 4.40–4.43 calculate a static control law, i.e., a selection of reference values I_d, I_q as a function of torque and speed, such that the stator current intensity is minimized. Do the calculation in the case (i) $I_s \leq I_{max}$, $U_s \leq U_{max} = mU_m$ (maximum torque region) and (ii) when the voltage constraint is active (flux weakening region). Neglect the stator resistance R_s and consider a machine with $p = 1$. The stator current and voltage intensities are defined as

$$I_s^2 = I_q^2 + I_d^2, \quad U_s^2 = U_q^2 + U_d^2.$$

Evaluate the base speed. *(Solution: $I_q = \frac{T_m'}{\varphi_m}$, (i) $I_d = 0$;*
(ii) $I_d = \sqrt{\left(\frac{U_{max}}{L_s\omega_m}\right)^2 - \left(\frac{T_m'}{\varphi_m}\right)^2} - \frac{\varphi_m}{L_s}$, with $T_m' \triangleq \frac{2}{3}T_m$. Base speed: $\omega_b = \frac{U_{max}}{\sqrt{\varphi_m^2 + (L_s \cdot I_{max})^2}}$).

Problem 4.12. Evaluate the torque limit curve for a PMSM both (i) in the maximum torque region and (ii) in the flux weakening region, see Problem 4.11, assuming $R_s = 0$. Evaluate the transition curve between these two regions.

Assume that $\varphi_m > L_s I_{max}$ (why is that important?). *(Solution: (i) T_{max} =*
$\frac{3}{2}\varphi_m I_{max}$, *(ii)* $T_{max}(\omega_m) = \frac{3}{2}\varphi_m \sqrt{I_{max}^2 - \frac{1}{4\varphi_m^2 L_s^2}\left(\left(\frac{U_{max}}{\omega_m}\right)^2 - \varphi_m^2 - (L_s I_{max})^2\right)^2}$.

$\omega_{max} = \frac{U_{max}}{\varphi_m - L_s I_{max}}$. *Transition curve:* $T_m(\omega_m) = \frac{3}{2}\frac{\varphi_m}{L_s}\sqrt{\left(\frac{U_{max}}{\omega_m}\right)^2 - \varphi_m^2}$ *for*
$\omega_b \leq \omega_m \leq \frac{U_{max}}{\varphi_m})$.

Problem 4.13. Using the same assumptions as in Problem 4.11, evaluate
the maximim power curve as a function of speed. *(Solution: For $\omega_m < \omega_b$,*
$P_{max} = \varphi_m I_{max}$. For $\omega_m < \omega_b$, $P_{max} = P_{max}(\omega_m) \leq U_{max} I_{max}$. There is
one value of speed $\omega_P = \frac{U_{max}}{\sqrt{\phi_m^2 - L_s^2 I_{max}^2}}$ for which $P_{max} = U_{max} I_{max}$).

Problem 4.14. Equation 4.42 is only valid when $L_d = L_q = L_s$. In the
general case in which $L_d \neq L_q$, the correct equation is

$$T_m = \frac{3}{2}\cdot p \cdot I_q \cdot (\varphi_m - p \cdot (L_q - L_d) \cdot I_d).$$

Consider again Problem 4.11 and derive a static control law I_d, I_q that minimizes the current intensity (MTPA), assuming that the constraints over current and voltage are not active, for a motor where $p = 4$, $R_s = 0.07\,\Omega$,
$L_q = 5.4 \cdot 10^{-3}\,\mathrm{H}$, $L_d = 1.9 \cdot 10^{-3}\,\mathrm{H}$, $\varphi_m = 0.185\,\mathrm{Wb}$ and $T = 50\,\mathrm{Nm}$. Then
compare the result with that obtained for $\Delta L = L_q - L_d = 0$. *(Solution: (i)*
$I_q = -17\,A$, $I_q = 34\,A$, (ii) for $\Delta L = 0$, $I_d = 0$, $I_q = 45\,A$).

Problem 4.15. Calculate the torque characteristic curve $T_m(\omega_m, U_s)$ of a
PMSM having the following characteristics: $R_s = 0.2\,\Omega$, $L = 0.003\,\mathrm{H}$, $\frac{3}{2}\varphi_m =$
$0.89\,\mathrm{Wb}$, $p = 1$, for a voltage intensity (see definition in Problem 4.11) $U_s =$
$30\,\mathrm{V}$. Derive an affine approximation of the DC-motor type, $T_{m,lin}(\omega_m, U_s)$.
Evaluate the torque error $\varepsilon(\omega_m) \triangleq U_s^2(T_m) - U_s^2(T_{m,lin})$ and calculate its
maximum value. *(Solution: breakaway torque $= \frac{U_s \varphi_m}{R_s}$, zero-torque speed =*
$\frac{U_s}{\varphi_m}$, $\varepsilon_{max} = 14.4\ V^2$ or 1.6%).

Problem 4.16. A simple thermal model of an electric machine reads

$$C_{t,m}\cdot\frac{d}{dt}\vartheta_m(t) = P_l(t) - \frac{\vartheta_m(t) - \vartheta_a}{R_{th}},$$

where $C_{t,m}$ is an equivalent thermal capacity, R_{th} is an equivalent thermal
resistance, $\vartheta_m(t)$ is the relevant motor temperature, and ϑ_a is the external
temperature. Derive the current limitation to I_{max} from thermal considerations using the models of Sect. 4.3.3 (DC motor). How would the result
change if other losses of the type $\beta \cdot \omega_m$ (iron losses, mechanical losses)
were taken into account? *(Solution: (i) $I_{max} = \sqrt{\frac{\vartheta_{max} - \vartheta_a}{R_{th} R_a}}$, (ii) $I_{max}(\omega_m) =$*
$\sqrt{\frac{\vartheta_{max} - \vartheta_a - \beta R_{th}\omega_m}{R_a R_{th}}}$).

Problem 4.17. Derive a (simplified) rule to express peak torque limits of an electric machine as a function of application time. Use the result of Problem 4.16. *(Solution: $T_{max}(t) = \kappa_a \sqrt{\frac{\vartheta_{max} - \vartheta_a}{R_{th} R_a \cdot \left(1 - e^{-\frac{t}{\tau}}\right)}}$).*

Problem 4.18. One PMSM has the following design parameter: external diameter $d_1 = 0.145$ m, weight $m_1 = 14$ kg, length $l_1 = 0.06$ m. At 5500 rpm it delivers a maximum torque $T_1 = 12$ Nm. Predict the power P_2 and the weight m_2 for a similarly designed machine with a diameter $d_2 = 0.2$ m and a length $l_2 = 0.2$ m. Compare the cases in which the design is made (i) on the basis of constant tangential stress and peripheral speed, or (ii) of constant speed. *(Solution: (i) $P_2 = 31.5$ kW, $m_2 = 89$ kg, (ii) $P_2 = 43$ kW, $m_2 = 89$ kg).*

Problem 4.19. Evaluate the specific power of a motor and inverter assembly, knowing that $\left(\frac{P}{m}\right)_{motor} = 1.2$ kW/kg and $\left(\frac{P}{m}\right)_{inverter} = 11$ kW/kg. *(Solution: $\left(\frac{P}{m}\right) = 1.08$ kW/kg).*

Range Extenders

Problem 4.20. Consider an APU for a series hybrid. Given the engine model

$$P_f = \frac{P_e + P_0}{e},$$

$$\frac{1}{e} = 5.07 - 0.0117 \cdot \omega_e + 1.50 \cdot 10^{-5} \cdot \omega_e^2 = a_1 + b_1 \cdot \omega_e + c_1 \cdot \omega_e^2$$

$$\frac{P_0}{e} = -1.22 \cdot 10^3 + 31.7 \cdot \omega_e + 0.421 \cdot \omega_e^2 = a_2 + b_2 \cdot \omega_e + c_2 \cdot \omega_e^2$$

$$T_{max} = 96.9 + 1.35 \cdot \omega_e - 0.0031 \cdot \omega_e^2 = h \cdot \omega_e^2 + g \cdot \omega_e + f,$$

and a generator model with constant efficiency $\eta_g = 0.92$, derive an OOL structure $\hat{\omega}(P_g)$. Then calculate the operating points for (i) $P_g = 10$ kW, (ii) $P_g = 40$ kW, and (iii) $P_g = 60$ kW. *(Solution: (i) $\hat{\omega} = 104.7$ rad/s, (ii) $\hat{\omega} = 222$ rad/s, (iii) $\hat{\omega} = 279$ rad/s).*

Problem 4.21. For the APU model of Problem 4.20, find a piecewise affine approximation $P_f = a + b \cdot P_g$. Evaluate the error with respect to the nonlinear model, for (i) $P_g = 10$ kW, (ii) $P_g = 40$ kW, and (iii) $P_g = 60$ kW. *(Solution: The piecewise affine model is (i) $a = b = 0$ for $P_g = 0$; (ii) $a = 6.7 \cdot 10^3$, $b = 4.36$ for $0 < P_g \leq 12.9 \cdot 10^3$, (iii) $a = 15.8 \cdot 10^3$, $b = 3.66$ for $14 \cdot 10^3 < P_g < 62 \cdot 10^3$, (iv) $a = -14.7 \cdot 10^3$, $b = 4.19$ for $62 \cdot 10^3 < P_g < 68.3 \cdot 10^3$. The error is (i) 2%, (ii) 2.5 %, (iii) 1%).*

Problem 4.22. Propose an algorithm to calculate the OOL of an engine for a combined hybrid from the data w4x (APU speed breakpoint vector), T4x (APU torque breakpoint vector), Tmax(w4x) (APU maximum torque), and mfuel(w4x,T4x) (engine consumption map).

Problem 4.23. Find an optimal design for a range extender working at a stationary operating point, i.e., select optimal values for the displacement volume V_d and the speed ω_e of the engine, neglecting stop-and-start effects and making the following approximations. The IC engine is modeled with a Willans line

$$p_{me} = e \cdot p_{mf} - (p_{me0} + p_{me2} \cdot \omega_e^2)$$

with $e = 0.4$, $p_{me0} = 1.5 \cdot 10^5$ Pa, $p_{me2} = 1.4$ Pa·s^2. The generator is a DC machine with constant armature resistance $R_a = 0.2\,\Omega$ and $\kappa_a = 0.5$. The battery is modeled as an internal voltage source $U_{oc} = 180$ V with an internal resistance $R_b = 0.3\,\Omega$. The overall system efficiency

$$\eta_{ov} = \frac{U_{oc} \cdot I_a}{\overset{*}{m}_f \cdot H_l}$$

should be optimal. The nominal power P_e of the range extender should be greater than 30 kW and the brake mean effective pressure of the engine smaller than 9 bar. The design parameters can be chosen between the following boundaries: $V_d \in [0.5, 2]$ l, $\omega_e \in [U_{oc}/\kappa_a, 600]$ rad/s. *(Solution: $\omega_e = 484\,rad/s$, $V_d = 0.87\,l$, $\eta_{ov} = 0.19$).*

Problem 4.24. In Problem 4.23, set the engine displacement volume to $V_d = 1.6$ l and find the optimal operating speed that maximizes the overall efficiency while respecting the two constraints on P_e and p_{me}. *(Solution: $\omega_e = 503\,rad/s$, $\eta_{ov} = 0.15$).*

Batteries

Problem 4.25. For a battery pack having the following characteristics: $Q_{cell} = 5$ Ah, $U_{cell} = 3.14 + 1.10 \cdot \xi$ (V), $R_{cell} = 0.005 - 0.0016 \cdot \xi$ (Ω) under discharge and $R_{cell} = 0.0020 \cdot \xi^2 - 0.0020 \cdot \xi + 0.0041$ (Ω) under charge, $N = 96$, calculate the electrochemical power P_{ech} for an electric power demand of 15 kW in charge and discharge, respectively, and for 20% and 90% state of charge. *(Solution: For (i) $P_b = 15\,kW$ and $\xi = 0.2$, $P_{ech} = 16.1\,kW$, for (ii) $P_b = 15\,kW$ and $\xi = 0.9$, $P_{ech} = 15.4\,kW$, for (iii) $P_b = -15\,kW$ and $\xi = 0.2$, $P_{ech} = -14.3\,kW$, for (iv) $P_b = -15\,kW$ and $\xi = 0.9$, $P_{ech} = -14.7\,kW$).*

Problem 4.26. Find a quadratic approximation for the relationship between battery power P_b and electrochemical power P_{ech}. Compare the results with those of Problem 4.25. *(Solution: $P_{ech} \approx P_b + 2 \cdot \frac{R}{U_{oc}^2} \cdot P_b^2$).*

Problem 4.27. One couple of electrodes for a lithium cell has the following characteristics: Negative electrode (graphite): capacity $q_{rev,n} = 340$ mAh/g, potential $U_n = 0.25$ V, density $\rho_n = 2.2$ g/cm^3, thickness $s_n \leq 80\,\mu$m. Positive electrode (LiCoO$_2$): capacity $q_{rev,p} = 140$ mAh/g, potential $U_p = 3.85$ V, density $\rho_p = 4$ g/cm^3, thickness $s_p \leq 80\,\mu$m. Separator, collector: surface density 0.047 g/cm^2. Calculate the cell voltage and the cell specific energy. *(Solution: $U_{cell} = 3.6$ V, $q_{rev} = 356$ Wh/kg, $q_{cell} = 176$ Wh/kg).*

Problem 4.28. Develop an equation for the battery apparent capacity as a function of the current for constant current discharge using the battery modeling equations of Section 4.5.2. Then evaluate the apparent capacity of a battery with nominal capacity $Q_0 = 72$ Ah, $\kappa_4 = -0.005$, $\kappa_2 = 1.22$, at $C/10$, C_1, and C_{10} discharge. *(Solution: (i) at $C/10$, $Q_0 = 70$ Ah, (ii) at C_1, $Q_0 = 56$ Ah, (iii) at C_{10}, $Q_0 = 18$ Ah).*

Problem 4.29. Verify that the round-trip efficiency of a battery under constant current discharge-charge and for varying parameters U_{oc}, R_i as described in Section 4.5.2 has the same form as (4.82) but with U_{oc} and R_i calculated for $\xi = 0.5$.

Problem 4.30. Verify that (4.111) simplifies to (4.110) when $\eta_z = 0$ and $\partial \phi_z / \partial \vartheta_b = 0$. *(Hint: Use the charge balance equations (4.103)–(4.104) and the voltage equation (4.63) with $U_{oc} = \Delta \Phi_{pos} - \Delta \Phi_{neg}$).*

Supercapacitors

Problem 4.31. Derive Equation (4.117).

Problem 4.32. Derive an analytical solution for the discharge of a supercapacitor with maximum power. Then verify the solution with the data of Figure 4.42. *(Solution: $Q_{sc}(t) = Q_0 \cdot e^{-\frac{t}{2 \cdot C_{sc} \cdot R_{sc}}}$, $I_{sc}(t) = \frac{Q_0}{2 \cdot C_{sc} \cdot R_{sc}} \cdot e^{-\frac{t}{2 \cdot C_{sc} \cdot R_{sc}}}$, $U_{sc}(t) = \frac{Q_0}{2 \cdot C_{sc}} \cdot e^{-\frac{t}{2 \cdot C_{sc} \cdot R_{sc}}}$, $P_{sc}(t) = \frac{Q_0^2}{4 \cdot R_{sc} \cdot C_{sc}^2} \cdot e^{-\frac{t}{C_{sc} \cdot R_{sc}}}$).*

Problem 4.33. Yet another definition of supercapacitor efficiency that is sometimes found (e.g., in [222]) is

$$\eta_{sc,d} = 1 - 2\frac{\tau}{t_f},$$

during discharge at constant current, and

$$\eta_{sc,c} = \frac{1}{1 + 2\frac{\tau}{t_f}},$$

during charge at constant current, where $\tau = C_{sc} \cdot R_{sc}$ and t_f has the same meaning as in the text. Explain this definition. *(Hint: Use the definition (4.129)).*

Electric Power Links

Problem 4.34. Consider again Problem 4.9 and account for a battery internal resistance of $0.025\,\Omega$. *(Solution: (i) $\alpha = 57\%$, (ii) $\alpha = 100\%$ and $\kappa_a = 0.14$).*

Problem 4.35. For an electric drive including a battery, a step-down DC–DC converter (chopper), and a DC motor, calculate the duty cycle that maximizes the regenerated power during braking. Calculate the corresponding motor current, using the data presented in Problem 4.9. *(Solution: $\alpha_{max} = 76\%$, $I_{a,min} = -250\,A$).*

Problem 4.36. Consider an electric drive including a battery, a boost DC–DC converter and a motor. Derive a relationship between the maximum torque curve of the motor and the converter ratio, assuming that for $\omega_m > \omega_b$, $P_{max} \approx U_{max} \cdot I_{max}$. Conceive a strategy to perform quasistatic simulations in this case.

Problem 4.37. Consider a semi-active power link with a battery, a super-capacitor, an electric motor, and a DC–DC converter on the supercapacitor branch. Derive an analytical relationship between the control factor u (4.132) and the DC–DC converter voltage ratio R. Calculate the values of R to obtain a pure battery supply ($u = 0$) or a pure supercapacitor supply ($u = 1$). Describe the supercapacitor on a quasistatic basis, i.e.,

$$C_{sc} \cdot \frac{U_{s0} - U_{sc}}{\tau} = -I_{sc},$$

where τ is the time step. *(Solution: $R(u)$ is the solution of the quadratic equation $\tau \cdot I_m \cdot u \cdot R^2 - C_{sc} \cdot U_{s0} \cdot R + C_{sc} \cdot (R_b \cdot I_m \cdot u - R_b \cdot I_m + U_{oc}) = 0$ and it depends on I_m, in agreement with the physical causality representation. For $u = 0$, $R = \frac{U_{oc} - R_b I_m}{U_{s0}}$).*

Torque Couplers

Problem 4.38. Consider a through-the-road parallel hybrid. The torque coupling has the following characteristics: transmission ratio between the rear-axle motor and the wheels $\gamma_m = 11$, transmission ratios between the front-axle engine and the wheels $\gamma_e = \{15.02, 8.09, 5.33, 3.93, 3.13, 2.59\}$, wheel radius $r_{wh} = 31.7\,\text{cm}$. Moreover, $T_{m,max} = -T_{m,min} = 140\,\text{Nm}$, $P_{m,max} = -P_{m,min} = 42\,\text{kW}$, engine speed limited to $\omega_{e,max} = 4500\,\text{rpm}$, engine maximum torque $T_{e,max} = 1.34 \cdot 10^{-5} \cdot \omega_e^3 - 0.0149 \cdot \omega_e^2 + 4.945 \cdot \omega_e - 243\,(\text{Nm})$, engine minimum torque $T_{e,min} = -20\,\text{Nm}$. For a driving situation with $V = 30\,\text{m/s}$ and $T_{wh} = 675\,\text{Nm}$, determine the limits imposed by the motor operation on the engine operation. *(Solution: For gears 3rd to 6th, $T_e \leq \{242, 270, 255, 225\}$ and $T_e \geq \{1.6, 2.1, 2.7, 3.2\}$).*

Power Split Devices

Problem 4.39. Consider a power-split combined hybrid powertrain with a planetary gear set linking the engine, the generator, and the output shafts with the following Willis relation

$$\omega_g = 3.6 \cdot \omega_e - 2.6 \cdot \omega_f$$

The second electric machine is mounted directly on the output shaft without any reduction gear. The generator has the following characteristics (both in motor and in generator modes): maximum torque $= 160$ Nm, maximum power $= 25$ kW, maximum speed $= 1200$ rad/s. The motor has the following characteristics (both in motoring and in generating mode): maximum torque $= 400$ Nm, maximum power $= 25$ kW, and maximum speed $= 700$ rad/s. The S.I. engine has the following characteristics: maximum torque curve $= \{0\ 87\ 110\ 107\}$ Nm @ $\{400\ 1300\ 3700\ 5000\}$ rpm. The maximum battery power is 35 kW. Consider a driving situation in which the speed of the output shaft is 343 rad/s and the required torque is 129 Nm. Supposing that the decision variables of the energy management strategy are engine speed and torque, evaluate the admissible range of these variables. *(Solution: Admissible range limited by curves $\mathcal{A}$, $\mathcal{B}$, $\mathcal{C}$, and $\mathcal{D}$, where $\mathcal{A}$: engine max torque curve, $\mathcal{B}: T_e = 28$ Nm is the motor max torque curve, $\mathcal{C}: \frac{T_e}{3.6} = \frac{25000}{3.6 \cdot \omega_e - 892}$ is the generator max power curve, $\mathcal{D}: \omega_e = 524\ rad/s$).*

Problem 4.40. Derive the coupling matrix for the four torque levels of a PSD from the elements of the kinematic matrix in the case of quasistatic modeling. Do the same in the case of forward modeling, i.e., derive (4.166). *(Hint: Use power balance and speed relationships, then equate the factors of the input speed levels).*

Problem 4.41. For the compound power split device architecture shown in Fig. 4.56, derive the kinematic matrix $\mathbf{M}$ and the values of the two kinematic nodes. *(Solution: $\mathbf{M}$ as in (4.165) with $A = 1/(1 + z_2)$, $B = z_2/(1 + z_2)$, $C = (z_1 + z_2 + z_1 z_2)/z_1(1 + z_2)$, $D = -1/z_1(1 + z_2)$. The two nodes are calculated (see Problem 4.42) as $K_1 = 1/(1 + z_1)$ and $K_2 = 1 + z_2$).*

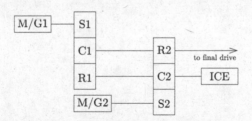

Fig. 4.56 Compound power-split configuration for Problem 4.41

Problem 4.42. Derive equation (4.167) for K_v as a function of K and equation (4.168) for r as a function of K, including the definitions of K_1 and K_2. *(Hint: Use the definition of K in speed equations, then in the definition of K_v. Use speed and torque equations in r both defined as P_g/P_e and as P_m/P_e, then equate the two equations to solve for the unknown variable T_m/T_e).*

Problem 4.43. Derive (4.163)-(4.164). *(Hint: proceed as in Problem 4.40 while taking into account inertia terms).*

Problem 4.44. Extend equations (4.149)-(4.150) to the case where there are losses in the planetary gearset. *(Hint: Define efficiencies of the contacts carrier-sun and carrier-ring and rewrite power balance).*

5

Non-electric Hybrid Propulsion Systems

The introductory section of this chapter describes several devices that may all be classified as short-term storage systems. The use of short-term storage systems in powertrain applications and the possible powertrain configurations are discussed first.

The next sections analyze the modeling approaches of hybrid-inertial, hybrid-pneumatic, and hybrid-hydraulic powertrains. Specific components such as flywheel accumulators, continuously-variable transmissions, pneumatic and hydraulic accumulators, pumps/motors are all described in accordance with the quasistatic and the dynamic modeling approach.

5.1 Short-Term Storage Systems

Aside from the energy carriers encountered in Chap. 3 (fossil fuels) and in Chap. 4 (electrochemical batteries), other methods are suitable for storing energy on-board. However, due to their low specific energy — they are referred to as short-term storage systems (3S) —, these systems are not suitable as the only on-board energy carrier. Instead, they may be used in hybrid vehicles in combination with a prime mover, such as an internal combustion engine or a fuel cell, with two main goals. On the one hand, due to the high specific power of 3S, such vehicles are very efficient in recuperating brake energy.[1] On the other hand, 3S-based architectures have a degree of freedom in load leveling the prime mover and shifting its operating points towards high efficiency. For instance, they enable the implementation of cyclic operation (duty-cycle operation, DCO) with the prime mover that either operates in a high-efficiency point or is turned off. In the engine-on phase, the short-term storage system is recharged, while in the engine-off phase it provides the energy for traction. However, due to the low energy density of 3S, the zero-emission driving range

---

[1] Typically, the energy available for recuperation, $E_v = \frac{1}{2} m_v v^2$ (order of magnitude, 10–100 Wh), is commonly stored at a power of 10–50 kW.

is limited to a few hundred meters. Of course, the overall benefit obtained with a 3S-based hybridization is partially overcome by the additional mass installed on-board, which therefore has to be carefully limited at a reasonable fraction (e.g., 10%) of the vehicle mass, see Chap. 2.

In general, several principles are conceivable for 3S-based power trains, for instance:

1. electrochemical, generator/motor and battery;
2. electrostatic, generator/motor and supercapacitor;
3. electromagnetic, generator/motor and superconductor coil;
4. inertial, CVT and flywheel;
5. potential, CVT and torsion spring;
6. pneumatic, pneumatic pump/motor and accumulator; and
7. hydraulic, hydraulic pump/motor and accumulator.

Technically, only the solutions (1), (2), (4) and (7) are currently employed, though also (3) has been proposed [306]. The solution (6) has been analyzed using simulations [137] and is the topic of ongoing research. The first two electric solutions are used in hybrid-electric vehicles and have been discussed already in Chap. 4, while solutions (4), (6), and (7) will be treated in more detail below.

Figure 5.1 shows the typical ranges of specific energy and specific power of the most common short-term storage systems (Ragone plot). Each device is located in a characteristic region of the power–energy plane. As a general rule, the specific energy that can be delivered by a specific device decreases when its specific power is increased. This behavior can be explained mainly by the internal losses of the device, which increase with the power. Instead of showing a single curve, the plot shows a region for each device since the characteristics of energy storage systems vary substantially with the materials used, the manufacturing processes, etc.

Ragone curves of batteries and supercapacitors have been described in Chapter 4. For batteries, this analysis yields (4.83)–(4.85) and is repeated here as

$$E_b(P_b) = \frac{2 \cdot R_i \cdot Q_0 \cdot P_b}{U_{oc} - \sqrt{U_{oc}^2 - 4 \cdot R_i \cdot P_b}}, \tag{5.1}$$

which clearly is a monotonously decreasing curve due to the presence of the internal resistance R_i. Maximum power is retrieved as being (4.74), while maximum energy is $E_{b,max} = Q_0 \cdot U_{oc}$. If battery self-discharge is considered [57], the Ragone curve presents a maximum.

For supercapacitors, the Ragone curve is obtained from (4.118),

$$E_{sc}(P_{sc}) = \frac{C_{sc}}{2} \cdot \left(-R_{sc} \cdot P_{sc} \cdot \ln\left(\frac{U_0(P_{sc})^2}{R_{sc} \cdot P_{sc}}\right) + U_0(P_{sc})^2 - R_{sc} \cdot P_{sc} \right), \tag{5.2}$$

with

$$U_0(P_{sc}) = \frac{Q_0}{2 \cdot C_{sc}} + \sqrt{\frac{Q_0^2}{4 \cdot C_{sc}^2} - P_{sc} \cdot R_{sc}}.$$ (5.3)

This curve is decreasing as well, again due to the presence of the internal resistance R_{sc}. The maximum power is given by (4.130) and the maximum energy by (4.129).

Similar curves will be obtained in the following sections for other short-term storage systems.

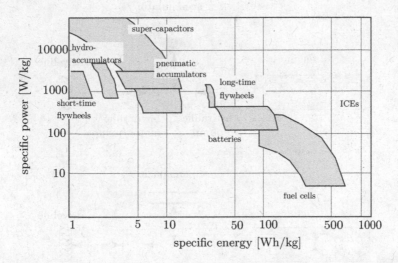

Fig. 5.1 Specific power versus specific energy for various short-term energy storage systems

The vehicular concepts with short-term storage systems can be series or parallel hybrids, in accordance with the classification of Chap. 4. Series hybrid system (see Fig. 5.2) have three power converters and two main energy conversion steps. In engine–supercapacitor systems the linking energy is electrical, while in engine–accumulator systems it is hydraulic energy or enthalpy. Parallel concepts are used in micro or mild hybrids with a hydraulic accumulator or a supercapacitor. In engine–flywheel systems the linking energy is mechanical, so there is no difference in principle between the parallel and the series configuration.

The design of models of hybrid-hydraulic and hybrid-inertial systems is based on the basic ideas introduced in the previous chapters. In particular, the concepts of modularity and of quasistatic versus dynamic models are still valid. Figures 5.3 and 5.4 show the flow of power factors in quasistatic and dynamic simulations of a hybrid-inertial and a hybrid-hydraulic vehicle. Notice that the power flow of Fig. 5.3 also describes configurations in which the flywheel is mounted on the engine transmission shaft, as shown in Fig. 5.2.

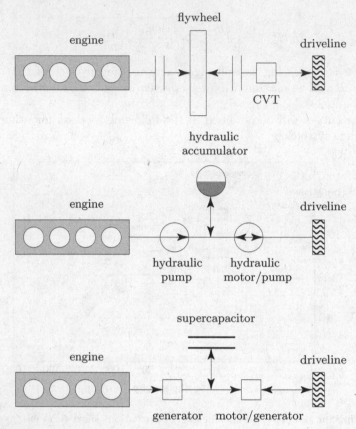

Fig. 5.2 Series hybrid concepts with short-term energy storage systems. For a description of a pneumatic hybrid, see Fig. 5.24.

5.2 Flywheels

5.2.1 Introduction

Low-speed flywheels have been used in various forms for centuries, and they have a long history of use in automotive applications. Early passenger cars featured a hand crank connected to a flywheel to start the engine, whereas all of today's internal combustion engines use flywheels to smooth the output power despite the reciprocating nature of the combustion torque.

In hybrid-inertial concepts a flywheel is mostly used as a kinetic energy recovery system (KERS) that allows regenerating braking energy [322, 67, 346]. During deceleration, kinetic energy is transferred from the wheels to the flywheel through mechanical connections, and the flywheel accelerates. Later, the stored kinetic energy can be transferred back to the wheels, thus helping the engine during vehicle launch. Additionally, the energy in the flywheel can

(a) quasistatic approach

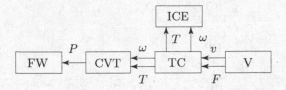

(b) dynamic approach

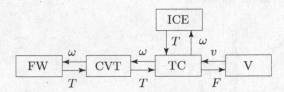

Fig. 5.3 Flow of power factors for a hybrid-inertial propulsion system, with the quasistatic approach (a) and the dynamic approach (b). Same block nomenclature as in Fig. 4.3 (CVT: continuously-variable transmission, FW: flywheel), F: force, P: mechanical power, T: torque, v: speed, ω: rotational speed.

be used to propel the vehicle, solely or in assistance to the engine, once it reaches cruising speed. Despite the short duration of the energy storage, benefits in fuel economy can be effective, particularly during driving that includes repeated stops and starts. In order to allow for such operation, the flywheel rotational speed must vary independently of the vehicle speed. Therefore a continuously variable transmission (CVT) system with a very wide range is necessary between the flywheel shaft and the drive train. Section 5.3 treats CVT systems in more detail. In principle, the flywheel can be not only accelerated during vehicle deceleration but also by the engine, if the latter provides more power than necessary during traction. Cycles of discharge (boost) and recharge (regenerative braking and recharge by the engine) may be operated, similarly to the duty-cycle operation of series HEVs. The case study in Sect. A.3 describes the optimization of the duty-cycle parameters.

Mixed inertial/electrical hybrid concepts employ a high-speed flywheel[2] to load-level electrochemical batteries [356] or as the only energy storage system. The latter concept has found application as a KERS in racing cars, where it provides regenerative braking as well as boosting capability, e.g., when accelerating out of a bend or when overtaking [353]. An electric motor/generator is mounted on the rotor shaft both to accelerate (charge, increase kinetic energy) and decelerate (discharge, decrease kinetic energy) the rotor, while traction power is provided by a second electric machine or the engine directly.

---

[2] Often referred to as "electromechanical batteries" in these applications.

(a) quasistatic approach

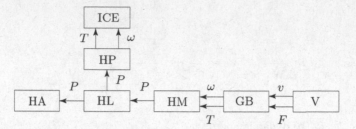

(b) dynamic approach

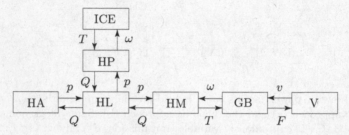

Fig. 5.4 Flow of power factors for a hybrid-hydraulic propulsion system, with the quasistatic approach (a) and the dynamic approach (b). Same block nomenclature as in Fig. 4.2 (HA: hydraulic accumulator, HM: hydraulic motor, HP: hydraulic pump, HL: hydraulic line), F: force, p: pressure, P: mechanical power, Q: flow rate, T: torque, v: speed, ω: rotational speed.

Combined architectures with a mechanical connection of the two electric machines through a power split device have been also proposed [142]. Duty-cycle operation is possible in both cases, but usually applied to road vehicles only [67].

Compared to electrochemical batteries, flywheels have a lower specific energy, but a higher specific power (up to 10 times higher). The advantages of flywheels are that they contain no acids or other potentially hazardous materials, that they are not affected by extreme temperatures, and that they usually exhibit a longer life. So far, flywheels have only been used in busses and racing cars. Their use in passenger cars has been advocated as well [271] but to be successful, flywheel technology would probably need to provide a specific energy higher than the levels currently available. In addition, there are some concerns regarding the complexity and the safety of flywheels. The latter aspect strongly limits the flywheel mass that can be installed on board, thus limiting maximum power and energy capacity.

Flywheels store kinetic energy within a rapidly spinning wheel-like rotor or disk. In order to achieve a sufficient amount of specific energy, modern flywheel rotors must be constructed from materials of high specific strength.

That requirement led to the selection of composite materials employing carbon fibers, while metals (steel) can be used for the internal hub that yields a smaller contribution to the inertia.[3] With this type of construction, rotor speed ranges from typical engine speeds up to very high speeds. Typical values are between 100 rev/s (road applications) and 1000 rev/s (racing cars). The reduction of the aerodynamic losses associated with such high rotor speeds requires the rotor to be placed in a vacuum chamber. This in turn leads to an additional vaccum pump and vacuum management system, and additional design requirements for the rotor bearing. These bearings must have low losses and must be stiff to adequately constrain the rotor and stabilize the shaft. These requirements frequently lead to choosing magnetic bearings with losses of kinetic energy in the order of 2% per hour. In order to reduce the gyroscopic forces transmitted to the magnetic bearings during pitching and rolling motions of the vehicle, a gimbal mount is often adopted [135]. The alternative of using counter-rotating rotors, which do not transmit any gyroscopic forces to the outside, is burdened by the fact that internally they transmit very large forces that are not easily supported by the magnetic bearings.

The specific energy of flywheel systems is limited by several factors, such as the maximum allowable stress and the maximum allowable rotor speed.[4] Small units that are technically feasible today reach 30 Wh/kg, including housing, electronics, possibly vacuum pump, etc. Much higher values of up to 140 Wh/kg are predicted by some authors for advanced rotor materials [333]. In contrast, KERS applications may have a high specific power of 2.4 kW/kg, but the system specific energy can be as low as 5 Wh/kg [67].

5.2.2 *Quasistatic Modeling of Flywheel Accumulators*

The causality representation of a flywheel accumulator in quasistatic simulations is sketched in Fig. 5.5. The input variable is the power $P_f(t)$ required at the output shaft. A positive value of $P_f(t)$ discharges the flywheel, while a negative value of $P_f(t)$ charges it. The output variable is the flywheel speed $\omega_f(t)$, sometimes regarded as the "state of charge" of the flywheel.

$$\xrightarrow{\quad P_f \quad} \boxed{\text{FW}} \xrightarrow{\quad \omega_f \quad}$$

Fig. 5.5 Flywheel accumulators: causality representation for quasistatic modeling

Models of flywheel accumulators may be derived on the basis of Newton's second law for a rotational system. The resulting equation is

---

[3] In integrated "electromechanical" batteries, the composite flywheel rotor integrates the magnets of the electric machine.

[4] This is usually limited by the first critical speed of the flywheel [333].

$$\Theta_f \cdot \omega_f(t) \cdot \frac{d}{dt}\omega_f(t) = -P_f(t) - P_l(t), \qquad (5.4)$$

with Θ_f being the moment of inertia of the rotor, from which the flywheel speed can be calculated. The term P_l describes the power losses.

For flywheel accumulators, two main loss contributions are usually considered, namely air resistance and bearing losses, $P_l(t) = P_{l,a}(t) + P_{l,b}(t)$. Both terms are functions of $\omega_f(t)$, i.e., of the peripheral velocity at the outer radius $u(t) = \omega_f(t) \cdot d/2$. The general expression for the air resistance force is proportional to the air density ρ_a, to $u^2(t)$, and to the frontal wheel area. The proportionality coefficient is a function of the Reynolds number and of the geometric ratio $\beta = b/d$, where b is the axial width and d the wheel diameter, see Fig. 5.7. For Reynolds numbers above $3 \cdot 10^5$, an expression for the power losses due to air resistance [94] is

$$P_{l,a}(t) = 0.04 \cdot \rho_a^{0.8} \cdot \eta_a^{0.2} \cdot u^{2.8}(t) \cdot d^{1.8} \cdot (\beta + 0.33), \qquad (5.5)$$

where η_a is the dynamic viscosity of air.

For the bearing losses, a general expression frequently used [94] is

$$P_{l,b}(t) = \mu \cdot k \cdot \frac{d_w}{d} \cdot m_f \cdot g \cdot u(t), \qquad (5.6)$$

where the physical quantities involved are a friction coefficient μ, a corrective force factor k that models unbalance and gyroscopic forces, etc., and the ratio of the shaft diameter d_w to the wheel diameter d. The variable m_f stands for the rotor mass.

A first estimation of the power losses may be obtained using the values for the physical parameters listed in Table 5.1. Figure 5.6 shows the variation of the bearing losses and the air resistance losses as a function of the flywheel rotational speed for an optimized flywheel construction [94]. The figure clearly shows the dependency of the power losses on the speed as given by (5.5). With the same data it is possible to obtain a value for the time range of the flywheel, i.e., the time that the flywheel speed remains above a certain threshold without any external torque. Typical values for road applications are about 10 min.

Charge and round-trip efficiency of flywheels under "Ragone" conditions, i.e., at constant power, are discussed in Problems 5.1–5.3.

Specific Energy of Flywheels

The specific energy of a flywheel is calculated as the ratio of the energy stored E_f to the flywheel mass m_f. The kinetic energy stored is $E_f = \frac{1}{2} \cdot \Theta_f \cdot \omega_f^2$, with Θ_f being the moment of inertia of the flywheel and ω_f its rotational speed. Using the notation of Fig. 5.7 the moment of inertia is calculated as [94]

$$\Theta_f = \rho \cdot b \cdot \int r^2 \cdot 2 \cdot \pi \cdot r\, dr = 2 \cdot \pi \cdot \rho \cdot b \cdot \left. \frac{r^4}{4} \right|_{q \cdot d/2}^{d/2} = \frac{\pi}{2} \cdot \rho \cdot b \cdot \frac{d^4}{16} \cdot (1 - q^4), \quad (5.7)$$

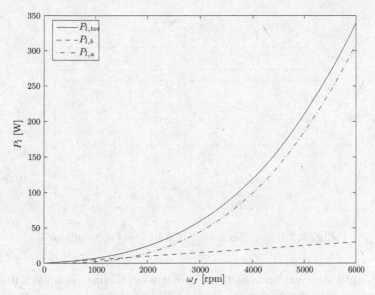

Fig. 5.6 Power losses of a flywheel as a function of the rotational speed. Flywheel data: $d = 0.36$ m, $b = 0.108$ m, $\beta = 0.3$, $q = 0.6$, $m_f = 56.33$ kg.

where ρ is the material's density and q the ratio between the inner and the outer flywheel ring. The flywheel mass is given by

$$m_f = \pi \cdot \rho \cdot b \cdot \frac{d^2}{4} \cdot \left(1 - q^2\right). \tag{5.8}$$

Consequently, the energy-to-mass ratio is evaluated as

$$\frac{E_f}{m_f} = \frac{d^2}{16} \cdot \left(1 + q^2\right) \cdot \omega_f^2 = \frac{u^2}{4} \cdot \left(1 + q^2\right). \tag{5.9}$$

Equation (5.9) may be written in a more compact way, as

$$\frac{E_f}{m_f} = k_f \cdot u^2, \tag{5.10}$$

where the coefficient k_f typically ranges from 0.5 to 5, according to the type of construction.

When designing flywheels, the main task consists of assigning values to the flywheel dimensions b, d in order to obtain the desired kinetic energy at a given rotational speed (thus, a given moment of inertia), while minimizing weight and power losses. A typical value for the stored energy may be estimated considering the recuperation energy of a braking maneuver from

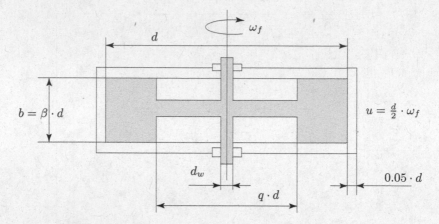

Fig. 5.7 Rotor dimensions of a flywheel accumulator

80 km/h until stop, with a vehicle mass of 910 kg. Having assumed a first tentative value of the flywheel mass equal to 10% of the vehicle mass, the energy to be recuperated is 247 kJ. Further assuming that the maximum rotational speed of the flywheel is 100 rev/s yields a moment of inertia of 1.24 kg m^2. The rotor dimensions can be calculated using equations (5.7)–(5.8). Figure 5.8 shows that the width b of the flywheel increases as a quadratic function of m_f, while the diameter d is proportional to $1/\sqrt{m_f}$. Consequently, the power losses and the peripheral speed u decrease as well. A possible compromise between flywheel weight and power losses can be obtained with a flywheel mass of about 50 kg. KERS applications need much lighter constructions (an example is discussed in Problem 5.2).

Table 5.1 Model parameters for a steel flywheel rotating in air

μ	$1.5 \cdot 10^{-3}$
k	4
d_w/d	0.08
ρ_a	$1.3 \, \text{kg/m}^3$
η_a	$1.72 \cdot 10^{-5} \, \text{Pa s}$
ρ	$8000 \, \text{kg/m}^3$

5.2.3 Dynamic Modeling of Flywheel Accumulators

The physical causality representation of a flywheel accumulator is sketched in Fig. 5.9. The model input variable is the rotational speed at the shaft, $\omega_f(t)$. The model output variable is the torque at the output shaft, $T_f(t)$.

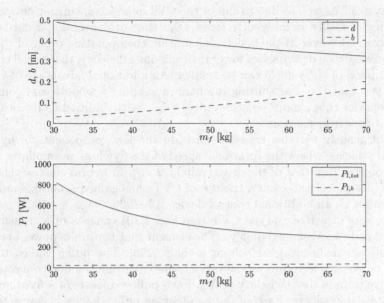

Fig. 5.8 Dimensions and losses of a flywheel as a function of its mass, for a given energy $E_f = 247\,\mathrm{kJ}$, speed $n = 100\,\mathrm{rev/s}$, and thus a moment of inertia $J_f = 1.24\,\mathrm{kg\,m^2}$

$$\xrightarrow{\omega_f} \boxed{\text{FW}} \xrightarrow{T_f}$$

Fig. 5.9 Flywheel accumulators: physical causality for dynamic modeling

The equations developed in the previous section are suitable also for dynamic modeling. The only dynamic term is related to the derivative of the flywheel speed. It can be estimated easily since $\omega_f(t)$ is an input variable.

5.3 Continuously Variable Transmissions

5.3.1 Introduction

The use of continuously variable transmissions (CVT) is a promising technology to increase the energy efficiency of conventional and hybrid vehicles. In these vehicles, a CVT allows the engine speed to be decoupled from the vehicle speed, such that the operating point of the engine can be optimized. In contrast, gearbox transmissions allow only for a fixed number of speed ratios (see Chap. 3). Due to higher transmission losses in the hydraulic part of the

CVT system[5] as well as due to slip in the CVT power transmitter device, the efficiency of CVTs is inherently lower than that of conventional fixed-ratio gear boxes. However, if the benefits of shifting the operating point of the engine outweigh the drawback of lower transmission efficiency, the overall energy consumption of the vehicle can be smaller. An additional advantage of CVTs is the fact that the gear shifting mechanism exhibits a smooth and comfortable behavior that cannot be achieved – or only with additional clutches – by conventional transmissions.

Continuously variable transmissions are the key components in hybrid-inertial vehicles, where the rotational speed of the flywheel accumulator must be decoupled from that of the drive train [322, 67]. In hybrid-electric vehicles, though not strictly necessary, the use of CVTs has gained a notable amount of attention as an additional control device [140, 229, 300].

The core of a Reeves-type CVT (see Fig. 5.10) consists of a transmitter element and two V-shaped pulleys. The element that transmits power between the pulleys can be a metal belt or a chain [250]. The pulley linked to the prime mover is referred to as the primary pulley, while the one connected to the drive train is the secondary pulley. Each pulley consists of a fixed and an axially slidable sheave. Each of the two moving pulley halves is connected to a hydraulic actuation system, consisting of a hydraulic cylinder and piston. In the simplest hydraulic configuration, the pressure on the secondary side is the hydraulic supply pressure. The pressure on the primary side is determined by one or more hydraulic valves, actuated by a solenoidal (electromagnetic) valve, which connect the two circuits with the pump return line [346].

To actuate a higher ratio and reduce the engine speed, the solenoidal valve moves the hydraulic valve and lets some pump pressure enter the primary pulley actuator. The additional pressure causes the pulley halves to squeeze together and the belt to ride toward the outer diameter, thus simulating a larger gear. Simultaneously, as the secondary pressure is discharged, the pulley actuator allows the pulley halves to spread and the belt to ride toward its inner diameter. The opposite operation occurs when shifting to lower ratios.

In the i^2-CVT used in the ETH-III propulsion system [81], the range of the CVT alone (typically 1:5) is substantially increased by combining the CVT core (a chain converter) with a two-ratio gear box. In the "slow" gear arrangement, the overall transmission ratio is given by the product of the CVT ratio and the gear ratio. In the "fast" gear arrangement, the power flow through the CVT is inverted, thus the overall transmission ratio is proportional to the reciprocal of the CVT ratio. This allows for reaching transmission ratios higher than 1:20.

When the transmission ratio exceeds the capacity of the CVT, clutches must be operated. As an example, in hybrid-inertial applications the transmission ratio could be too low (moving vehicle, stationary flywheel), thus

---

[5] For this reason CVTs actuated by electromechanical devices [338] are being studied, with the goal of reducing the power losses by 25%.

requiring the slipping of a clutch downstream of the CVT. If the transmission ratio is too high (stationary vehicle, rotating flywheel), a clutch betwheen the flywheel and the CVT is slipped [67].

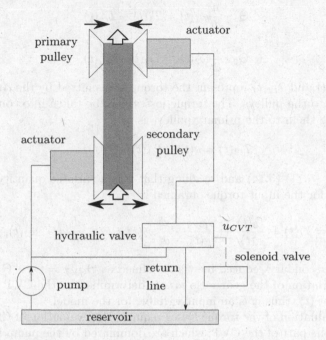

Fig. 5.10 Schematic diagram of a CVT system. The arrows illustrate the movements of the pulley halves and the belt during an upshift.

5.3.2 *Quasistatic Modeling of CVTs*

The causality representation of CVTs in quasistatic simulations is sketched in Fig. 5.11. The input variables are the torque $T_2(t)$ and the speed $\omega_2(t)$ at the downstream shaft. The output variables are the torque $T_1(t)$ and the speed $\omega_1(t)$ required at the input shaft.

Fig. 5.11 CVTs: causality representation for quasistatic modeling

A simple quasistatic model of a CVT can be derived neglecting slip at the transmitter but considering torque losses [129, 300, 250, 305]. The definition of the transmission (speed) ratio ν implies that

$$\omega_1(t) = \nu(t) \cdot \omega_2(t). \tag{5.11}$$

Newton's second law applied to the two pulleys yields

$$\Theta_1 \cdot \frac{d}{dt}\omega_1(t) = T_1(t) - T_{t1}(t), \tag{5.12}$$

$$\Theta_2 \cdot \frac{d}{dt}\omega_2(t) = T_{t2}(t) - T_2(t), \tag{5.13}$$

where $T_{t1}(t)$ and $T_{t2}(t)$ represent the torque transmitted by the chain or the metal belt to the pulleys. The torque losses may be taken into consideration by applying them to the primary pulley, as

$$T_{t2}(t) = \nu(t) \cdot (T_{t1}(t) - T_l(t)). \tag{5.14}$$

Combining (5.11)-(5.14) and recalling that ν is a variable quantity, the final expression for the input torque obtained is

$$T_1(t) = T_l(t) + \frac{T_2(t)}{\nu(t)} + \frac{\Theta_{CVT}}{\nu(t)} \cdot \frac{d}{dt}\omega_2(t) + \Theta_1 \cdot \frac{d}{dt}\nu(t) \cdot \omega_2(t), \tag{5.15}$$

where the secondary reduced inertia is defined as $\Theta_{CVT} = \Theta_2 + \Theta_1 \cdot \nu^2(t)$.

The variation of the gear ratio $\nu(t)$ is determined by the CVT controller, $\dot{\nu}(t) = u_{CVT}(t)$, thus it is an input variable for the model.

The evaluation of the torque losses requires the evaluation of the losses in the hydraulic part of the CVT, which are dominated by the pump losses, and the friction losses at the mechanical contacts between various CVT components. Analytical expressions for $T_l(t)$ are derived by fitting experimental data, but due to the complexity of the processes involved, at least a second-order dependency of $T_l(t)$ on $\nu(t)$, $\omega_1(t)$ and on the input power $P_1(t)$ is required [347, 81].

An alternative approach is based on the definition of the CVT efficiency $\eta_{CVT}(\omega_2, T_2, \nu)$, according to which the torque required at the input shaft is evaluated as

$$T_1(t) = \frac{T_2(t)}{\nu(t) \cdot \eta_{CVT}(\omega_2(t), T_2(t), \nu(t))}, \quad T_2(t) > 0, \tag{5.16}$$

$$T_1(t) = \frac{T_2(t)}{\nu(t)} \cdot \eta_{CVT}(\omega_2(t), T_2(t), \nu(t)), \quad T_2(t) < 0. \tag{5.17}$$

The typical dependency of η_{CVT} on output speed and torque as well as on the transmission ratio is depicted in Fig. 5.12. The efficiency of the CVT increases with torque at constant speed and ν, exhibiting a maximum that is more pronounced at higher speeds. Higher transmission ratios also favorably affect the efficiency. Values around 90% can be reached for high-load,

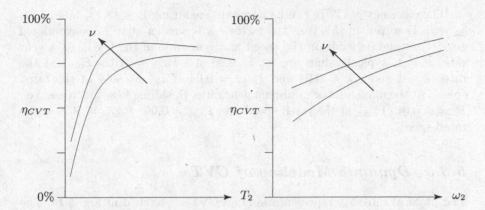

Fig. 5.12 Qualitative dependency of CVT overall efficiency on torque, speed, and transmission ratio

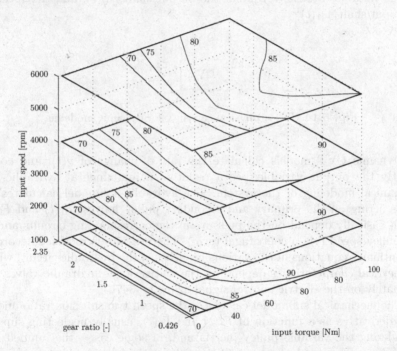

Fig. 5.13 Efficiency of a push-belt CVT as a function of input speed ω_e, input torque T_1, and gear ratio ν [249]

low-speed conditions. Values lower than 70% are typical during low-load operation [347, 250].

As an example, Fig. 5.13 shows the result of a detailed model of a push-belt CVT [249] of the type discussed in the next section.

The efficiency of CVTs can be approximated using Eqs. (3.13) and (3.14) as well. However, in this case the factor e_{gb} is smaller than in conventional gear boxes and depends on the speed and gear ratio of the CVT. At a gear ratio $\nu = 1$, typical values are $e_{gb} = 0.96$ and $P_{0,gb} = 0.02 \cdot P_{max}$ at the rated speed and $e_{gb} = 0.94$ and $P_{0,gb} = 0.04 \cdot P_{max}$ at 50% of the rated speed. At the maximum or minimum gear ratio, the idling losses increase, i.e., $P_{0,gb} = 0.04 \cdot P_{max}$ at the rated speed and $P_{0,gb} = 0.06 \cdot P_{max}$ at 50% of the rated speed.

5.3.3 Dynamic Modeling of CVTs

The physical causality representation of a CVT is sketched in Fig. 5.14. The model input variables are the rotational speed at the output or downstream shaft, $\omega_2(t)$, and the torque at the input or upstream shaft, $T_1(t)$. The model output variables are the torque at the output shaft, $T_2(t)$, and the speed at the input shaft, $\omega_1(t)$.

$$\xleftarrow{\omega_1} \boxed{\text{CVT}} \xleftarrow{\omega_2}$$
$$\xrightarrow{T_1} \qquad \xrightarrow{T_2}$$

Fig. 5.14 CVTs: physical causality for dynamic modeling

Dynamic CVT models calculate the rate of change of $\nu(t)$, and consequently the output variables, as a result of a fundamental hydraulic and mechanical modeling of the system. The hydraulic submodel calculates the forces acting on the primary and secondary pulley halves, $F_1(t)$ and $F_2(t)$, which basically depend on the pressure in the corresponding circuits and on the pulley speed (via a centrifugal term). The pressure is modeled according to continuity equations in the various branches of the hydraulic circuit, whose geometrical characteristics depend on the status of the hydraulic valves and ultimately on the electric control signal $u_{CVT}(t)$ [347].

The mechanical submodel calculates the speed transmission ratio and its time derivative as a function of $F_1(t)$ and $F_2(t)$, usually neglecting slip but considering the variable pulley inertia and, in some cases, the lumped belt mass. A model of the latter type [301] results in the following equation for the rate of change of the transmission ratio [347],

$$\frac{d}{dt}\nu(t) = k(\nu(t)) \cdot (F_1(t) - \kappa(\nu(t), T_1(t), F_2(t)) \cdot F_2(t)), \qquad (5.18)$$

where $F_1(t)$ and $F_2(t)$ are the forces acting on the primary and the secondary pulley, respectively, while k and the so-called pulley thrust ratio κ are functions that have to be determined experimentally. Other models, mainly proposed for metal-belt CVTs, although based on a different derivation, show a final dependency similar to that of (5.18) [148, 125].

Once $\dot{\nu}(t)$ is calculated or assumed, the output model variables are evaluated using the equations derived in the previous section. In particular (5.11) can be used to evaluate $\omega_1(t)$ and (5.15) to evaluate $T_2(t)$, since both $\omega_2(t)$ and its derivative are known as input variables.

5.4 Hydraulic Accumulators

5.4.1 Introduction

Besides supercapacitors, flywheels, and pneumatic systems, a fourth type of short-term storage system is represented by hydraulic accumulators. Like the other concepts mentioned, hydraulic accumulators are characterized by a higher power density and a lower energy density than electrochemical batteries. Due to the high power flow that can be recuperated during deceleration and then made available, the application of hydraulic hybrid concepts has been of interest so far for use in heavy vehicles, i.e., sport utility vehicles, urban delivery vehicles, trucks, etc. As the energy storage system, the hydraulic accumulator has the ability to accept both high frequencies and high rates of charging/discharging, both of which are not possible for electrochemical batteries. However, the relatively low energy density of the hydraulic accumulator requires a carefully designed control strategy if the fuel economy potential is to be realized to its fullest.

In this context, both series and parallel hybrid architectures have been proposed. The parallel architecture (Eaton's Hydraulic Launch Assist [89], Ford's Hydraulic Power Assist [102]) includes an engine (often a Diesel engine) and a hydraulic reversible machine (pump/motor). Besides regenerative braking, typically the hydraulic motor is used alone at low loads (stop-and-go) and for the first few seconds during acceleration. The hydraulic motor also provides additional torque during hill climbing and heavy acceleration, thus covering many of the possibilities already illustrated for hybrid-electric vehicles. Expected benefits are in terms of fuel consumption in stop-and-go operation (20–30%), exhaust emissions, and acceleration time. The series hybrid configuration (EPA [161], Parker [244]) requires two hydraulic pumps/motors. It is often referred to as "full hybrid," since larger degrees of hybridization are used, with larger improvements (30–50%) in reducing fuel consumption than with parallel "mild" concepts.

All types of hybrid-hydraulic propulsion systems include a high-pressure accumulator and a low-pressure reservoir. The accumulator contains the hydraulic fluid and a gas such as nitrogen (N_2) or methane (CH_4), separated by a membrane (see Fig. 5.15). During the charge phase, the hydraulic fluid is extracted from the reservoir by the pump. As the fluid flows into the accumulator, the gas is compressed. During the discharge phase, the fluid flows out through the hydraulic motor and then into the reservoir. The state of charge is defined as the ratio of instantaneous gas volume in the accumulator, V_g, to its maximum capacity. Typical data of an accumulator designed for a truck are listed in Table 5.2. The reservoir can be regarded as an accumulator working at a much lower pressure, e.g., 8.5–12.5 bar.

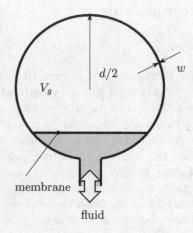

Fig. 5.15 Schematic of a hydraulic accumulator

Table 5.2 Typical data of a bladder-type hydraulic accumulator

fluid capacity	50 l
maximum gas volume	100 l
minimum gas volume	50 l
pre-charge pressure (at 320 K)	125 bar
maximum pressure	360 bar

5.4.2 Quasistatic Modeling of Hydraulic Accumulators

The causality diagram of a simple hydraulic accumulator used for quasistatic simulations is sketched in Fig. 5.16. The input variable is the hydraulic power $P_h(t) = p_h(t) \cdot Q_h(t)$ at the inlet. Here $p_h(t)$ is the fluid pressure and $Q_h(t)$ is the volumetric flow from the accumulator. The output variable is the state of charge of the accumulator, i.e., the volume occupied by the gas, $V_g(t)$.

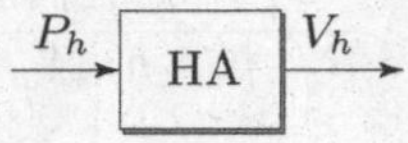

$$\xrightarrow{P_h} \boxed{\text{HA}} \xrightarrow{V_h}$$

Fig. 5.16 Hydraulic accumulators: causality representation for quasistatic modeling

Thermodynamic Model

A basic physical model of a hydraulic accumulator can be derived from the mass and energy conservation laws and the ideal gas state equation for the charge gas [258, 100]. The resulting set of nonlinear equations includes an energy balance

$$m_g \cdot c_v \cdot \frac{d}{dt}\vartheta_g(t) = -p_g(t) \cdot \frac{d}{dt}V_g(t) - h \cdot A_w \cdot (\vartheta_g(t) - \vartheta_w), \qquad (5.19)$$

a mass balance

$$\frac{d}{dt}V_g(t) = Q_h(t), \qquad (5.20)$$

and the ideal gas law

$$p_g(t) = \frac{m_g \cdot R_g \cdot \vartheta_g(t)}{V_g(t)}, \qquad (5.21)$$

where A_w is the effective accumulator wall area for heat convection, h is the effective heat transfer coefficient, m_g is the gas mass, c_v is the constant-volume specific heat of the gas, $\vartheta_g(t)$ is the gas temperature, $p_g(t)$ the gas pressure, $V_g(t)$ the gas volume, ϑ_w the wall temperature, and R_g is the gas constant. More detailed models [258] may describe the charge gas with a nonlinear state equation (Benedict–Webb–Rubin equation of state) and may take into account the heat exchange between the charge gas and the elastomeric foam that is usually inserted on the gas side of the accumulator to reduce the thermal loss to the accumulator walls.

In a first approximation the gas pressure $p_g(t)$ equals the fluid pressure at the accumulator inlet, $p_h(t)$. Frictional losses caused by flow entrance effects, viscous shear, and piston-seal friction or bladder hysteresis can hardly be accounted for in an analytical fashion. Usually the pressure loss term is evaluated as a fraction of $p_h(t)$ (e.g., 2%) [258].

A relationship between fluid flow rate, charge volume, and power can be derived by solving (5.19)–(5.21) at steady state. Assuming that $p_g(t) = p_h(t)$, i.e., no pressure losses at the accumulator inlet, the resulting equation for fluid pressure is written as

$$p_h(t) = \frac{h \cdot A_w \cdot \vartheta_w \cdot m_g \cdot R_g}{V_g(t) \cdot h \cdot A_w + m_g \cdot R_g \cdot Q_h(t)}. \qquad (5.22)$$

Combining (5.22) with the definition of output power, the fluid flow rate is obtained,

$$Q_h(t) = \frac{V_g(t)}{m_g} \cdot \frac{h \cdot A_w \cdot P_h(t)}{R_g \cdot \vartheta_w \cdot h \cdot A_w - R_g \cdot P_h(t)}. \tag{5.23}$$

The state of charge is thus given by integrating (5.20). With only one state variable, this quasi-stationary model has a complexity that is equivalent to that of most battery models. Thus it can easily be embedded, for instance, in a dynamic programming algorithm [359] and also generally in quasistatic algorithms.

Accumulator Efficiency

The efficiency of a hydraulic accumulator may be defined as the ratio of the total energy delivered during a complete discharge to the energy that is necessary to charge the device. This definition is conceptually identical to the "global efficiency" introduced in Chap. 4 for electrochemical batteries. The energy spent to charge the accumulator depends on the charge/discharge cycle. Similarly to electrochemical batteries, constant flow rate or constant power cycles are possible. For hydraulic accumulators, however, the usual definition of efficiency is based on a reference cycle consisting of isentropic compression and expansion from the maximum to the minimum volume and vice versa.

The thermodynamic transformations followed by the gas in the reference cycle are represented in the temperature/entropy (ϑ–s) diagram of Fig. 5.17. The transformation AB is an isentropic compression from an initial state A, with the gas occupying the whole volume of the accumulator at a temperature that equals that of the surrounding ambient ($\vartheta_A = \vartheta_w$). The transformation BC is an isochoric cooling of the gas, which loses thermal energy[6] to the ambient until $\vartheta_C = \vartheta_A$. The transformation CD is an isentropic expansion that ends when $p_D = p_A$. The reference cycle followed by the gas is thus ABCDA. As a basis of comparison, a second reference cycle ABCEA can be defined by introducing the transformation DE, a further expansion that ends when $V_E = V_A$.

By definition, the isentropic transformation yields the following equations

$$\frac{p_B}{p_A} = \left(\frac{V_A}{V_B}\right)^{\gamma} = r^{-\gamma}, \quad \frac{\vartheta_B}{\vartheta_A} = \left(\frac{p_B}{p_A}\right)^{\frac{\gamma-1}{\gamma}} = r^{1-\gamma}, \tag{5.24}$$

where r is the expansion ratio and γ is the ratio of the specific heats. The compression work is calculated along the path AB. For a closed, adiabatic system, the work exchanged equals the variations of the internal energy. Thus the work W_{AB} is calculated as

$$W_{AB} = m_g \cdot c_{v,g} \cdot \vartheta_A \cdot \left(\frac{\vartheta_B}{\vartheta_A} - 1\right) = m_g \cdot c_{v,g} \cdot \vartheta_A \cdot \left(r^{1-\gamma} - 1\right). \tag{5.25}$$

---

[6] This corresponds to a worst-case scenario in which the device is left to cool for a long time.

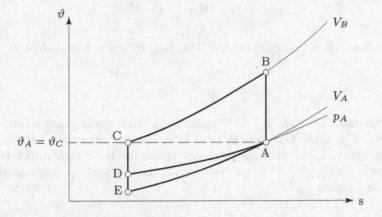

Fig. 5.17 Temperature–entropy (ϑ–s) diagram of the accumulation process

The transformation BC is isochoric, i.e., the volume V_C is equal to V_B. Moreover, since $\vartheta_C = \vartheta_A$, the pressure can be calculated as

$$p_C = \frac{m_g \cdot R_g \cdot \vartheta_C}{V_C} = \frac{m_g \cdot R_g \cdot \vartheta_A}{V_B} = \frac{p_A \cdot V_A}{V_B} = \frac{p_A}{r}. \tag{5.26}$$

For the isobaric reference cycle ABCDA, the expansion ends at the state D, which is characterized by a pressure $p_D = p_A$, and a temperature ϑ_D evaluated as

$$\frac{\vartheta_D}{\vartheta_C} = \left(\frac{p_D}{p_C}\right)^{\frac{\gamma-1}{\gamma}} = \left(\frac{p_A}{p_A}r\right)^{\frac{\gamma-1}{\gamma}} = r^{\frac{\gamma-1}{\gamma}}. \tag{5.27}$$

The expansion work is thus calculated as

$$W_{CD} = m_g \cdot c_{v,g} \cdot \vartheta_C \cdot \left(\frac{\vartheta_D}{\vartheta_C} - 1\right) - m_g \cdot c_{v,g} \cdot \vartheta_A \cdot \left(r^{\frac{\gamma-1}{\gamma}} - 1\right). \tag{5.28}$$

For the isochoric reference cycle ABCEA, the expansion ends at the state E, which is characterized by a volume $V_E = V_A$, and a temperature evaluated as

$$\frac{\vartheta_E}{\vartheta_C} = \left(\frac{V_E}{V_C}\right)^{1-\gamma} = r^{\gamma-1}. \tag{5.29}$$

The expansion work is thus

$$W_{CE} = m_g \cdot c_{v,g} \cdot \vartheta_C \cdot \left(\frac{\vartheta_E}{\vartheta_C} - 1\right) = m_g \cdot c_{v,g} \cdot \vartheta_A \cdot \left(r^{\gamma-1} - 1\right). \tag{5.30}$$

The accumulator efficiency is the ratio between the discharge energy and the charge energy. For the isochoric reference cycle ABCEA, the efficiency is evaluated as

$$\eta_{ha,V} = \frac{-W_{CE}}{W_{AB}} = \frac{1 - r^{\gamma-1}}{r^{1-\gamma} - 1}. \tag{5.31}$$

For the isobaric reference cycle ABCDA, the efficiency is evaluated as

$$\eta_{ha,P} = \frac{-W_{CD}}{W_{AB}} = \frac{1 - r^{\frac{\gamma-1}{\gamma}}}{r^{1-\gamma} - 1}. \tag{5.32}$$

The variations of W_{AB}, W_{CD}, W_{DE} and η_{ha} with the expansion ratio r are shown in Fig. 5.18. The plots clearly show that the efficiency of the isobaric cycle is always lower than the efficiency of the isochoric cycle which reaches a value of 100% for $r = 1$.[7] In both cases, the efficiency is a monotonically increasing function of r.

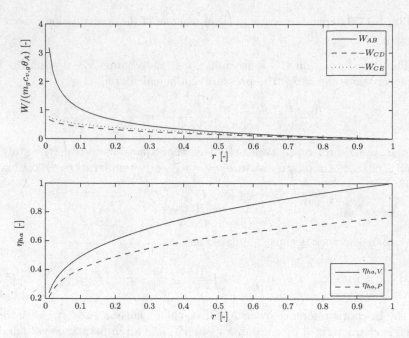

Fig. 5.18 Compression work W_{AB}, expansion works W_{CD}, W_{CE} (top), and efficiency η_{ha} (bottom) as a function of the expansion ratio r

Specific Energy of Accumulators

The results of the analysis developed in the previous section allow for the evaluation of the specific energy of a hydraulic accumulator. The compression work given by (5.25) must be divided by the accumulator mass. This is given

---

[7] Of course, the isochoric reference cycle implies maintaining pressures below ambient pressure.

by the sum of three terms, *viz.* the mass of the charge gas, the mass of the hydraulic fluid, and the mass of the housing. The latter term may be evaluated as the mass of a spherical shell designed to bear a pressure p_B, the maximum gas pressure. The thickness of the housing (see Fig. 5.15) may be evaluated using the formula:

$$w = \frac{p_B \cdot d}{4 \cdot \sigma},\tag{5.33}$$

where d is the diameter of the accumulator shell and σ is the maximum tensile stress. The housing mass is thus

$$m_m = \rho_m \cdot \pi \cdot d^2 \cdot w = \frac{3}{2} \cdot V_A \cdot p_B \cdot \frac{\rho_m}{\sigma},\tag{5.34}$$

since V_A, the maximum volume occupied by the gas, by definition is the accumulator volume. The fluid mass to be considered in the calculation is the mass that occupies the maximum volume left by the gas, i.e., the volume $V_A - V_B$. The mass of the charge gas is the constant m_g already introduced in the previous section. The orders of magnitude of these masses are rather different. Evaluating the gas mass as $m_g = V_A \cdot p_A/(R_g \cdot \vartheta_A)$, the ratios of fluid to gas mass and of housing to gas mass are

$$\frac{m_h}{m_g} = \rho_o \cdot \frac{R_g \cdot \vartheta_A}{p_A} \cdot (1 - r), \quad \frac{m_m}{m_g} = \frac{3}{2} \cdot R_g \cdot \vartheta_A \cdot r^{-\gamma} \cdot \frac{\rho_m}{\sigma}.\tag{5.35}$$

Using typical data for the materials, such as the ones listed in Table 5.3, clearly shows that m_g can be neglected when compared with m_h and m_m. For higher pressures, even m_h can be neglected when compared with m_m.

Table 5.3 Typical parameter values for hydraulic accumulator models (steel construction, methane as compressed gas)

ρ_m/σ	$10^{-5}\,\mathrm{kg/J}$
ρ_o	$900\,\mathrm{kg/m^3}$
ϑ_A	$300\,\mathrm{K}$
R_g	$520\,\mathrm{J/(kg\,K)}$
p_B	400–$800\,\mathrm{bar}$

In the latter case, the expression for the specific energy is written as

$$\frac{E_{ha}}{m_{ha}} \approx \frac{W_{AB}}{m_m} = \alpha \cdot r^\gamma \cdot \left(r^{1-\gamma} - 1\right),\tag{5.36}$$

with $\alpha = 2 \cdot c_{v,g} \cdot \sigma/(3 \cdot \rho_m \cdot R_g) = (2/3) \cdot (\sigma/\rho_m)/(\gamma - 1)$. The variation of the specific energy as a function of the compression ratio r is shown in Fig. 5.19.

The specific energy of (5.36) can be maximized with respect to r by setting the relevant derivative to zero. This condition yields the optimal compression ratio

$$r_{opt} = \left(\frac{1}{\gamma}\right)^{\frac{1}{\gamma-1}}. \tag{5.37}$$

With $\gamma = 1.31$ (methane), the optimal compression ratio is $r_{opt} = 0.42$, the corresponding efficiency is $\eta_{ha,P} = 0.60$, the corresponding specific energy is around 6 Wh/kg. With $\gamma = 1.4$ (nitrogen), lower specific energy values are obtained.

For modern lightweight accumulators, carbon-fiber shells are used. That increases the maximum pressure up to 500 bar, although for safety concerns and risks of leakages, practical values are often limited to 350–400 bar. The lower values of ρ_m that characterize these materials with respect to steel increase the specific energy to 10–15 Wh/kg and more [101, 193].

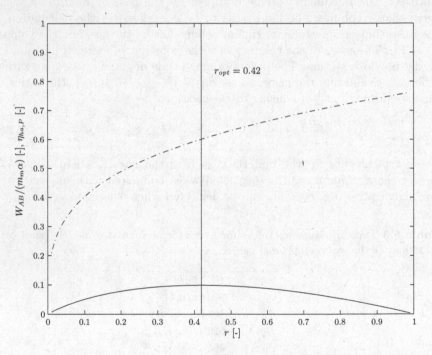

Fig. 5.19 Specific energy (solid curve) and efficiency (dashed curve) of a hydraulic accumulator as a function of the compression ratio with methane as a hydraulic fluid

5.4.3 Dynamic Modeling of Hydraulic Accumulators

The physical causality representation of a hydraulic accumulator is sketched in Fig. 5.20. The model input variable is the flow rate of the hydraulic fluid $Q_h(t)$. A positive value of $Q_h(t)$ discharges the accumulator, while a negative

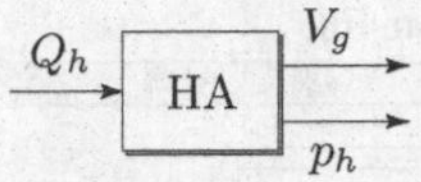

Fig. 5.20 Hydraulic accumulators: physical causality for dynamic modeling

value of $Q_h(t)$ charges it. The model output variables are the hydraulic fluid pressure $p_h(t)$ and the state of charge $V_g(t)$.

The state of charge is calculated by directly integrating (5.20), while in a first approximation the fluid pressure equals the gas pressure $p_g(t)$, which may be integrated using (5.19) and (5.21).

5.5 Hydraulic Pumps/Motors

5.5.1 Introduction

In hybrid-hydraulic propulsion systems, hydraulic motors convert the hydraulic energy of a fluid into mechanical energy available at the motor shaft. Conversely, the machine can act as a pump, converting mechanical energy back into hydraulic energy.

Hydraulic motors are rotary or reciprocating volumetric machines with a fixed or a variable displacement volume. Different applications may require different types of machines, according to the required maximum pressure (100–600 bar) and the fluid flow rate. Rotary machines handle the fluid in chambers whose volume cyclically changes in accordance with the design of the walls (vane machines) or the action of teeth (gear machines). However, most types have a fixed displacement volume. Reciprocating machines are characterized by the cyclic variation of the chamber volume for the action of a piston. Variable displacement operation is usually possible by changing the geometry of the driving mechanism. Reciprocating machines are further classified into radial piston and axial piston machines.

In the former type, the cylinder axes are arranged perpendicular to the shaft. In pump operation, as the cylinder block rotates, the pistons are pressed against the rotor and are forced in and out of the cylinders, thereby receiving fluid and pushing it out into the system. The motor operation is the reverse.

In axial piston machines, the cylinder axes are arranged parallel to the shaft (see Fig. 5.21). In the swash-plate type, the reciprocating motion is created by a plate mounted on the shaft at a fixed or variable angle. One end of each piston rod is held in contact with the plate as the cylinder block and piston assembly rotates with the drive shaft. This causes the pistons to reciprocate within the cylinders. In bent-axis units, it is the pistons that are bent with respect to the shaft and plate axes.

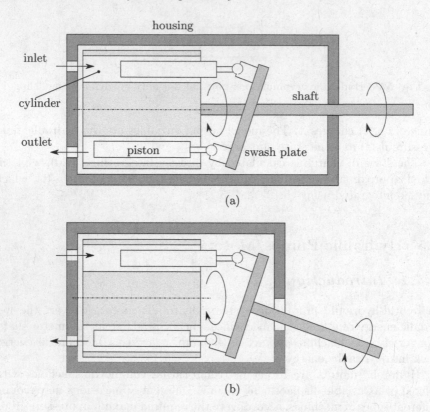

Fig. 5.21 Schematics of piston pumps/motors: swash-plate (a), bent-axis (b). The arrows indicate shaft rotation and fluid pumping.

In automotive applications, the faster axial piston machines of the bent-axis or swash-plate types, with typical speeds of 300–3000 rpm, are preferred to slower radial piston machines, which are used instead for very high pressure applications. The displacement per revolution of the machine is adjusted to control the power delivered or absorbed, thus letting the machine operate as a pump or as a motor. Both in bent-axis units and in swash-plate units, the quantity adjusted is the swivel angle, i.e., the angle between the pistons' axes and the connecting plate axis. Typical values of this angle are 20–25°.

5.5.2 *Quasistatic Modeling of Hydraulic Pumps/Motors*

The causality representation of a hydraulic pump/motor in quasistatic simulations is sketched in Fig. 5.22. The input variables are the torque $T_h(t)$ and speed $\omega_h(t)$ required at the shaft. The ouput variable is the hydraulic power $P_h(t) = p_h(t) \cdot Q_h(t)$.

Fig. 5.22 Hydraulic pumps/motors: causality representation for quasistatic modeling

The relationship between $T_h(t) \cdot \omega_h(t)$ and $P_h(t)$ may be described by defining a pump/motor efficiency $\eta_{hm}(\omega_h, T_h)$ such that

$$P_h(t) = \frac{T_h(t) \cdot \omega_h(t)}{\eta_{hm}(\omega_h(t), T_h(t))}, \quad T_h(t) > 0, \tag{5.38}$$

$$P_h(t) = T_h(t) \cdot \omega_h(t) \cdot \eta_{hm}(\omega_h(t), T_h(t)), \quad T_h(t) < 0. \tag{5.39}$$

Of course, as for other energy converters, an alternative quasistatic description for hydraulic pumps/motors is the Willans approach

$$P_h(t) = \frac{T_h(t) \cdot \omega_h(t) + P_0}{e}, \tag{5.40}$$

where e is the efficiency of the energy conversion process and P_0 represents the friction losses of the pump/motor.

A basic physical model of variable-displacement piston pumps/motors can be derived by following the classic Wilson's approach [258]. The volumetric flow rate of the hydraulic fluid through a pump is given by

$$Q_h(t) = -\eta_v(t) \cdot x(t) \cdot D \cdot \omega_h(t), \quad T_h(t) < 0, \tag{5.41}$$

where D is the maximum displacement and x is the fraction of D available. For a bent-axis unit, $x(t)$ is related to the swivel angle, i.e., the control input of the machine. The negative sign in (5.41) is necessary because usually for pumps $x(t)$ is considered to be a positive quantity. The volumetric efficiency $\eta_v(t)$ accounts for various phenomena, including leakage and fluid compressibility [258]. For motors, (5.41) is modified as

$$Q_h(t) = -\frac{x(t) \cdot D \cdot \omega_h(t)}{\eta_v(t)}, \quad T_h(t) > 0, \tag{5.42}$$

with $x(t)$ that is now a negative quantity.

The fluid pressure in pump mode $p_h(t)$ is defined here as the differential between fluid pressure in the machine and in the reservoir and is calculated from a basic energy balance

$$p_h(t) = -\frac{\eta_t(t) \cdot T_h(t)}{x(t) \cdot D}, \quad T_h(t) < 0. \tag{5.43}$$

The mechanical efficiency $\eta_t(t)$ accounts for viscous, frictional, and hydrodynamic torque losses. The corresponding equation for the motor mode is

$$p_h(t) = -\frac{T_h(t)}{\eta_t(t) \cdot x(t) \cdot D}, \quad T_h(t) > 0. \tag{5.44}$$

A semi-physical approach to evaluate the volumetric and the torque efficiency is now illustrated [258]. The volumetric efficiency in the pump mode is evaluated as

$$\eta_v(t) = 1 - \frac{C_s}{x(t) \cdot S(t)} - \frac{p_h(t)}{\beta} - \frac{C_{st}}{x(t) \cdot \sigma(t)}, \quad x(t) > 0, \tag{5.45}$$

where C_s and C_{st} are the laminar and the turbulent leakage coefficients, respectively, and β is the bulk modulus of elasticity (1660 MPa for most hydraulic fluids). Cavitation losses, small in most modern pumps, are neglected in this model. The variables $S(t)$ and $\sigma(t)$ are defined by

$$S(t) = \frac{\mu_o \cdot \omega_h(t)}{p_h(t)}, \tag{5.46}$$

$$\sigma(t) = \frac{\omega_h(t) \cdot D^{1/3} \cdot \rho_o^{1/2}}{(2 \cdot p_h(t))^{1/2}}, \tag{5.47}$$

where μ_o is the fluid viscosity and ρ_o its density. The volumetric efficiency in the motor mode is calculated from simple considerations as

$$\eta_v(t) = \frac{1}{1 - \frac{C_s}{x(t) \cdot S(t)} + \frac{p_h(t)}{\beta} - \frac{C_{st}}{x(t) \cdot \sigma(t)}}, \quad x(t) < 0. \tag{5.48}$$

By accounting for the viscous torque, the frictional torque, and the hydrodynamic loss with the three coefficients C_v, C_f and C_h, respectively, it can be shown that for a motor the mechanical efficiency is

$$\eta_t(t) = 1 + \frac{C_v \cdot S(t)}{x(t)} + \frac{C_f}{x(t)} - C_h \cdot x^2(t) \cdot \sigma^2(t), \quad x(t) < 0. \tag{5.49}$$

In the pump mode, the following relationship applies,

$$\eta_t(t) = \frac{1}{1 + \frac{C_v \cdot S(t)}{x(t)} + \frac{C_f}{x(t)} + C_h \cdot x^2(t) \cdot \sigma^2(t)}, \quad x(t) > 0. \tag{5.50}$$

The overall machine efficiency is the product of the volumetric efficiency and the torque efficiency

$$\eta_{hm}(t) = \eta_v(t) \cdot \eta_t(t). \tag{5.51}$$

Thus (5.48)–(5.50) can be used to estimate the motor efficiency from pump data and the pump efficiency from motor data, at least for normal operation. At very low loads, the difference between the loss coefficients of the same unit operating as a motor and a pump may be substantial [258].

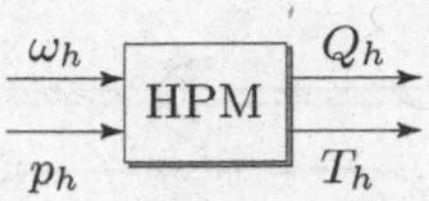

Fig. 5.23 Hydraulic pumps/motors: physical causality for dynamic modeling

5.5.3 Dynamic Modeling of Hydraulic Pumps/Motors

The physical causality representation of a hydraulic pump/motor is sketched in Fig. 5.23. The model input variables are the rotational speed $\omega_h(t)$ and the fluid pressure $p_h(t)$. The model output variables are the shaft torque $T_h(t)$ and the fluid flow rate $Q_h(t)$.

The quasistatic model derived in the previous section can easily be used also for dynamic simulations. The output torque is calculated from (5.43)–(5.44) as a function of the fluid pressure, while (5.41)–(5.42) serve to calculate the fluid flow rate. The behavior of the machine as a pump ($T_h(t) < 0$) or as a motor ($T_h(t) > 0$) is determined by the sign of the fractional capacity $x(t)$, which is the control input to the machine.

5.6 Pneumatic Hybrid Engine Systems

5.6.1 Introduction

The combination of a conventional IC engine and a pneumatic short-term storage system is an interesting approach to achieve lower fuel consumption. Instead of using a battery, a hybrid pneumatic vehicle uses a robust and inexpensive air pressure tank for energy storage. Moreover, the IC engine is able to run in purely pneumatic modes, acting as a pneumatic pump or motor, without fuel injection. The resulting concept is called Hybrid Pneumatic Engine (HPE) [286, 137, 85]. This concept allows recuperating some of the energy that is otherwise lost when braking and the elimination of the most inefficient engine operating points, i.e., idling and very low loads. Moreover, it ideally complements a downsized and supercharged engine [131, 336].

In fact, turbocharged engines usually have a reduced drivability due to the relatively slow acceleration of the compressor-turbine during load steps. This leads to the choice of small turbines, which minimize the delays but have a rather low efficiency. In HPE systems, the air available in the pressure tank can be used for supercharging the engine in heavy transients and, therefore, the turbines can be designed for optimal fuel economy. Since the air is provided to the cylinder by a fully variable charge valve, the torque can be raised from idling to full-load from one engine cycle to the next, i.e., in the shortest time possible. Moreover, the combustion with the larger amount of fresh air and fuel accelerates the turbocharger much faster, resulting in a fast pressure rise in the

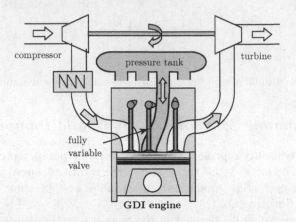

Fig. 5.24 Hybrid pnuematic engine system

intake manifold. Therefore, the additional air from the pressure tank is needed only for a very short time such that relatively small air tanks can be used.

The hardware configuration necessary for a *directly-connected* HPE includes an additional valve in the cylinder head, which is connected to the pressure tank [286, 137, 85, 331, 328]. A fully variable actuation of this valve is mandatory. To achieve maximum efficiency in the pneumatic modes, fully variable intake and exhaust valves are beneficial as well. An illustration of the hardware configuration which corresponds to the discussed system is shown in Fig. 5.24. Indirectly-connected HPE concepts have been proposed as well, where the connection between the cylinders and the air tank is via either an intake or an exhaust receiver [187].

The combination of mainly mechanical components makes the HPE system a very cost-efficient alternative to hybrid-electric propulsion systems. The additional weight can be kept very small since only a low-pressure air tank is used. Of course the fuel economy potential offered by the additional degrees of freedom can only be realized by a well-designed supervisory control system. In Chap. A.8 a case study is presented in which this problem is analyzed.

Specific Energy

The energy stored in a pressure tank of fixed volume is the internal energy of the air contained in the tank, $U_t = m_t c_v \vartheta_t$, where m_t is the air mass, c_v its specific heat, and ϑ_t its temperature. Assuming and ideal gas, energy can be expressed as a function of pressure as

$$U_t = \frac{p_t V_t}{\gamma - 1}, \tag{5.52}$$

where V_t is the fixed tank volume and γ the ratio of the specific heats (approximately 1.4 for air). The maximum energy content is thus obtained at the maximum pressure level $p_{t,max}$ that the compression device used for pumping air into the tank can achieve. Since in an HPE the compression device is the IC engine used as a pneumatic pump, the maximum tank pressure is generally given by

$$p_{t,max} = p_{amb} \cdot r^n, \tag{5.53}$$

where r is the engine's compression ratio and n the polytropic coefficient $(1 < n < \gamma)$.

Using the two expressions above, the energy density is easily calculated. Assuming a compression ratio of $r = 10$ and $n = 1.4$, an energy density value of 6.28 kJ/l for compressed air results. Compared to the energy density of gasoline, this value is about 5000 times smaller [85]. Therefore, compressed air as the main or sole energy source for propulsion, an old concept that periodically regains some interest, is not expected to be succesfully implemented in a significant number of vehicles due to the very low range that these vehicles would have.

5.6.2 Description of Operation Modes

The directly-connected HPE enables several operation modes that can be classified in two groups: (i) combustion modes, comprising the conventional ICE mode, the pneumatic supercharged mode, and the pneumatic undercharged mode, and (ii) purely pneumatic modes, comprising the pneumatic pump mode and the pneumatic motor mode. The latter can be based on a two-stroke or four-stroke cycle and require fuel cut off. Two-stroke modes further require a fully variable control of the intake and exhaust valves. In contrast, for the four-stroke modes, the intake and exhaust valves can remain camshaft-driven, which reduces complexity and cost of this configuration.

All the modes can be best explained by plotting the thermodynamic cycle inside the cylinders in a p–V diagram [168]. Since for any polytropic expansion $p \cdot V^n$ is constant, a double logarithmic p–V diagram is used in Figs. 5.25 – 5.28. In these figures, all polytropic processes are represented by a straight line. The dashed lines represent operating cycles with a higher torque than those depicted with solid lines.

Conventional ICE Mode

The cyclic operation in conventional ICE mode (throttled operation) is represented in Fig. 5.25 as a succession of eight ideal transformations: (i) a polytropic compression from bottom dead center (BDC) to top dead center (TDC), (ii) an isochoric combustion at TDC, followed by (iii) an isobaric part of the combustion, (iv) a polytropic expansion from end of combustion to BDC, (v) an isochoric expansion at BDC from cylinder pressure to exhaust pressure,

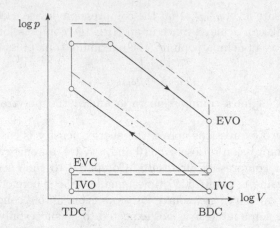

Fig. 5.25 Double logarithmic p–V diagram of the conventional engine mode. IVO: intake valve opening, IVC: intake valve closing, EVO: exhaust valve opening, EVC: exhaust valve closing. Dashed line: cycle with higher torque.

(vi) an isobaric exhaust stroke from BDC to TDC, (vii) an isochoric expansion at TDC from exhaust pressure to intake manifold pressure, and (viii) an isobaric intake stroke from TDC to BDC.

The manifold pressure varies from levels below ambient to higher than ambient if the turbocharger supplies enough fresh air. For a comparison with a typical efficiency map of the conventional combustion mode the reader is referred to Fig. 3.2.

Pneumatic Supercharged Mode

The pneumatic supercharged mode is used for operating points at which the turbocharger does not yet provide enough air. As schematically shown in Fig. 5.26, during compression the charge valve is opened (CVO) and closed again (CVC), at the latest when the tank pressure is reached. The tank pressure and the timing of this short opening determines the amount of additional air in the cylinder. The amount of fuel in the cylinder has to be metered to match the amount of air coming both from the intake manifold and from the tank.

Pneumatic Undercharged Mode

For a demanded engine torque lower than that achieved at naturally aspirated full load conditions, the excess air can be used to fill the tank. As shown schematically in Fig. 5.27, during the compression stroke the charge-valve is opened at a crank angle where the cylinder pressure is equal to the tank pressure. The duration of the opening determines the amount of air remaining

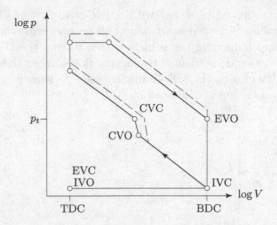

Fig. 5.26 Double logarithmic p–V diagram of the pneumatic supercharged mode. Nomenclature as in Fig. 5.25, plus CVO: charge valve opening, CVC: charge valve closing.

in the cylinder. For this mode, a direct fuel injection system is crucial because for safety reasons no fuel is allowed to enter the air tank. Accordingly, fuel injection can start only after CVC.

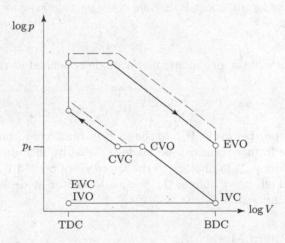

Fig. 5.27 Double logarithmic p–V diagram of the pneumatic undercharged mode. Nomenclature as in Fig. 5.26.

Pneumatic Motor Mode

In this mode, a positive torque is generated at the expenses of pressurized air injected from the tank, while the fuel injection is cut off. High and smooth torque transients can be provided with a two-stroke cycle. The pneumatic motor mode enables fast stop–start, stop-and-go operations without causing any emissions and launching the vehicle without using the clutch [337]. The load control in this mode is effected by varying the charge valve closing (CVC) and exhaust valve closing (EVC)[8] simultaneously, as shown in Fig. 5.28. The efficiency-optimal timing is discussed in [314].

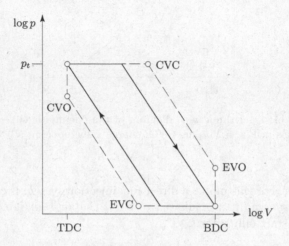

Fig. 5.28 Double logarithmic p–V diagram of the pneumatic motor mode. Nomenclature as in Fig. 5.26.

The efficiency of the pneumatic motor mode is defined as the ratio

$$\eta_{pm} = \frac{W_o}{W_t} \tag{5.54}$$

of the work output per cycle, W_o, to the energy transferred from the tank per cycle, W_t. The former term can be made scalable by introducing the *mean indicated pressure* p_{mi}, in analogy to the approach of Sect. 3.1.2. Figure. 5.29 shows a typical efficiency map of the pneumatic motor mode as a function of tank pressure and p_{mi}.

Pneumatic Pump Mode

The pneumatic pump mode is used for recuperating energy during braking. A two-stroke operation, illustrated in Fig. 5.30, yields the highest braking torque. The pressure tank is provided with pressurized air which can be used later

---

[8] The exhaust stroke can be carried out using the exhaust valves or the intake valves, the latter method being actually favorable if oxygen flooding of tree-way catalyst has to be avoided.

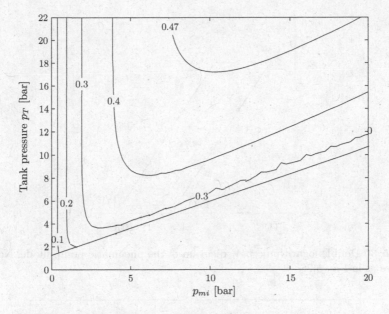

Fig. 5.29 Typical efficiency map for the pneumatic motor mode

in the pneumatic motor mode, or for supercharging the engine. The braking torque can be varied by changing the timing of the charge valve and of the intake valve simultaneously. For instance, the dashed-line cycle of Fig. 5.30 provides a less negative torque at the crankshaft than the solid-line cycle, thus a smaller braking torque results.[9] Starting with ambient pressure at BDC, a polytropic compression brings the cylinder pressure up to the pressure of the tank. Then, the charge valve opens and an enthalpy flow into the tank starts. The timing of the charge valve closing (CVC) determines the braking torque. The highest braking torque is reached if CVC is at TDC. Leaving the charge valve open for a longer period reduces the braking torque because the tank pressure acts on the piston down-stroke again. When the cylinder pressure again reaches manifold pressure, the intake valve remains open until it closes at BDC.

To describe the efficiency of the pneumatic pump mode, it is reasonable to look at the ratio of transferred tank energy per cycle $W_t < 0$ to the work output per cycle $W_o < 0$. However, since the energy from the intake is not considered in this definition (since it is "for free"), values higher than 1 can be achieved. For this reason, the term *coefficient of performance* (COP) is used for this ratio,

$$COP_{pp} = \frac{W_t}{W_o}. \tag{5.55}$$

---

[9] Larger negative torque values can be obtained following a different p–V cycle [85].

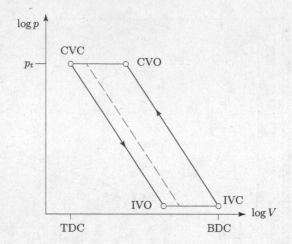

Fig. 5.30 Double logarithmic p–V diagram of the pneumatic pump mode. Nomenclature as in Fig. 5.26.

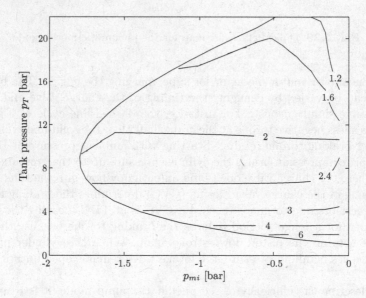

Fig. 5.31 Typical COP map for the pneumatic pump mode

This definition corresponds to that used for, e.g., heat pumps. Similarly to the motor mode, the work output per cycle W_o can be normalized as p_{mi} (now a negative quantity).

Figure 5.31 shows typical values of the COP of the pneumatic pump mode as a function of p_{mi} (i.e., indicated torque). For a given tank pressure, only a limited range of brake torques can be applied in an efficient way. Higher torques would be possible, but then the additional work would not lead to a higher enthalpy flow to the tank, but rather would be lost by throttling, backflow, or heat transfer. Depending on the temperature of the engine, it might thus be preferable to use the conventional brakes to provide this additional braking torque. Highest levels of COP can be achieved when the tank is empty.

5.7 Problems

Hybrid-Inertial Powertrains

Problem 5.1. Derive a Ragone curve similar to (5.1) and (5.2) for a flywheel battery. Then evaluate the maximum energy and power. Use a simplified expression for the loss power of the type $P_l = R \cdot \omega_f^2$. *(Solution:* $E_f(P_f) = P_f \cdot \tau \cdot \ln(1 + \frac{R\omega_0^2}{P})$, *where* $\tau = \frac{\Theta_f}{2R}$*).*

Problem 5.2. Dimension a flywheel for the following application (F1 KERS): $P_{max} = 60\,\text{kW}$, $E_{max} = 500\,\text{kJ}$, $\omega_{max} = 60000\,\text{rpm}$. Assume $\beta = 0.5$, $\rho = 1658\,\text{kg/m}^3$ (carbon fiber), $\rho_a^{0.8}\eta_a^{0.2} = 0.48$ (sealed flywheel). *(Solution:* $d = 20\,cm$, $b = 10\,cm$, $m_f = 5\,kg$, $\Theta_f = 0.026\,kgm^2$*).*

Problem 5.3. Evaluate the charging efficiency of the flywheel of Problem 5.2 for a braking at maximum power for 2 s. Evaluate the round-trip efficiency. *(Solution:* $\eta_c = 0.89$, $\eta_f = 0.81$*).*

Problem 5.4. Evaluate the CVT ratio during a deceleration of a vehicle equipped with a flywheel-based KERS and the opening time of the clutch. Use the flywheel data of Problem 5.3. The flywheel is connected to the input stage of the CVT through a fixed-reduction gear with ratio 8.33. Final drive and wheel ratio $\left(\frac{\gamma_{fd}}{r_w}\right) = 13$. Initial conditions: $v(0) = 80\,\text{km/h}$, $\omega_f(0) = 10000\,\text{rpm}$, $m_v = 600\,\text{kg}$, braking time 2 s (assume a constant braking power), CVT range $\nu_{max}/\nu_{min} = 6$. *(Solution:* $t_{oc} = 1.9\,s$*).*

Hybrid-Hydraulic Powertrains

Problem 5.5. Derive a Ragone curve similar to (5.1) and (5.2) for a hydraulic accumulator. Show that for high power this definition is equivalent to that adopted in the text. *(Solution:* $E_{ha}(P_h) = P_h \cdot \tau \cdot \ln\left(1 + \frac{W_{AB}}{\tau \cdot P_h}\right)$, *where* $\tau = m_g \cdot c_{v,g}/(h \cdot A_w)$. *For* $P_h \to \infty$, $E_{ha} \to W_{AB}$. *The definition (5.36) is thus correct)).*

Problem 5.6. Derive (5.48) and (5.50).

6

Fuel-Cell Propulsion Systems

This chapter contains three sections. The first section briefly describes the application of fuel-cell systems as stand-alone energy sources for powertrains or, in combination with a storage system, as fuel-cell hybrid powertrains. The second section introduces some thermodynamic and electrochemical models of fuel cells, as well as some fluid dynamic models of the complete fuel-cell system. The last section introduces the on-board production of hydrogen through fuel reforming and presents some system-level models of methanol reformers.

The main objective of this chapter is to introduce models that are useful in the context of energy management. Readers interested in the low-level control of fuel-cell systems are referred to [266].

6.1 Fuel-Cell Electric and Fuel-Cell Hybrid Vehicles

6.1.1 Introduction

Fuel cells are electrochemical devices that convert chemical energy directly into electrical energy. In contrast to internal combustion engines, there is no intermediate conversion into thermal energy. The efficiency of a fuel cell is thus not limited by the Carnot efficiency, and its work output theoretically can reach values that are higher than the lower heating value of the fuel.

Fuel cells deliver electrical energy, i.e., pure exergy, without any combustion products. This is the reason why, along with purely electric vehicles, fuel-cell vehicles are classified as zero-emission vehicles. As shown in Chap. 1, the specific energy of hydrogen as a fuel is substantially better than that of electrochemical batteries. Therefore, fuel-cell vehicles seem to be able to combine the best features of EVs and of ICE-based vehicles, namely zero local emissions, high efficiency, and a reasonable range. However, the operation of the on-board auxiliaries and all the conversion steps required to obtain and store the fuel (see Chap. 1) strongly affect the performance of fuel-cell systems. These shortcomings thus must be carefully taken into account when comparisons with other propulsion systems are made.

In principle, fuel cells can convert all fluid oxidizable substances. However, from a technical point of view, only hydrogen, natural gas, and methanol are currently of practical use.[1] Other liquid fossil fuels, such as ethanol and gasoline, can be used as energy carriers that are converted ("reformed") on-board into hydrogen. Hydrogen has the decisive advantage that the reaction product is pure water. Table 6.1 presents a comparison of the most important data for hydrogen and gasoline.

Table 6.1 Comparison of fuel data for hydrogen and gasoline

	Hydrogen	Gasoline
Molecular weight (g/mol)	2.016	107
Boiling point (K)	13.308	310–478
Vapor density at normal conditions (g/m^3)	84	4400
Liquid density (kg/m^3)	70.8	700
Higher heating value (MJ/kg)	141.86	48
Lower heating value (MJ/kg)	119.93	43.5
Gas constant (J/kg/K)	4124	78.0
Flammability limit in air (vol. %)	4.0–75	1.0–7.6
Detonation limit in air (vol. %)	18.3–59.0	1.1–3.3
Auto-ignition temperature (K)	858	501–744
Adiabatic flame temperature in air (K)	2318	2470
Flame front speed in air (cm/s)	≈ 300	37–43

The integration of a fuel cell in a propulsion system seems simple in principle, and various prototypes have already been developed. Since the fuel cell delivers energy in electrical form, the final power train is reduced to that of an electric vehicle (fuel-cell electric vehicles, FCEV). However, the resulting system is rather complex due to the multiple energy storage systems and load-levelers, various auxiliary devices, etc. that are required.

The combined problems of on-board hydrogen storage and of the lack of a hydrogen refuelling infrastructure so far have represented an impediment to the wide-scale adoption of FCEVs. On-board fuel processors that generate hydrogen from on-board hydrocarbons (methanol, gasoline, diesel, or natural gas) have therefore been proposed as an alternative hydrogen source for FCEVs. Although the energy density is much higher than for hydrogen storage, these systems are rather complex and introduce additional efficiency

---

[1] Metal/air fuel cells are one of the most promising alternatives, particularly for the ease of storing the anode fuel. In zinc–air systems (ZAFC) the anode is provided with zinc pellets [303, 282].

losses related to the various stages of fuel preparation, reforming, and hydrogen cleaning. Moreover, they produce CO_2 emissions and a poor dynamic response that makes control a difficult task.

The drivability and the power performance of an FCEV may be improved by the addition of a short-term storage system, resulting in a fuel-cell hybrid vehicle (FCHEV). The standard option consists of using electrochemical batteries, the alternative being represented by supercapacitors. Supercapacitors (see Sect. 4.6) have an extremely high power density and a higher efficiency than batteries for energy charge and discharge. They are used to cover power peaks, typically during accelerations, and for regenerative braking, while the fuel cell is operated almost stationarily. There are many advantages associated with this operation. First, the fuel cell can be downsized with respect to the peak power and thus it can be a smaller and cheaper unit. Secondly, the stationary operation increases not only the efficiency of the fuel cell but also its lifetime.

6.1.2 Concepts Realized

In the 1990s and 2000s many prototypes of passenger cars and other types of vehicles equipped with a fuel cell have been demonstrated by research institutions and major car manufacturers. However, none of these concepts have gone beyond the demonstration or pre-commercial stage. Concerning the propulsion architecture, purely fuel-cell vehicles must be distinguished from fuel-cell hybrid-electric vehicles. After pioneering attempts in the preceding decades, the development of FCEVs received a renewed impulse in the mid 1990s. Often the same manufacturer explored various hydrogen storage scenarios, i.e., gaseous, liquid, metal hydride, methanol, or gasoline reforming, in successive prototypes (e.g., Daimler Necar series [72], Ford Focus FCV/FC5 [106], Opel HydroGen3 series [237]), although pre-commercial projects mostly employed compressed gaseous hydrogen (GM's Hy-Wire [115]). Another concept used sodium borohydride as a hydrogen storage system (DaimlerChrysler Natrium [71]).

More recently, fuel-cell hybrids with a secondary battery as an energy storage system have been investigated with various solutions for hydrogen storage (Toyota FCHV series [323], Ford Focus FCV Hybrid [104], Nissan X-Trail FCV [231], Daimler F-Cell [215] and Citaro bus [70]). Hybrid fuel-cell systems using supercapacitors have also been introduced. From being the subject of research and development projects [275, 178], this solution has been explored by several car manufacturers (Mazda Demio FCEV [213], Honda FCX [141]). However, the introduction of lithium-based batteries has somehow diminished the interest in the supercapacitor-based FCHEV concept [146]. On the contrary, the role of fuel cells in FCHEVs seems to shift progressively towards the use as a range extender, similarly to series HEVs (Ford HySeries [354]).

6.2   Fuel Cells

6.2.1   Introduction

Hydrogen Storage Systems

As mentioned above, one of the main problems for the development of fuel cells as prime movers in passenger cars arises from the usage of hydrogen as a fuel. The lower energy density of hydrogen in comparison with gasoline makes its storage a difficult task. Various hydrogen storage technologies are the subject of research and testing. Example technologies are pressure vessels, cryogenic accumulators, and metal hydrides (chemical adsorption). These technologies lead to a specific energy which is significantly lower than that of gasoline. However, in consideration of the higher system efficiency of fuel-cell power sources, the range of such vehicles is comparable to that of ICE-based passenger cars. Additional disadvantages are generally higher cost and weight, aside from a more complicated fuel management.

Storage under pressures of up to 700 bar is achieved in conventional pressurized vessels. High-pressure tanks are complex systems that must be periodically tested and inspected to ensure their safety. However, this technology is widely developed, efficiently controllable, and relatively inexpensive. Therefore most of the current fuel-cell vehicle applications use compressed gaseous hydrogen. The specific energy of the fuel stored in a pressurized vessel is

$$\frac{E_{ht}}{m_{ht}} \approx \frac{E_{ht}}{m_m} = H_h \cdot \rho_h \cdot \gamma_{ht}, \tag{6.1}$$

where m_{ht} is the mass of the fully charged hydrogen tank, practically coincident with the mass of the vessel m_m, H_h is the lower heating value (120 MJ/kg or 33.3 kWh/kg) of hydrogen, ρ_h its density (depending on the pressure in the vessel), and γ_{ht} is the storage capacity of the vessel, i.e., the volume stored per unit mass. The energy density E_{ht}/V_{ht} is given by the product $H_h \cdot \rho_h$.

Hydrogen can be liquefied, but only at extremely low temperatures. Liquid hydrogen typically has to be stored at 20 K or –253°C. The storage tanks are insulated to preserve that low temperature and reinforced to store the liquid hydrogen under pressure. The energy density is still evaluated with (6.1). It is higher than with pressurized vessels, but energy losses of about 1% of the lower heating value are typical during vehicle operation, aside from those occurring in the liquefaction and compression process (typically, 30% of the lower heating value). Moreover, the cryogenic hydrogen must be heated before supplying the fuel cell, which causes additional losses and possibly further problems, especially in transient operation. However, the cryogenic storage has already found some application in fuel-cell vehicles [72].

A hybrid storage concept combining both high-pressure gaseous and cryogenic storage is being studied. These insulated pressure vessels (cryo-compressed tanks, "CcH2") are lighter and more compact than ambient-temperature, high-pressure vessels. On the other hand, because of the higher temperatures required (e.g., 77 K), there is a lower energy penalty for liquefaction, and smaller evaporative losses occur than with ambient-pressure liquid hydrogen storage [172].

Metal hydride storage systems represent a relatively new technology, although they have already been used in various fuel-cell vehicle prototypes [323, 213]. The structure of metal hydrides causes hydrogen molecules to be decomposed and hydrogen atoms to be incorporated in the interstices of specific combinations of metallic alloys. During vehicle operation, these hydrogen atoms are liberated through the addition of heat. The energy density is comparable to that of liquid hydrogen. However, the overall specific energy drops to lower values due to the additional weight of the metal hydride material. The main advantage of metal hydride storage is a relatively simple and safe fuel handling and delivery. This technology avoids the risks concomitant with high pressure or low temperature and, due to chemical bonds, it liberates only very little hydrogen in case of an accident. A serious drawback is the higher cost that originates on the one hand from the material and on the other hand from the complicated thermal management. The life of a metal hydride storage system is directly related to the purity of the hydrogen. The specific energy of metal hydride storage systems is calculated differently from (6.1), i.e.,

$$\frac{F_{ht}}{m_{ht}} \approx \frac{E_{ht}}{m_m} = H_h \cdot \xi_{ht}, \qquad (6.2)$$

where ξ_{ht} is the mass fraction of hydrogen stored in the system. The energy density of a metal hydride system is obtained from the specific energy by dividing it by γ_{ht}, the density of the material. Values of ξ_{ht} and γ_{ht} vary with the specific material used. The hydrogen storage capacity is typically 1–2%, though some alloys are theoretically capable of storing up to 11.5% hydrogen. Practical values can be reduced by a factor of 2 due to the porosity of the material. Density is generally higher for those materials that have a lower storage capacity. Currently, materials of interest are complex hydrides based on lithium, boron, aluminum, and magnesium [350].

A particular metal-hydride technology using sodium borohydride (NaBH$_4$) has found some vehicle application [71] but is now considered non-viable [234]. When NaBH$_4$ is combined with a specific catalyst, liquid borax and pure hydrogen gas are produced. The former subsequently can be recycled back into sodium borohydride. This technology is rather expensive due to the costs of the catalyst (ruthenium) and of the processes to produce sodium borohydride and to recycle borax. Moreover, the energy losses associated with the several conversion steps are substantial.

Alternative technologies using various physical storage concepts of hydrogen are being studied and tested. Storage efficiency claims of these methods (carbon nanotubes, glass microspheres, metal-organic frameworks and others) sometimes reach 10% or more.

Table 6.2 lists typical[2] values of the storage parameters of current storage technologies together with reference values for conventional gasoline storage. Also shown are the technical targets set by the US Department of Energy (DOE) for the year 2015 [332]. Note that these data do not include the conversion efficiencies as was the case in Fig. 1.7.

Table 6.2 Typical storage parameters of current storage technologies; [a]: for a steel tank, aluminum $1.5\,l/kg$, composite $3–4\,l/kg$ and more, [b]: $LiBH_4 + 1/2MgH_2$, optimized for specific energy [350], [c]: the mass of gasoline is not negligible when compared with that of the tank

	γ_{ht} (1/kg)	ρ_h (kg/m^3)	E_{ht}/m_{ht} (kWh/kg)	E_{ht}/V_{ht} (kWh/l)	ξ_{ht} (%)
Pressure vessel	1^a	15–30	0.5	0.5	
Cryogenic storage	1.7	71	4.0	2.4	
Metal-hydride storage^b	1.5	60	1.9	1.3	5.75
Gasoline	1.2^c	750	10.8	8.8	
DOE target, 2015			3.0	2.7	9.0

Types of Hydrogen Fuel Cells

The classification of hydrogen fuel cells follows the type of electrolyte. The main types are listed in Table 6.3.

Table 6.3 Types of fuel cells. See below for the definition of the fuel cell efficiency η_{fc}

Type	Electrolyte	$\eta_{fc}(\%)$	ϑ (°C)	Use
AFC	Alkaline (NaOH, KOH)	50–65	80–250	aerospace
PEM	Ionic membrane (Nafion)	50–60	40–100	automotive
DMFC	Ionic membrane	40	50–100	automotive
PAFC	Phosphoric acid (H$_3$PO$_4$)	35–45	160–220	power
MCFC	Molten carbonate (KLiCO$_3$)	40–60	600–650	power
SOFC	Solid oxides (ZrO$_2$, Y$_2$O$_3$)	50	850–1000	power

Alkaline fuel cells (AFC) use an aqueous solution of alkaline (e.g., potassium or sodium) hydroxide soaked in a matrix as the electrolyte. The cathode reaction is faster than in other electrolytes, which means a higher performance. In fact, AFCs yield the highest electrochemical efficiency levels, up to 65%.

---

[2] The research efforts in this field are currently very active such that a reasonable and updated estimation of the average values is very difficult.

Their operating temperature ranges from 80 to 250°C, although newer, low-temperature designs can operate below 80°C. They typically have an output of 300 to 5000 W. This technology is widely developed since it was initially used in aerospace applications as far back as the 1960s, to provide not only the power but also the drinking water for the astronauts. Although some experimental vehicles (e.g., ZEVCO London hybrid taxi, 1998, Lada Antel, 2001) were powered by AFCs, this technology is not considered to be suitable yet for automotive applications. On the one hand, the caustic electrolyte is highly corrosive and thus high standards are demanded from the material and the safety technology. On the other hand, this type of cell is very sensitive to contaminations in the supply gases, thus requiring very pure hydrogen to be used. Moreover, until recently AFCs were too costly for commercial applications.

Proton-exchange membrane fuel cells (PEM) use a thin layer of solid organic polymer[3] as the electrolyte. This ion-conductive membrane is coated on both sides with highly dispersed metal alloy particles (mostly platinum, an expensive material) that are active catalysts. The PEM fuel cell basically requires hydrogen and oxygen as reactants, though the oxidant may also be ambient air, and these gases must be humidified to prevent membrane dehydration. Hydrogen must be as pure as possible, since CO contained in alcohol or hydrocarbon fuels poisons platinum catalysts (up to 10 ppm of CO are tolerated). Methane and methanol reforming is thus possible only for low loads. However, research is ongoing on platinum/ruthenium catalysts that are more resistent to CO [52]. Because of the limitations imposed by the thermal properties of the membrane, PEM fuel cells operate at relatively low temperatures of about 80°C, which permits a quick start-up. Due to this fact and to other advantages such as higher power density and higher safety with the solid electrolyte, PEM fuel cells are particularly suitable for automotive applications. Further improvements are required in system efficiency, the goal being an efficiency level of 60%.

Direct methanol fuel cells (DMFC) are similar to the PEM fuel cells in that their electrolyte is also a polymer membrane. However, in the DMFC the anode catalyst itself draws the hydrogen from the liquid methanol, eliminating the need for a fuel reformer. That is quite an advantage in the automotive area where the storage or generation of hydrogen is one of the main obstacles for the introduction of fuel cells. Another field of application is in the very small power range, e.g., laptops. There are principal problems, including the lower electrochemical activity of the methanol as compared with hydrogen, giving rise to lower cell voltages and efficiency levels. Efficiencies of only about 40% may be expected from the DMFC, at a typical temperature of operation of 50–100°C. Also, DMFCs use expensive platinum as a catalyst and, since methanol is miscible in water, some of it is liable to cross the water-saturated membrane and cause corrosion and exhaust gas problems on the cathode side.

Phosphoric-acid fuel cells (PAFC) use liquid phosphoric acid soaked in a

---

[3] The most commonly used material is Nafion by DuPont.

matrix as the electrolyte. This type of fuel cell is the most commercially developed and is used particularly for power generation. Already various manufacturers are represented on the market with complete power plants that generate from 100 to 1000 kW. The efficiency of PAFCs is roughly 40%. The fact that they can use impure hydrogen as a fuel allows the possibility of reforming methane or alcohol fuels. Nevertheless, due to the higher operating temperatures of 160–220°C and the associated warm-up times, for automotive applications this type is suitable only for large vehicles such as buses.[4] Moreover, PAFCs use expensive platinum as a catalyst and their current and power density is small compared with other types of fuel cells.

Molten-carbonate fuel cells (MCFC) use a liquid solution of lithium, sodium, and/or potassium carbonates, soaked in a matrix as the electrolyte. This cell operates at a temperature of about 650°C, which is required to achieve a sufficiently high conductivity of the electrolyte. The higher operating temperature provides the opportunity for achieving higher overall system efficiencies (up to 60%) and greater flexibility in the use of available fuels and inexpensive catalysts, although it imposes constraints on choosing materials suitable for a long lifetime. MCFCs have been operated with various fuels in power plants ranging from 10 kW to 2 MW. The necessity for large amounts of ancillary equipment would render a small operation, such as an automotive application, uneconomic.

Solid-oxide fuel cells (SOFC) use solid, nonporous metal oxide electrolytes. The metal electrolyte normally used in manufacturing SOFCs is stabilized zirconia. This cell operates at a temperature of about 1000°C, allowing internal reforming and/or producing high-quality heat for cogeneration or bottoming cycles. Thus this type is used in large power plants of up to 100 kW, where it reaches efficiency levels of 60% or even as high as 80% for the combined cycle. However, high temperatures limit the use of SOFCs to stationary operation and impose severe requirements on the materials used. On the other hand, SOFCs do not need any expensive electrode materials. Moreover, various fuels can be used, from pure hydrogen to methane to carbon monoxide. Some developers are testing SOFC auxiliary power units for automotive applications [68].

Electrochemistry of Hydrogen Fuel Cells

In electrochemical cells, the reaction consists of two semi-reactions, which take place in two spatially separated sections. These two zones are connected by an electrolyte that conducts positive ions but not electrons. The electrons that are released by the semi-reaction of oxidation can arrive at the reduction electrode (cathode) only through an external electric circuit. This process yields an electric current, which is the useful output of the cell. The normal direction of the external current is from the reduction side (cathode, positive

---

[4] In 1994 a prototype PAFC bus was demonstrated by Georgetown University.

electrode) to the oxidation side (anode, negative electrode). In Fig. 6.1 a simple PEM fuel cell is depicted. It consists of a particular membrane that does not conduct electrons but rather lets ions pass. This membrane is impermeable for neutral gas and serves as an electrolyte. At both sides of the membrane porous electrodes are mounted. The electrodes allow for the gas diffusion, and they accomplish a triple contact gas–electrolyte–electrode. As described below, at the anode and the cathode sides of the membrane hydrogen and oxygen are supplied, respectively.

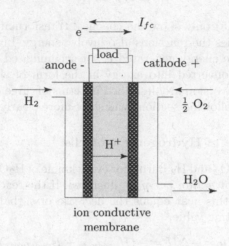

Fig. 6.1 Principle of operation of a fuel cell

At the anode, molecular hydrogen is in equilibrium with simple protons and electrons. The anode reaction is written as

$$H_2 \rightarrow 2H^+ + 2c^-. \tag{6.3}$$

Under standard operating conditions, the dissociation rate is small, such that most of the hydrogen is present in the form of electrically neutral molecules. The equilibrium can be modified with a change in the boundary conditions, e.g. the adoption of a catalyst or an increase in temperature. Since the H^+ ions are formed at the anode, a concentration gradient is established between the two ends of the membrane, generating an ion diffusion toward the cathode. The H^+ ion current transports a positive charge from the anode to the cathode. As a consequence, a difference of potential arises across the membrane, with the anode as a negative electrode and the cathode as a positive electrode. Since the free electrons cannot follow the electric field through the membrane, they flow through the external circuit.

The electron current dissipates energy across an external resistance. This energy is generated by the chemical reaction that takes place at the cathode, where electrons (from the external path), protons (through the membrane), and oxygen from an external source are combined to yield water. The cathode reaction is

$$2H^+ + 2e^- + \frac{1}{2}O_2 \rightarrow H_2O. \tag{6.4}$$

The water product is at a lower energetic state than the original combination of protons, electrons, and oxygen molecules. The difference is the energy delivered by the fuel cell. The overall reaction is the combination of (6.3) and (6.4),

$$H_2 + \frac{1}{2}O_2 \rightarrow H_2O. \tag{6.5}$$

Equations (6.3)–(6.5) only show the chemical transformations. But together with chemical species the reactions also involve energy. In contrast to a combustion process, the energy generated is not accumulated as thermal energy, but it is directly converted into exergy in the form of electric energy. The limits of such a conversion are described in terms of "free energy" (or Gibbs' potential), as the following sections will show more clearly.

Thermodynamics of Hydrogen Fuel Cells

When a mixture of O_2 and H_2 burns, for every kmole of H_2O a defined quantity of heat known as *formation energy* is liberated. If the reaction takes place at constant pressure, this heat equals the decrease of enthalpy H and is often referred to as "heating value,"

$$Q_H = -\Delta H = - \left(H_{products} - H_{reactants} \right). \tag{6.6}$$

The heating value is a function of the temperature and the pressure at which the reaction takes place. For a reaction such as that of (6.5), Q_H also depends on the state of the water in the products. If water occurs in the form of vapor, the lower heating value is obtained. Vice versa, condensated water frees an additional amount of energy that leads to the higher heating value. Since the enthalpy of all pure substances in their natural state is zero by definition, the enthalpies of H_2 and O_2 are zero, thus $H_{reactants} = 0$ in (6.6). The enthalpy of water thus corresponds to the heating value of the reaction. For vapor water at 1 atmosphere and 298.15 K (reference temperature and pressure, RTP), $\Delta H_H = -241.8\,\text{MJ/kmol}$. For liquid water at RTP, $\Delta H_l = -285.9\,\text{MJ/kmol}$. The difference corresponds to the enthalpy of the vaporization of water.

If the reaction were to take place in an isolated system, all the fuel heating value in principle could be converted into electrical work. In fact, in a system without any heat exchange with the surrounding ambient, the work not related to variations of volume equals the variations of enthalpy,

$$W_{id} = -\Delta H = Q_H. \tag{6.7}$$

However, it is theoretically impossible to collect all the work W_{id} from a fuel cell. One limiting condition playing a role similar to that of the Carnot efficiency for thermal systems arises when the entropy is taken into account.

Besides its internal energy each substance is also characterized by a certain entropy level that depends on the particular thermodynamic state. In a closed system and in the ideal case of a reversible reaction, the heat dissipated to the surrounding ambient equals the entropy variation,

$$Q_S = -\vartheta \cdot \Delta S, \tag{6.8}$$

where Q_S is positive if released and negative if absorbed. In the general case, (6.8) is transformed into the well-known Clausius inequality. For the substances involved in an H_2–O_2 fuel cell, the entropy values are listed in Table 6.4 [92]. If 1 kmol of H_2O is formed, 1 kmol of H_2 and 0.5 kmol of O_2 disappear with their entropy. In the balance, 44.4 kJ/(kmol K) are missing. Thus the fuel cell releases the heat $Q_S = 298 \cdot 44.4 = 13.2$ MJ/kmol to the ambient.

Table 6.4 Thermodynamic data for hydrogen fuel cells at RTP

H_2 (gaseous)	H = 0 MJ/kmol	S = 130.6 kJ/(kmol K)
O_2 (gaseous)	H = 0 MJ/kmol	S = 205.0 kJ/(kmol K)
H_2O (vapor)	H = –241.8 MJ/kmol	S = 188.7 kJ/(kmol K)

The fact that in association with the entropy variations a certain amount of heat is always liberated limits the useful work available. In fact, for a reversible reaction at a constant temperature ϑ, a simple energy balance yields

$$-W_{rev} - Q_S = \Delta H \rightarrow W_{rev} = -\Delta H + \Delta(\vartheta \cdot S) = -\Delta G, \tag{6.9}$$

where G is the state function known as "free energy"

$$G = H - \vartheta \cdot S. \tag{6.10}$$

The quantity W_{rev} in (6.9) is the maximum electrical work that can be obtained from a fuel cell operating at constant pressure and temperature. When non-reversible processes are taken into account, the value of (6.9) is further diminished by a quantity corresponding to the non-reversible entropy sources.

Equation (6.9) can also be written as

$$W_{rev} = Q_H - Q_S, \tag{6.11}$$

which clearly shows that not all the heating value can be converted into useful work. Thus the highest possible cell efficiency for an ideal electrochemical cell ("electrochemical Carnot efficiency") is given by the ratio between the maximum work available in the case of a reversible process and the heating value of the single energy carriers,

$$\eta_{id} = \frac{-\Delta G}{-\Delta H} = 1 - \frac{\vartheta \cdot \Delta S}{\Delta H} = 1 - \frac{Q_S}{Q_H}. \tag{6.12}$$

Table 6.5 Reversible voltage and efficiency for various fuel cell types; [a]: vapor water

Fuel	Reaction	U_{rev} (V)	η_{id}
Hydrogen	$H_2 + 0.5\ O_2 \rightarrow H_2O$	1.18	0.94[a]
Methane	$CH_4 + 2\ O_2 \rightarrow CO_2 + 2\ H_2O$	1.06	0.92
Methanol	$CH_3OH + 0.5\ O_2 \rightarrow CO_2 + 2\ H_2O$	1.21	0.97
Propane	$C_3H_8 + 5\ O_2 \rightarrow CO_2 + 2\ H_2O$	1.09	0.95
Carbon monoxide	$CO + 0.5\ O_2 \rightarrow CO_2$	1.07	0.91
Formaldehyde	$CH_2O + O_2 \rightarrow CO_2 + H_2O$	1.35	0.93
Formic acid	$HCOOH + 0.5\ O_2 \rightarrow CO_2 + H_2O$	1.48	1.06
Carbon	$C + 0.5\ O_2 \rightarrow CO$	0.71	1.24
	$C + O_2 \rightarrow CO_2$	1.02	1.00
Ammonia	$NH_3 + 0.75\ O_2 \rightarrow 0.5\ N_2 + 1.5\ H_2O$	1.17	0.88

Table 6.5 shows a collection of theoretical efficiency levels for various fuels that may be used in fuel cells.

The work available per unit quantity of reactants with the cell voltage is calculated as follows. Each kmol of hydrogen contains $N_0 = 6.022 \cdot 10^{26}$ molecules. Equation (6.3) shows that for each molecule of hydrogen two electrons circulate in the external circuit. In general, the useful work W can be expressed in terms of cell voltage U_{fc} as

$$W = n_e \cdot q \cdot N_0 \cdot U_{fc} = n_e \cdot F \cdot U_{fc}, \tag{6.13}$$

where n_e is the number of free electrons for every kmol of hydrogen, with $q = 1.6 \cdot 10^{-19}$ C being the charge of an electron and $F = q \cdot N_0 = 96.48 \cdot 10^6$ C/kmol being the Faraday constant. Assuming that the cells work in a reversible way, (6.13) can be written as

$$W_{rev} = n_e \cdot q \cdot N_0 \cdot U_{rev} = n_e \cdot F \cdot U_{rev} = -\Delta G, \tag{6.14}$$

where U_{rev} is the reversible cell voltage. Solving (6.14) for U_{rev}, the expression for the cell voltage obtained is

$$U_{rev} = -\frac{\Delta G}{n_e \cdot F}. \tag{6.15}$$

If water is in its vapor form, $W_{rev} = 228.6$ MJ/kmol and the cell voltage is $U_{rev} = 1.185$ V (see Table 6.5). If in contrast the fuel cell produces liquid water, the higher heating value leads to $W_{rev} = 237.2$ MJ/kmol and thus the ideal open-circuit voltage is 1.231 V.

Besides that of the reversible voltage U_{rev}, an important role for further considerations is played by the "caloric voltage" U_{id}. This voltage is defined by

$$U_{id} = -\frac{\Delta H}{n_e \cdot F} \tag{6.16}$$

and it measures the voltage (impossible to reach) that would be provided by a total conversion of enthalpy into electrical energy. From (6.12), $U_{rev} = \eta_{id} \cdot U_{id}$ follows.

Under load ($I_{fc}(t) > 0$), real cells deliver a voltage $U_{fc}(t)$, which is lower than U_{rev}. The power lost in the form of heat is evaluated as [226]

$$P_l(t) = I_{fc}(t) \cdot (U_{id} - U_{fc}(t)) \tag{6.17}$$

This heat has to be removed by means of a dedicated cooling device (see Sect. 6.2.2).

Fuel-Cell Systems

For typical automotive applications, the voltage of a single cell is too low. Higher voltages can be obtained by arranging several cells in series. This combination, often referred to as a "stack", is depicted in Fig. 6.2.

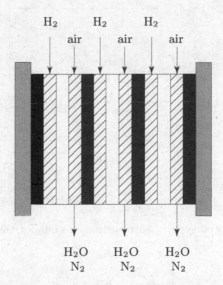

Fig. 6.2 Typical arrangement of a PEM fuel-cell stack. White fields represent the membrane–electrode arrangement (MEA), gray fields are the flow fields of reactant gases, black fields are bipolar plates.

The core of a single PEM cell consists of the membrane, the catalyst layers, and the porous electrodes (often referred to as "diffusion layers" or "backing layers"). Together, these layers form the membrane–electrode arrangement, MEA. Single cells are separated by flow fields that supply reactant gas to each electrode. Flow fields are arranged in bipolar plates as parallel flow channels (in coflow or counterflow form), a serpentine channel (one long channel with many passes over the diffusion layer), and interdigitated channels that force the flow through the diffusion layer. Bipolar plates also have to conduct electricity to the external circuit and allow for water flow to remove heat generated by the reaction. Research on materials for bipolar plates is ongoing since common graphite and metal plates suffer from some inconvenients (cost or corrosion).

A fuel-cell stack needs to be integrated with other components to form a fuel-cell system that is able to power a vehicle. Figure 6.3 shows a typical fuel-cell system with such components. These are grouped in four subsystems or circuits: (i) hydrogen circuit, (ii) air circuit, (iii) coolant circuit, and (iv) humidifier circuit. Most fuel cells use deionized water as a coolant, so that the subsystems (iii) and (iv) can be combined in a single water circuit [64]. The description in the following refers to a system with hydrogen storage. For systems fed by a fuel reformer, see Sect. 6.3.

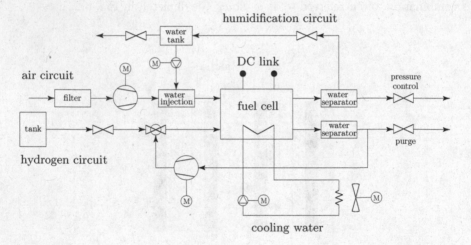

Fig. 6.3 Typical fuel-cell system, with distinct circuits for the humidification and the cooling

Hydrogen is supplied to the fuel cell anode from its storage system, e.g., a pressurized tank, through a regulating valve that adjusts the pressure to the fuel cell level. A pressure tap from the air circuit serves as the reference pressure for the regulator. In some applications the hydrogen flow is humidified (see below) in a chamber with a water injector. At the outlet of the fuel cell, hydrogen is recirculated by a pump to form a closed loop. This prevents the hydrogen from being consumed when the fuel cell is not absorbing current. In such a case the tank can be isolated by closing a shutoff valve located at the exit of the tank. A manual shutoff valve is also present in the recirculation loop to purge hydrogen during a shut-down.

The air supply system provides the fuel cell with clean air at a high relative humidity (>80%). A motor-driven volumetric (screw, scroll, rotary piston) or dynamic (centrifugal) compressor raises the air pressure typically by 70 kPa [352, 235]. The air is humidified (see below) and then fed to the fuel cell cathode. The exhaust air flow from the fuel cell outlet is regulated by a valve.

Due to the high flow rates and water carrying capacity of the reactants, current PEM fuel cells are quite sensitive to humidity. Humidification is accomplished by a controlled water injection system, for example. Liquid water accumulates inside the humidity chamber as the humidified air passes through it and cools down. A reservoir collects this condensate and supplies it to the injection system and the cooling circuit. The exhaust stream from the fuel cell contains air, water droplets, and steam. To keep the water level in the deionized water tank even, the water in the air exhaust stream must be captured. This is accomplished by using an air–water separator in the exhaust stream [235].

The coolant circuit provides cooling for the fuel cell and some auxiliaries, such as the motor that drives the compressor. The coolant circuit includes a radiator, often of the multi-stage type, a radiator fan, a coolant pump, a water reservoir, and often a deionizer. If deionized water is used, the coolant circuit is strongly integrated with the humidifier circuit, since both use the same water reservoir. Unlike most ICE-based vehicles, fuel-cell vehicles have an electrically driven coolant pump whose speed can be varied independently of the operating conditions of the prime mover. Strategies of "intelligent cooling" can be used to minimize power consumption [278].

A much simpler and less expensive circuitry is expected when using high-temperature PEM fuel cells, which are currently under development. These cells operate at 150 °C or more and consequently they do not require a liquid coolant. They only need a simple air cooling system, with a fan replacing radiator, pump, etc. Moreover, there is no need for a humidifier due to the particular composition of the MEA. The membrane is also more tolerant of CO and sulfur in the fuel supplied [214].

6.2.2 Quasistatic Modeling of Fuel Cells

The causality representation of a fuel cell in quasistatic simulations is sketched in Fig. 6.4. The input variable is the terminal power $P_{fcs}(t)$. The output variable is the hydrogen consumption $\overset{*}{m}_{h,c}(t)$.

First, the behavior of a single cell is discussed, then the arrangement of multiple cells in a stack and the related system. In the following, PEM fuel cell behavior is assumed where not explicitly stated otherwise. Fundamental fuel cell models that describe multi-phase, multi-dimensional flows in the electrodes, membrane, and catalyst layers, are not discussed here. The reader can find additional information in the bibliographies of the cited works, which deal with system-level, lumped-parameter modeling of PEM fuel cells.

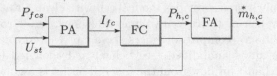

Fig. 6.4 Fuel cells: causality representation for quasistatic modeling

Cell Voltage

The behavior of a single cell is characterized in terms of cell voltage $U_{fc}(t)$ and current density, $i_{fc}(t)$, which is defined as the cell current per active area,

$$i_{fc}(t) = \frac{I_{fc}(t)}{A_{fc}}. \tag{6.18}$$

Figure 6.5 shows a typical polarization curve of a fuel cell, i.e., its static dependency between $U_{fc}(t)$ and $i_{fc}(t)$. The curve is depicted for given operating parameters such as partial pressure, humidity, and temperature. Increasing operating pressure would lift the curve up [126, 281, 204]. An increase of stack temperature or air humidity generally increases voltage also, but combined effects may be more difficult to predict [281, 273].

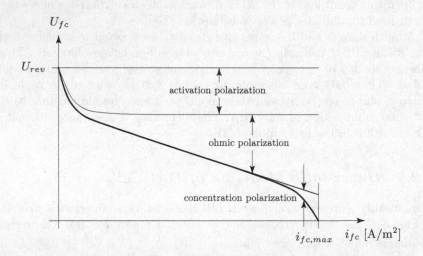

Fig. 6.5 Polarization curve of a hydrogen fuel cell with major loss contributions

The cell voltage is given by the equilibrium potential diminished by irreversible losses, often called overvoltages or polarizations. These may be due to three sources: (i) activation polarization, (ii) ohmic polarization, (iii) concentration polarization,

$$U_{fc}(t) = U_{rev} - U_{act}(t) - U_{ohm}(t) - U_{conc}(t). \qquad (6.19)$$

The reversible cell potential U_{rev} is a function of cell temperature and partial pressure of reactants and products, but not of cell current. The potential at standard conditions depends only on the reaction stoichiometry and is given in the previous section. The classical Nernst equation provides a correction for any temperature and pressure values different from the standard values [92]. Sometimes it is necessary to modify the theoretical coefficients of the Nernst equation to take into account other phenomena, e.g., the liquid water flooding the cathode [273].

The activation polarization $U_{act}(t)$ is a result of the energy required to initiate the reaction, which depends on the type of catalyst used. A better catalyst causes lower activation losses. The limiting reaction is that occurring at the cathode, which is inherently slower than the anode reaction. Activation losses increase as current density increases. The relation between $U_{act}(t)$ and $i_{fc}(t)$ is described by the semi-empirical Tafel equation [92, 204, 9, 209],

$$U_{act}(t) = c_0 + c_1 \cdot \ln(i_{fc}(t)), \qquad (6.20)$$

where the coefficients c_0 and c_1 depend on temperature (c_0 also depends on reactant partial pressure). However, such a model is not valid for very small current densities, i.e., where the influence of the activation polarization is dominant. Therefore, a convenient empirical approximation of the Tafel equation can be derived [126] as

$$U_{act}(t) = c_0 \cdot \left(1 - e^{-c_1 \cdot i_{fc}(t)}\right), \qquad (6.21)$$

where c_0 and c_1 depend on the partial pressure of the reactants and on the cell temperature. Equation (6.21) describes the fact that beyond a certain current density the activation polarization can be considered as a constant added to ohmic losses (see Fig. 6.5).

The ohmic losses $U_{ohm}(t)$ are due to the resistance to the flow of ions in the membrane and in the catalyst layers and of electrons through the electrodes, the former contributions being dominant. Assuming that both membrane and electrode behavior may be described by Ohm's law, the ohmic losses are expressed in terms of an overall resistance R_{fc} as

$$U_{ohm}(t) = i_{fc}(t) \cdot \tilde{R}_{fc}, \qquad (6.22)$$

where $\tilde{R}_{fc} = R_{fc} \cdot A_{fc}$. The ohmic resistance R_{fc} includes electronic, membrane (ionic), and contact resistance contributions. Usually, only the dominant membrane resistance is taken into account, which is related to the membrane conductivity [265]. This resistance depends strongly on the cell temperature and the membrane humidity. A minor nonlinear influence of current density may be included as well [204, 9], though it is often negligible [209].

The concentration polarization $U_{conc}(t)$ results from the change in the concentration of the reactants at the electrodes as they are consumed in the reaction. This loss becomes important only at high current densities. Its dependency on $i_{fc}(t)$ may be described [126] by the law

$$U_{conc}(t) = c_2 \cdot i_{fc}^{c_3}(t),$$ (6.23)

where c_2 and c_3 are complex functions of the temperature and the partial pressure of the reactants. A different formulation [185] expresses the concentration polarization as

$$U_{conc}(t) = c_2 \cdot e^{c_3 \cdot i_{fc}(t)}.$$ (6.24)

The measured polarization curve of a 100-cell stack is shown in Fig. 6.6. In the operating range of useful current densities, the figure clearly shows the effects of the activation losses and of the ohmic losses. With the exclusion of very low current densities ($i_{fc}(t) < 0.1\,\mathrm{A/cm^2}$), the curve can be conveniently linearized [126, 24], i.e., fitted by the equation

$$U_{fc}(t) = U_{oc} - R_{fc} \cdot I_{fc}(t).$$ (6.25)

In this equation, U_{oc} is the voltage at which the linearized curve crosses the y axis with $i_{fc} = 0$ (no-current state). Thus U_{oc} should not be confused with the reversible voltage U_{rev} that is the voltage value corresponding to the nonlinear curve at $i_{fc} = 0$. The "resistance" R_{fc} is constant for a given type of cell and at constant operating conditions (pressure, temperature, humidity). Equation (6.25) is equivalent to the electrical circuit depicted in Fig. 6.7, which will be used in the following to describe fuel cells.

Fuel-Cell System

The voltage of a stack is obtained by multiplying the cell voltage by the number N of single cells in series,

$$U_{st}(t) = N \cdot U_{fc}(t),$$ (6.26)

while the stack current equals the cell current $I_{fc}(t)$. The parallel arrangement of several stacks is also possible. In the model introduced above this arrangement corresponds to an increase of the cell surface, and therefore the output current increases.

The output power of a fuel-cell stack is

$$P_{st}(t) = I_{fc}(t) \cdot U_{fc}(t) \cdot N.$$ (6.27)

This power must cover the load demand $P_{fcs}(t)$ and the requirements of the auxiliaries $P_{aux}(t)$ (Fig. 6.3),

$$P_{fcs}(t) = P_{st}(t) - P_{aux}(t).$$ (6.28)

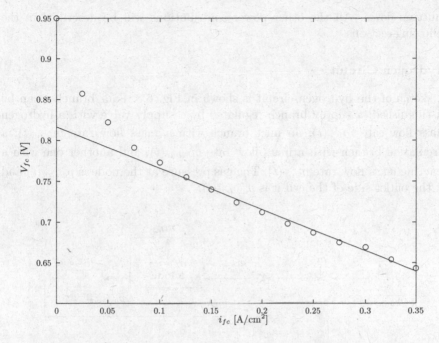

Fig. 6.6 Linear fitting of the polarization curve of a PEM fuel cell. Experimental data (circles) are from measurements taken on a stack of $N = 100$ cells with an active area of $A_{fc} = 200\,\mathrm{cm}^2$ [273]. Operating conditions: $\vartheta_{fc} = 60°\mathrm{C}$, $p_{ca,in} = 2.0\,\mathrm{bar}$, $p_{an,in} = 2.2\,\mathrm{bar}$, $\lambda_a = 2.2$. Voltage values: $U_{rev} = 0.95\,\mathrm{V}$, $U_{oc} = 0.82\,\mathrm{V}$.

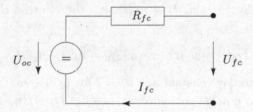

Fig. 6.7 Equivalent circuit of a fuel cell

The auxiliary power $P_{aux}(t)$ is the sum of various contributions,

$$P_{aux}(t) = P_0 + P_{em}(t) + P_{ahp}(t) + P_{hp}(t) + P_{cl}(t) + P_{cf}(t), \qquad (6.29)$$

where $P_{em}(t)$ is the power of the compressor motor, $P_{hp}(t)$ is the power of the hydrogen circulation pump, $P_{ahp}(t)$ is the power of the humidifier water circulation pump, $P_{cl}(t)$ is the power of the coolant pump, and $P_{cf}(t)$ is the power of the cooling fan motor. The term P_0 is the value of the bias power that covers the linkage current necessary to keep a minimum flow of reactants throughout the whole operating range of the cell in order to keep it from shutting down. All the other power contributions will be described in the following sections.

Hydrogen Circuit

A sketch of the hydrogen circuit is shown in Fig. 6.8. Four branches can be distinguished: a supply branch regulated by a supply valve with a hydrogen mass flow rate $\overset{*}{m}_{h,c}(t)$, an inlet branch with a mass flow rate $\overset{*}{m}_{h,in}(t)$, a circulation branch with a mass flow rate $\overset{*}{m}_{h,out}(t)$, and another one with a reacting mass flow rate $\overset{*}{m}_{h,r}(t)$. The gas pressure at the node is $p_{an,in}(t)$ and at the outlet side of the cell it is $p_{an,out}(t)$.

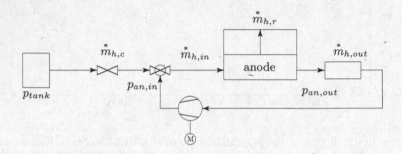

Fig. 6.8 Relevant variables in the hydrogen circuit

The electric current intensity $I_{fc}(t)$ is given by the product of the quantity flow rate of the electrons, $\overset{*}{n}_e(t) = \overset{*}{n}_h(t) \cdot n_e$, the electron charge q, and the Avogadro constant N_0,

$$I_{fc}(t) = \overset{*}{n}_h(t) \cdot n_e \cdot q \cdot N_0 = \overset{*}{n}_h(t) \cdot n_e \cdot F, \tag{6.30}$$

where F is the Faraday constant and $n_e = 2$ for hydrogen.

From the molar flow rate of hydrogen, $\overset{*}{n}_h(t) = I_{fc}(t)/(n_e \cdot F)$, the hydrogen mass flow rate $\overset{*}{m}_{h,r}(t) = M_h \cdot \overset{*}{n}_h(t)$ follows, where M_h is the hydrogen molar mass. By substitution, the relationship between the hydrogen mass flow rate and the fuel cell current for N cells is obtained as

$$\overset{*}{m}_{h,r}(t) = \frac{N \cdot I_{fc}(t) \cdot M_h}{n_e \cdot F}. \tag{6.31}$$

To obtain a uniform feeding of hydrogen on the cell surface, particularly in transient operation, the system must be fed with a hydrogen excess quantified by the variable $\lambda_h(t)$. The actual hydrogen entering the fuel cell is usually expressed as

$$\overset{*}{m}_{h,in}(t) = \lambda_h(t) \cdot \frac{N \cdot I_{fc}(t) \cdot M_h}{n_e \cdot F} = \lambda_h(t) \cdot \kappa_h \cdot I_{fc}(t). \tag{6.32}$$

The value of $\lambda_h(t)$ is regulated by adjusting the opening of the hydrogen supply valve, since the mass flow rate $\overset{*}{m}_{h,in}(t)$ is the sum of a quantity recirculated $\overset{*}{m}_{h,rec}(t)$ and a quantity extracted from the hydrogen tank,

$$\overset{*}{m}_{h,in}(t) = \overset{*}{m}_{h,rec}(t) + \overset{*}{m}_{h,c}(t). \tag{6.33}$$

The mass flow rate recirculated may differ from the anode output for a quantity that is periodically purged,[5]

$$\overset{*}{m}_{h,rec}(t) = \overset{*}{m}_{h,out}(t) - \overset{*}{m}_{h,pur}(t). \tag{6.34}$$

The output mass flow rate in turn is given by

$$\overset{*}{m}_{h,out}(t) = \overset{*}{m}_{h,in}(t) - \overset{*}{m}_{h,r}(t) = (\lambda_h(t) - 1) \cdot \kappa_h \cdot I_{fc}(t), \tag{6.35}$$

such that, if there is no purge, the flow rate of the mass extracted must equal the flow rate of the mass consumed in the cell reaction.

The hydrogen mass flow rate is related to the pressure difference between the cell inlet and outlet. Such a relationship depends on the type of flow fields that distribute the reactant gas to the electrodes. Graphite or carbon composite plates with small grooves (dimensions of ≈ 0.5 mm) for gas flow are extensively used in commercial fuel-cell stacks. For these flow fields, assuming laminar gas flow with continuous, uniform subtraction of mass due to the cell reaction, it is possible to write [126, 319]

$$\overset{*}{m}_{h,in}(t) = K_h \cdot (p_{an,in}(t) - p_{an,out}(t)) + \frac{1}{2} \cdot \overset{*}{m}_{h,r}(t), \tag{6.36}$$

that is, to assume a linear dependency between mass flow rate and pressure drop. For other types of flow fields, e.g., porous plates, a quadratic dependency described by Darcy's law arises [319, 139].

The mass extracted from the tank flow rate is related to the hydrogen tank pressure and the cell pressure through the opening of the supply control valve. Such a dependency can be simplified as

$$p_{an,in}(t) = p_{tank} - \xi_h(t) \cdot \overset{*}{m}_{h,c}^{2}(t), \tag{6.37}$$

where $\xi_h(t)$ describes the supply valve opening.

The circulation mass flow rate $\overset{*}{m}_{h,rec}(t)$ as well as the pressure levels $p_{an,out}(t)$ and $p_{an,in}(t)$ are related to the operation of the hydrogen circulation pump. The performance of the hydrogen pump is expressed by quasistatic characteristic curves, which represent the dependency

---

[5] Purging prevents the accumulation of impurities in the hydrogen feed and N_2 accumulation due to diffusion from the cathode and, if done properly [276], removes excess water.

$$f_{hp}\left(\omega_{hp}(t), \overset{*}{m}_{h,rec}(t), \Pi_{hp}(t)\right) = 0, \tag{6.38}$$

where $\Pi_{hp}(t) = p_{an,in}(t)/p_{an,out}(t)$ is the pump pressure ratio. The power absorbed by the motor-driven hydrogen pump is calculated as

$$P_{hp}(t) = \overset{*}{m}_{h,rec}(t) \cdot c_{p,h} \cdot \vartheta_{fc} \cdot \left(\Pi_{hp}(t)^{\frac{\gamma_h - 1}{\gamma_h}} - 1\right) \cdot \frac{1}{\eta_{hp} \cdot \eta_{em}}, \tag{6.39}$$

where η_{em} is the efficiency of the drive motor and η_{hp} is the efficiency of the pump. Equation (6.39) can be approximated for small pressure drops as

$$P_{hp}(t) \approx \frac{\overset{*}{m}_{h,rec}(t) \cdot (p_{an,in}(t) - p_{an,out}(t))}{\rho_h \cdot \eta_{hp} \cdot \eta_{em}}, \tag{6.40}$$

where ρ_h is the average hydrogen mass density.

The system represented by equations (6.31)–(6.39) is a system of nine equations with the thirteen unknowns $\overset{*}{m}_{h,r}(t)$, $\overset{*}{m}_{h,in}(t)$, $\overset{*}{m}_{h,out}(t)$, $\overset{*}{m}_{h,rec}(t)$, $\overset{*}{m}_{h,pur}(t)$, $\overset{*}{m}_{h,c}(t)$, $I_{fc}(t)$, $p_{an,in}(t)$, $p_{an,out}(t)$, $\lambda_h(t)$, $\xi_h(t)$, $P_{hp}(t)$, and $\omega_{hp}(t)$. Since the cell current $I_{fc}(t)$ is the independent variable, three other variables must be specified for the system equations to be solved. In quasistatic simulations, system outputs are prescribed and control variables are calculated therefrom.

The hydrogen circuit control system is usually designed to keep the anode pressure $p_{an,in}(t)$ and the hydrogen circulation ratio $\lambda_h(t)$ at prescribed values. In particular, the anode pressure is prescribed as a function of the cathode pressure $p_{ca,in}(t)$, usually with a constant pressure difference (e.g., 0.2 bar) between the two sides. To meet the control requirements, the control system acts on the power of the hydrogen pump motor $P_{hp}(t)$, on the supply valve opening $\xi_h(t)$, and, if necessary, on the purge mass flow rate $\overset{*}{m}_{h,pur}(t)$.

Based on (6.36) and (6.32), the pressure drop $p_{an,in}(t) - p_{an,out}(t)$ is proportional to the reacting mass $\overset{*}{m}_{h,r}(t)$. As given by (6.35) the hydrogen circulating mass flow rate $\overset{*}{m}_{h,out}(t)$ is also proportional to $\overset{*}{m}_{h,r}(t)$. The pump power calculated with the approximation of (6.40) may thus be expressed as

$$P_{hp}(t) = \kappa_{hp} \cdot I_{fc}^2(t). \tag{6.41}$$

Air Circuit

The flow rate of the oxygen mass that reacts with hydrogen is calculated from the cell electrochemistry. For each mole of hydrogen converted, 0.5 mol of oxygen is needed. Thus the molar flow rate of air that is strictly necessary for the reaction is given by

$$\overset{*}{n}_o(t) = \frac{1}{2} \cdot \overset{*}{n}_h(t). \tag{6.42}$$

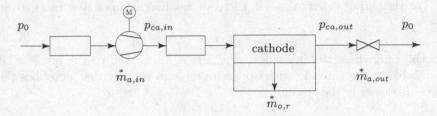

Fig. 6.9 Relevant variables in the air circuit

The reacting air mass flow rate thus can be calculated for the whole stack as a function of the fuel cell current,

$$\overset{*}{m}_{o,r}(t) = \frac{N \cdot I_{fc}(t) \cdot M_o}{2 \cdot n_e \cdot F}, \tag{6.43}$$

where M_o is the molar mass of oxygen.

For a stack that operates with excess air quantified by the variable $\lambda_a(t) > 1$, the effective air mass flow rate needed is given by

$$\overset{*}{m}_{a,in}(t) = \lambda_a(t) \cdot \frac{N \cdot I_{fc}(t) \cdot M_o}{2 \cdot n_e \cdot F} \frac{100}{21} = \lambda_a(t) \cdot \kappa_a \cdot I_{fc}(t), \tag{6.44}$$

assuming the mass fraction of oxygen in dry inlet air to be 21%. The relationship between mass flow rate and pressure drop in the air circuit (see Fig. 6.9) is similar to (6.36) for laminar flow in the cathode channel,

$$\overset{*}{m}_{a,in}(t) = K_a \cdot (p_{ca,in}(t) - p_{ca,out}(t)) + \frac{1}{2} \cdot \overset{*}{m}_{o,r}(t). \tag{6.45}$$

This relationship can be properly adapted to represent a nozzle-type behavior (see below).

The mass flow rate $\overset{*}{m}_{a,in}(t)$ and the supply manifold pressure $p_{ca,in}(t)$ are related to the air compressor operation. In fact, the performance of the air compressor is expressed by quasistatic characteristic curves, which represent the dependency

$$f_{cp}\left(\omega_{cp}(t), \overset{*}{m}_{a,in}(t), \Pi_{cp}(t)\right) = 0, \tag{6.46}$$

where $\Pi_{cp}(t) = p_{ca,in}(t)/p_0$ is the compressor pressure ratio. Various approaches are available in the literature to fit steady-state compressor data and to derive the analytical expression of (6.46) [319]. The power absorbed by a motor-driven air compressor is calculated as

$$P_{cp}(t) = \overset{*}{m}_{a,in}(t) \cdot c_{p,a} \cdot \vartheta_a \cdot \left(\Pi_{cp}(t)^{\frac{\gamma_a - 1}{\gamma_a}} - 1\right) \cdot \frac{1}{\eta_{cp}}, \tag{6.47}$$

where ϑ_a is the supply air temperature and η_{cp} is the compressor efficiency.

At the output side of the cell, there are residual air mass flow rates, given by

$$\overset{*}{m}_{a,out}(t) = \overset{*}{m}_{a,in}(t) - \overset{*}{m}_{o,r}(t). \tag{6.48}$$

In the case where the mass flow $\overset{*}{m}_{a,out}(t)$ is discharged to the ambient, it is related to the cathode outlet pressure through the law that describes the pressure losses in the circuit,

$$p_{ca,out}(t) - p_0 = \xi_a(t) \cdot \overset{*}{m}_{a,out}^2(t), \tag{6.49}$$

where ξ_a describes the opening of the pressure-control valve.

In the presence of an expander to recuperate part of the enthalpy contained in the exhaust air, (6.49) is substituted by the dependency expressed by the expander characteristic curves,

$$f_{ex}\left(\omega_{cp}(t), \overset{*}{m}_{a,out}(t), \Pi_{ex}(t)\right) = 0, \tag{6.50}$$

where $\Pi_{ex}(t) = p_{ca,out}(t)/p_0$ is the expansion ratio. The power recuperated through the expander is given by

$$P_{ex}(t) = \overset{*}{m}_{a,out}(t) \cdot c_{p,a} \cdot \vartheta_{fc} \cdot \left(1 - \Pi_{ex}(t)^{-\frac{\gamma_a-1}{\gamma_a}}\right) \cdot \eta_{ex}, \tag{6.51}$$

where η_{ex} is the expander efficiency. Unfortunately, expanders often have a rather small efficiency. Moreover, the available enthalpy is small as well. For these reasons, expanders are rarely installed, and controllable outlet valves (acting on $\xi_a(t)$) are used instead. Notice that the composition of the outlet air is different from that of the inlet air, since oxygen is consumed in the fuel cell. In some cases, it may be necessary to take this effect into account by properly modifying the specific heat $c_{p,a}$.

At steady state, a relationship exists between the compressor power, the expander power (if any), and any additional power provided by an external source, e.g., an electric motor,

$$-P_{cp}(t) + P_{ex}(t) + P_{em}(t) \cdot \eta_{em} = 0, \tag{6.52}$$

where η_{em} is the efficiency of the electric motor.

The system represented by (6.43)–(6.52)[6] is a system of nine equations with the twelve unknowns $\omega_{cp}(t)$, $p_{ca,in}(t)$, $p_{ca,out}(t)$, $P_{cp}(t)$, $P_{ex}(t)$, $\overset{*}{m}_{a,in}(t)$, $\overset{*}{m}_{a,out}(t)$, $\overset{*}{m}_{o,r}(t)$, $\lambda_a(t)$, $I_{fc}(t)$, $\xi_a(t)$, and $P_{em}(t)$. Since the cell current $I_{fc}(t)$ is the independent variable, two other variables must be specified for the system equations to be solved. In quasistatic simulations, system outputs are prescribed and control variables are calculated therefrom.

The air circuit control system is usually designed such as to keep the excess air $\lambda_a(t)$ and the cathode pressure $p_{ca,in}(t)$ – and thus $\Pi_{cp}(t)$ – at prescribed constant values, acting on the purge valve opening $\xi_a(t)$ and on the

---

[6] With (6.49) in alternative to (6.50).

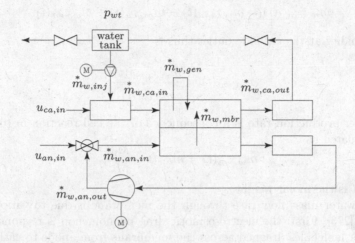

Fig. 6.10 Relevant variables in the humidification water circuit

compressor motor power $P_{em}(t)$. Based on (6.52) and (6.47), with constant motor efficiency η_{em} and omitting the possibility of an expander, it is possible to express $P_{em}(t)$ as

$$P_{em}(t) = \kappa_{cp} \cdot I_{fc}(t). \tag{6.53}$$

Water Circuit

To prevent the dehydration of the fuel cell membrane, the air flow and in some cases also the hydrogen flow are humidified by the injection of water. The humidifier can act also as an air cooler, to reduce the temperature of the air entering the cell. Even if the hydrogen is not humidified, there still may be some water content at the anode outlet, due to the water diffusion through the fuel cell membrane. In the following, the presence of liquid water flooding the electrodes will not be taken into consideration for simplicity, though its influence is treated in the literature [273].

The water mass entering the cathode is calculated as the sum (see Fig. 6.10)

$$\overset{*}{m}_{w,ca,in}(t) = u_0 \cdot \overset{*}{m}_{a,in}(t) + \overset{*}{m}_{w,inj}(t), \tag{6.54}$$

where u_0 is the humidity mass fraction in the ambient air and $\overset{*}{m}_{w,inj}(t)$ is the injected water mass flow rate. This is regulated by the water pump to achieve a specified humidity $u_{ca,in}(t)$ at the cathode inlet. This humidity is given by

$$u_{ca,in}(t) = \frac{\overset{*}{m}_{w,ca,in}(t)}{\overset{*}{m}_{a,in}(t)}. \tag{6.55}$$

At the cathode outlet, the water mass flow rate is increased by the water production in the reaction and by the water flow rate across the membrane,

$$\overset{*}{m}_{w,ca,out}(t) = \overset{*}{m}_{w,ca,in}(t) + \overset{*}{m}_{w,gen}(t) + \overset{*}{m}_{w,mbr}(t). \tag{6.56}$$

The humidity at the cathode outlet thus is

$$u_{ca,out}(t) = \frac{\overset{*}{m}_{w,ca,out}(t)}{\overset{*}{m}_{a,out}(t)}. \tag{6.57}$$

The water production rate is a consequence of the cell reaction in (6.4) and it is evaluated as

$$\overset{*}{m}_{w,gen}(t) = \overset{*}{m}_{h,r}(t) \cdot \frac{M_w}{M_h}, \tag{6.58}$$

where M is the molar weight.

The water mass flow rate through the membrane is due to various phenomena [273]. First, the electro-osmotic drag phenomenon is responsible for the water molecules dragged across the membrane from anode to cathode by the hydrogen ions. Second, the gradient of the water concentration across the membrane causes a back diffusion of water from cathode to anode. Another source of back diffusion is the presence of a gradient in the partial pressure of water across the membrane. Since the membrane is usually very thin (approximately 50 μm) the gradients can be approximated as constant. Neglecting the third term, which depends on the concentration of water in the membrane [265], the water flux is evaluated as

$$\overset{*}{m}_{w,mbr} = M_w \cdot N \cdot \left(n_d \cdot \frac{I_{fc}(t)}{F} - A_{fc} \cdot D_w \cdot \frac{\phi_{ca}(t) - \phi_{an}(t)}{\delta_{mbr}} \right), \tag{6.59}$$

where δ_{mbr} is the membrane thickness, n_d is the electro-osmotic drag coefficient, and D_w is the diffusion coefficient, the latter two being functions of the membrane water content and of the stack temperature [273]. The membrane water content (related to the concentration of water in the membrane) in turn can be expressed as a function of $\phi_{ca}(t)$ and $\phi_{an}(t)$, which are the relative humidities of the cathode and of the anode, respectively. The relationship between the u's (absolute humidity) and the ϕ's (relative humidity) is generally

$$u = \frac{M_w}{M_a} \cdot \frac{\phi \cdot p_{sat}(\vartheta)}{p - \phi \cdot p_{sat}(\vartheta)}. \tag{6.60}$$

It is possible to approximate $u_{ca}(t)$ as an average of the values $u_{ca,in}(t)$ and $u_{ca,out}(t)$, from which $\phi_{ca}(t)$ to be used in (6.59) follows. The evaluation of $\phi_{an}(t)$ for the anode side is analogous.

The mass flow $\overset{*}{m}_{w,ca,out}(t)$ is purged and recuperated in the air/water separator. An additional mass flow to compensate inevitable losses is provided by a water pump. The power for the air humidifier water pump is given by

$$P_{ahp}(t) = \frac{\overset{*}{m}_{w,inj}(t) \cdot \Delta p_{ahp}(t)}{\rho_w \cdot \eta_{ahp}} = \kappa_{ahp} \cdot \overset{*}{m}_{w,inj}(t), \tag{6.61}$$

where $\Delta p_{ahp}(t) = p_{ca,in}(t) - p_{wt}$ is the pressure rise of the air humidifier pump, p_{wt} is the constant pressure in the water tank, and η_{ahp} is the efficiency of the pump.

If there is no hydrogen humidifier, at the anode inlet only the vapor in the supply hydrogen is present. The water mass flow rate entering the anode is given by

$$\overset{*}{m}_{w,an,in}(t) = \overset{*}{m}_{h,c}(t) \cdot u_{an,c} + \overset{*}{m}_{w,an,out}(t), \tag{6.62}$$

with $u_{an,c}$ usually assumed to correspond to a relative humidity of 100%. At the anode outlet, the water flow rate across the membrane is lost. Thus the remaining flow rate, assuming no additional water purge, is given by

$$\overset{*}{m}_{w,an,out}(t) = \overset{*}{m}_{w,an,in}(t) - \overset{*}{m}_{w,mbr}(t). \tag{6.63}$$

The humidity levels at the anode inlet and outlet stages, useful to evaluate the quantity $\overset{*}{m}_{w,mbr}(t)$, are

$$u_{an,in}(t) = \frac{\overset{*}{m}_{w,an,in}(t)}{\overset{*}{m}_{h,in}(t)}, \tag{6.64}$$

$$u_{an,out}(t) = \frac{\overset{*}{m}_{w,an,out}(t)}{\overset{*}{m}_{h,out}(t)}. \tag{6.65}$$

The system of (6.54)–(6.65) is a system of eleven equations with the twelve unknowns $\overset{*}{m}_{w,ca,in}(t)$, $\overset{*}{m}_{w,ca,out}(t)$, $\overset{*}{m}_{w,inj}(t)$, $\overset{*}{m}_{w,gen}(t)$, $\overset{*}{m}_{w,mbr}(t)$, $u_{ca,in}(t)$, $u_{ca,out}(t)$, $P_{ahp}(t)$, $\overset{*}{m}_{w,an,out}(t)$, $\overset{*}{m}_{w,an,in}(t)$, $u_{an,in}(t)$, and $u_{an,out}(t)$. Thus, one of the variables must be specified for the system to be solved. In quasistatic simulations, system outputs are prescribed and control variables are calculated therefrom.

The water circuit control system is designed such as to keep the humidity of the inlet air $u_{cu,in}(t)$ at a prescribed constant value, acting on the water injected mass flow rate $\overset{*}{m}_{w,inj}(t)$. Based on (6.55) and (6.61), $\overset{*}{m}_{w,inj}(t)$ must be proportional to the inlet air mass flow rate. This is in turn related to the cell current, so that it is possible to express $P_{ahp}(t)$ as

$$P_{ahp}(t) = \kappa_{ahp} \cdot I_{fc}(t). \tag{6.66}$$

Coolant Circuit

The power for the coolant pump is evaluated as

$$P_{cl}(t) = \frac{\overset{*}{m}_{cl}(t) \cdot \Delta p_{cl}}{\rho_{cl} \cdot \eta_{cl}}. \tag{6.67}$$

The coolant mass flow rate is calculated as a function of the heat to be removed, as

$$\overset{*}{m}_{cl}(t) = \frac{P_{l,st}(t)}{c_{p,cl} \cdot \Delta\vartheta_{cl}}, \tag{6.68}$$

where $c_{p,cl}$ and $\Delta\vartheta_{cl}$ are the specific heat and the temperature rise of the coolant.

The heat power to be removed by the coolant, $P_{l,st}(t)$, is practically coincident with the amount of heat generated by the cells, since exhaust gas contains only a very limited amount of enthalpy. Thus the evaluation of $P_{l,st}(t)$ for a stack may be derived from (6.17) for a single cell,

$$P_{l,st}(t) = (U_{id} - U_{fc}(t)) \cdot N \cdot I_{fc}(t). \tag{6.69}$$

The temperature difference in (6.68) cannot exceed an admissible value which is a characteristic value of the cooling system. Equation (6.67) may thus be written as $P_{cl}(t) = \kappa_{cl} \cdot P_{l,st}(t)$.

The thermal power must be removed from the water with the help of cooling fans, usually capable of an on/off operation, such as in internal combustion engines. The power required for the cooling fan $P_{cf}(t)$ is proportional to the mass flow rate of the cooling air, by an expression similar to (6.47). In turn, the cooling air mass flow rate is proportional to the coolant mass flow rate, so that it is possible to write $P_{cf}(t) = \kappa'_{cf} \cdot \overset{*}{m}_{cl}(t) = \kappa_{cf} \cdot P_{l,st}(t)$.

The coolant control system regulates the coolant mass flow rate and the cooling fan operation in order to remove the heat generated by the fuel cell. If (6.69) is combined with (6.25), it is possible to express the power requirement of the cooling system as

$$P_{cl}(t) + P_{cf}(t) = \kappa_{cl,1} \cdot I_{fc}(t) + \kappa_{cl,2} \cdot I_{fc}^2(t). \tag{6.70}$$

Overall Model

Now the total auxiliary power can be evaluated as a function of the fuel cell current. Using the simplified expressions derived above, i.e., (6.41), (6.53), (6.66), and (6.70), equation (6.29) may be written as

$$P_{aux}(t) = P_0 + \kappa_{hp} \cdot I_{fc}^2(t) + \kappa_{ahp} \cdot I_{fc}(t) + \kappa_{cp} \cdot I_{fc}(t) + \kappa_{cl,1} \cdot I_{fc}(t) +$$
$$+ \kappa_{cl,2} \cdot I_{fc}^2(t) = P_0 + \kappa_1 \cdot I_{fc}(t) + \kappa_2 \cdot I_{fc}^2(t). \tag{6.71}$$

On the other hand, the stack output power $P_{st}(t)$ has a quadratic dependency on $I_{fc}(t)$, as described by (6.25) and (6.27). Thus, the result of (6.71) agrees with semi-empirical data that suggest a linear dependency between auxiliary power and stack output power [321]. For a larger output power the dependency may be less than linear, thus a nonlinear approximation should be used, e.g., exponential [297]. Other formulations were derived leading to a quadratic dependency between $P_{aux}(t)$ and $P_{st}(t)$ [24].

In a first approximation, (6.71) can be further simplified using a linearized accessory power [127],

$$P_{aux}(t) = P_0 + N \cdot \kappa_{aux} \cdot I_{fc}(t), \tag{6.72}$$

which explicitly takes into account multiple cells.

Equation (6.72) can be combined with (6.28), (6.27), and (6.25), or generally with (6.19), to obtain the quadratic expression for $I_{fc}(t)$

$$N \cdot R_{fc} \cdot I_{fc}^2(t) - (N \cdot U_{oc} - N \cdot \kappa_{aux}) \cdot I_{fc}(t) + P_{fcs}(t) + P_0 = 0, \tag{6.73}$$

from which the final equation for the cell current may be written as [127]

$$I_{fc}(t) = \frac{N \cdot (U_{oc} - \kappa_{aux}) - \sqrt{N^2 \cdot (U_{oc} - \kappa_{aux})^2 - 4 \cdot N \cdot R_{fc} \cdot (P_{fcs}(t) + P_0)}}{2 \cdot N \cdot R_{fc}}. \tag{6.74}$$

Once the cell current is known, the hydrogen consumption can be calculated using (6.33),

$$\overset{*}{m}_{h,c}(t) = N \cdot \frac{I_{fc}(t) \cdot M_h}{n_e \cdot F} = k_h \cdot I_{fc}(t). \tag{6.75}$$

Fuel Cell Efficiency

Besides the electrochemical efficiency $\eta_{id} = U_{rev}/U_{id}$, various other efficiencies can be defined for fuel-cell stacks. The voltage efficiency for a single cell is defined as

$$\eta_V(I_{fc}) = \frac{U_{fc}(I_{fc})}{U_{rev}}. \tag{6.76}$$

In the affine approximation (6.25), the voltage efficiency also is an affine function of the cell current. It decreases linearly as $I_{fc}(t)$ increases, showing its higher values at low loads. Another possible efficiency is the current or Faradaic efficiency that compares the effective current with the current theoretically delivered,

$$\eta_I(\lambda_h) = \frac{I_{fc}}{I_{th}} = \frac{1}{\lambda_h}. \tag{6.77}$$

The global efficiency of a fuel cell $\eta_{fc}(I_{fc}) = \eta_V(I_{fc}) \cdot \eta_I$ has the same dependency as $\eta_V(I_{fc})$ on the cell current, since η_I typically is a constant.

The system efficiency takes into account also the auxiliary power. The effective power delivered is compared with the theoretically deliverable power,

$$\eta_{st}(I_{fc}) = \frac{P_{fcs}(I_{fc})}{N \cdot U_{id} \cdot I_{fc}}, \tag{6.78}$$

thus the total stack efficiency is $\eta_{st,tot}(I_{fc}) = \eta_{st}(I_{fc}) \cdot \eta_I$. From (6.73) and (6.12), an expression for the efficiency may be derived

$$\eta_{st}(I_{fc}) = \eta_{id} \cdot \frac{U_{oc}}{U_{rev}} \cdot \left(1 - \frac{R_{fc} \cdot I_{fc}}{U_{oc}} - \frac{P_0}{U_{oc} \cdot I_{fc} \cdot N} - \frac{\kappa_{aux}}{U_{oc}}\right). \qquad (6.79)$$

Obviously, this expression has a maximum for η_{st} at $I_{fc,\eta} = \sqrt{P_0/(N \cdot R_{fc})}$.

Figure 6.11 illustrates the dependency of various power terms and efficiency values on the fuel cell current.

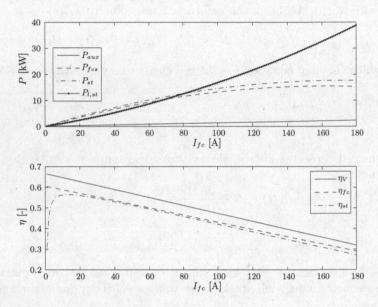

Fig. 6.11 Fuel cell power (top) and efficiency (bottom) as a function of fuel cell current. Fuel cell data: $N = 250$, $U_{rev} = 1.23\,\text{V}$, $U_{oc} = 0.82\,\text{V}$, $A_{fc} = 200\,\text{cm}^2$, $R_{fc} = 0.0024\,\Omega$, $\lambda_h = 1.1$, $P_0 = 100\,\text{W}$, $\kappa_{aux} = 0.05\,\text{V}$.

6.2.3 Dynamic Modeling of Fuel Cells

The physical causality representation of a fuel cell is sketched in Fig. 6.12. The model input variable is the stack current $I_{fc}(t) = P_{fcs}(t)/U_{st}(t)$. The model output variables are the stack voltage $U_{st}(t)$ and the fuel consumption $\overset{*}{m}_{h,c}(t)$.

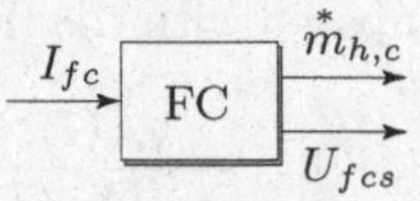

$$I_{fc} \rightarrow \boxed{\text{FC}} \begin{array}{l} \overset{*}{m}_{h,c} \\[4pt] U_{fcs} \end{array}$$

Fig. 6.12 Fuel cells: physical causality for dynamic modeling

The dynamic model of a PEM fuel cell that will be discussed here is a system-level, lumped-parameter model based on the ideas presented in the previous section. The dynamic effects include in principle electrochemical, fluid dynamic, and thermal effects. However, the order of magnitude of the relevant time constants of such processes is quite different [126],

- Hydrogen and air manifolds: $O(10^{-1}$ s)
- Membrane water content: $O(10^0$ s)
- Control system: $O(10^0$ s)
- Stack temperature: $O(10^2$ s)

where $O(\cdot)$ denotes the order of magnitude.

The fastest transient phenomena are the electrochemical ones. These dynamics are due to charge double layers at the membrane/electrode interfaces. The ion/electron charge separation at these interfaces creates a charge storage that can be described by a double-layer capacitance C_{dl} [245]

$$\frac{d}{dt}U_{act}(t) = \frac{i_{fc}(t)}{C_{dl}} - \frac{U_{act}(t)}{R_{act} \cdot C_{dl}}, \tag{6.80}$$

where R_{act} is defined as the ratio of the steady-state activation polarization, given by (6.20) or (6.21) and the current density $i_{fc}(t)$. The equivalent circuit of a cell is modified accordingly, as depicted in Fig. 6.13. For simplicity, the concentration polarization is not considered. However, the resulting time constants $R_{act} \cdot C_{dl}$ are so small that this class of dynamic effects may be neglected without causing any substantial errors.

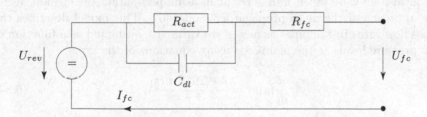

Fig. 6.13 Equivalent dynamic electric circuit for a fuel cell

The fluid dynamic transient effects taken into consideration are usually of the capacitive type, i.e., they describe the variation of gas pressure in a reservoir as a consequence of mass flows entering and leaving the reservoir. The models available in the literature differ in the number of pressure levels introduced.

The most immediate extension of the quasistatic equations presented in the previous section consists of keeping the same pressure levels, i.e., $p_{ca,in}(t)$ and $p_{ca,out}(t)$ for the cathode channel, $p_{an,in}(t)$ and $p_{an,out}(t)$ for the anode channel, but letting them vary as a function of the mass flow rates $\overset{*}{m}_{a,in}(t)$,

$\overset{*}{m}_{a,out}(t)$, $\overset{*}{m}_{h,in}(t)$, $\overset{*}{m}_{h,out}(t)$, etc. Assuming that the supply manifold temperature is kept close to the operating stack temperature ϑ_{st} by means of a cooling device,[7] the state equation for $p_{ca,in}(t)$ is

$$\frac{d}{dt}p_{ca,in}(t) = \frac{R_a \cdot \vartheta_{st}}{V_{a,sm}} \cdot \left(\overset{*}{m}_{a,cp}(t) - \overset{*}{m}_{a,in}(t)\right),$$ (6.81)

where R_a is the air gas constant and $V_{a,sm}$ is the volume of the air supply manifold. Analogously, the state equation for $p_{ca,out}(t)$ is

$$\frac{d}{dt}p_{ca,out}(t) = \frac{R_a \cdot \vartheta_{st}}{V_{a,rm}} \cdot \left(\overset{*}{m}_{a,out}(t) - \overset{*}{m}_{a,rm}(t)\right),$$ (6.82)

where $V_{a,rm}$ denotes the return manifold volume.

Equations (6.81)–(6.82) introduce two additional variables, namely $\overset{*}{m}_{a,cp}(t)$ and $\overset{*}{m}_{a,rm}(t)$, representing the air mass flow rate delivered by the compressor and the one discharged through the control valve or the expander. These quantities replace $\overset{*}{m}_{a,in}(t)$ and $\overset{*}{m}_{a,out}(t)$ in (6.46)–(6.47) and (6.49)–(6.51), respectively.

Substituting its quasistatic counterpart (6.52), an additional state equation can be written for the compressor speed as

$$\frac{d}{dt}\omega_{cp}(t) = \frac{P_{em}(t) \cdot \eta_{em} - P_{cp}(t) + P_{ex}(t)}{\Theta_{cp} \cdot \omega_{cp}(t)}.$$ (6.83)

Inertial effects in the connecting pipes, particularly at the compressor output [273], also may be taken into account in a lumped-parameter fashion, using the approach of the quasi-propagatory model [62]. This model describes the mass flow rates in the pipes as new state variables, evaluated as a function of the pressure levels at the manifolds using equations of the type

$$\frac{d}{dt}\overset{*}{m}(t) = \frac{\kappa \cdot \Delta p - \overset{*}{m}(t)}{\tau},$$ (6.84)

where $\kappa \cdot \Delta p$ is the steady-state mass flow rate and τ is the time constant of the process that depends primarily on the pressure levels and on the pipe lengths [62].

Compared with the quasistatic case, if inertial effects are not modeled there are two more variables, namely $\overset{*}{m}_{a,cp}(t)$, and $\overset{*}{m}_{a,rm}(t)$, balanced by two additional equations. The cell current $I_{fc}(t)$ is still the independent variable, while $P_{em}(t)$ is the main control variable of the air circuit. An additional control input that may contribute to the regulation of the cathode pressure is $\xi_a(t)$, although in most applications this valve opening is kept constant. Consequently, all the remaining quantities are determined, in particular the excess air $\lambda_a(t)$ by (6.44).

---

[7] The humidifier can perform this function.

A similar modeling approach can be used for the anode channel. The state equation for $p_{an,in}(t)$ is

$$\frac{d}{dt}p_{an,in}(t) = \frac{R_h \cdot \vartheta_{st}}{V_{h,sm}} \cdot \left(\overset{*}{m}_{h,c}(t) - \overset{*}{m}_{h,in}(t) - \overset{*}{m}_{h,rec}(t) \right), \qquad (6.85)$$

with obvious meaning of the variables. The state equation for $p_{an,out}(t)$ similarly is

$$\frac{d}{dt}p_{an,out}(t) = \frac{R_h \cdot \vartheta_{st}}{V_{h,rm}} \cdot \left(\overset{*}{m}_{h,out}(t) - \overset{*}{m}_{h,rec}(t) - \overset{*}{m}_{h,pur}(t) \right). \qquad (6.86)$$

As in the quasistatic case, the number of variables exceeds the number of relationships by four. The cell current $I_{fc}(t)$ is still the independent variable, while $P_{hp}(t)$, $\xi_h(t)$, and $\overset{*}{m}_{h,pur}(t)$ are the control variables of the hydrogen circuit. Therefore, all the remaining quantities are determined, in particular the excess hydrogen $\lambda_h(t)$ by (6.32) and the hydrogen consumption $\overset{*}{m}_{h,c}(t)$ by (6.33).

As for the water circuit, usually no dynamic effects are introduced in the manifolds. Therefore, as in the quasistatic case, the number of variables exceeds the number of relationships available by one. Since the control variable is the injected mass flow rate $\overset{*}{m}_{w,inj}(t)$, all the remaining quantities are determined, in particular the inlet air humidity $u_{ca,in}(t)$ by (6.55).

A different method of lumping the capacitive properties may lead to recognizing distinct pressure levels for the anode and the cathode channels, $p_{an}(t)$ and $p_{ca}(t)$, respectively [361, 245, 265, 267]. Mass flow rates entering and leaving the channels are evaluated according to some "nozzle-type" equation as a function of the pressure difference between the channels and the supply and return manifolds introduced above, which are characterized by their own pressure levels as described earlier.

This model structure allows a generalization of the equations above to the non-isothermal case. The energy conservation law must be invoked together with the mass conservation law to derive at least two additional equations (one for the cathode, one for the anode) at the two additional variable temperature levels $\vartheta_{an}(t)$ and $\vartheta_{ca}(t)$. Energy fluxes to be considered are the enthalpy flows associated with mass fluxes and heat exchanged with the solid walls facing the gas flows. An additional state equation may be written for the solid body temperature $\vartheta_{st}(t)$, which may vary due to heat exchanged with fluids, heat produced by the electrochemical reaction, or converted electric power [361, 245, 11].

Another model refinement possible consists of introducing two more state variables representing the vapor water masses $m_{w,ca}(t)$ and $m_{w,an}(t)$ in the cathode and in the anode, respectively. The corresponding state equations express the conservation of water mass in the electrode channels – thus replacing (6.56) and (6.63) – including the terms due to membrane transport and reaction. This allows a direct evaluation of the absolute humidity $u_{ca}(t)$, $u_{an}(t)$ to be used, e.g., in (6.59) [265, 267].

In summary, in dynamic simulation $U_{st}(t)$ is calculated using (6.19), possibly combined with (6.80) and (6.26). The hydrogen consumption rate is calculated by solving the coupled system of differential equations for the anode, the cathode, and the water circuit, taking the various control variables as inputs ($P_{hp}(t)$, $\xi_h(t)$, $\overset{*}{m}_{w,inj}(t)$, $P_{em}(t)$, and $\xi_a(t)$).

6.3 Reformers

6.3.1 Introduction

In view of the problems concerning the hydrogen storage, the use of different energy carriers that are easier to handle is a significant and still open research task. The main interest of current research efforts for automotive applications is concentrated on liquid hydrocarbons,[8] although they require an additional on-board process to extract hydrogen from the supply fuel to operate the fuel cell ("reforming"). The advantages are that (i) there is no need for special storage systems, (ii) until the production and distribution of hydrogen is better established, the existing infrastructure for fossil fuels can be used, and (iii) the consumer acceptance is likely to be higher. The refueling operation does not change for the user from how it is today. This would clearly favor the adoption of this new technology. Disadvantages are that (i) the resulting vehicle is not a zero-emission vehicle (CO_2 emissions), (ii) the propulsion system is more complex and more expensive, (iii) the tank-to-wheel efficiency is lower since the reforming requires energy, (iv) the fuel cell is likely to have a shorter life span due to the impurities in the reformer gas, and especially (v) the system exhibits poor response times, which makes the use of reformers critical during transient operation. These major drawbacks limit the application of fuel reforming in vehicles, although some prototypes of fuel-cell vehicles have adopted this technology [72]. The use of reforming-based fuel-cell systems as small, stationary auxiliary power units for trucks and camper vans, where efficiency and response time is not an issue, seems to be more promising.

Among liquid hydrocarbons, gasoline, diesel, and methanol are the most common reforming fuels for vehicles. Gasoline or diesel is often a simpler option since it benefits of an existing infrastructure [344]. On the other hand, the advantages of methanol (methyl alcohol), CH_3OH, are that (i) methanol can be obtained from various renewable resources (e.g., biomass), (ii) the conversion of natural gas into methanol allows the use of remote natural gas sources, (iii) due to its simpler molecular structure, it is technically easier to reform, which simplifies the hydrogen production and yields a very high H_2/CO_2 ratio

---

[8] There is also a deep interest in methane reforming, however, not for mobile applications.

for liquid fuel, and (iv) there already exist prototypes of fuel cells (DMFCs) that allow for a direct electrochemical conversion of methanol and thus render an upstream on-board reformer superfluous. However, methanol needs a dedicated distribution infrastructure and corrosion-resistant refuelling and storage equipment, it is poisonous if swallowed, and it burns with an invisible flame. Methanol is also water soluble, which makes it more dangerous.

For the on-board production of hydrogen from methanol[9] there are basically three methods, the "steam reforming," the partial oxidation (POx), and the methanol scission. Steam reforming is generally used for methanol reforming. Together with carbon dioxide, hydrogen is produced from methanol and water vapor. The overall reaction can be written as

$$CH_3OH + H_2O \rightarrow CO_2 + 3H_2, \quad \Delta h_R = 58.4\,kJ/mol. \tag{6.87}$$

In real reactions another product is carbon monoxide CO. The CO formation is due to the direct decomposition of methanol. Its concentration is affected by the water-gas shift reaction

$$CH_3OH \rightarrow CO + 2H_2, \quad \Delta h_R = 97.8\,kJ/mol, \tag{6.88}$$

$$CO + H_2O \rightarrow CO_2 + H_2 \quad \Delta h_R = -39.4\,kJ/mol. \tag{6.89}$$

However, even this model is a very simplified approximation of the reality. Reforming is a complex mechanism with many side reactions.

With the partial oxidation method the methanol is directly oxidized with the aid of oxygen. The overall reaction can be written as

$$CH_3OH + 1/2O_2 \rightarrow CO_2 + 2H_2, \quad \Delta h_R = -193\,kJ/mol, \tag{6.90}$$

and it can be regarded as the result of the two partial reactions

$$CH_3OH + 1/2O_2 \rightarrow CO + H_2 + H_2O, \quad \Delta h_R = -153.6\,kJ/mol, \tag{6.91}$$

$$CO + H_2O \rightarrow CO_2 + H_2, \quad \Delta h_R = -39.4\,kJ/mol, \tag{6.92}$$

with the possible formation of formaldehyde as another byproduct. Although partial oxidation allows an exothermal reaction, for proper methanol–oxygen ratios, the quality of the exhaust gas is not suitable for low-temperature fuel cells. Lower hydrogen concentrations, a higher carbon monoxide content, and an incomplete methanol conversion make a complex aftertreatment necessary. The same problems arise also with the methanol scission method, in which basically a thermal cracking takes place. Combinations of steam reforming and POx are also studied, since in this way autothermal reformers can be obtained.

Figure 6.14 shows a schematic of the methanol steam reforming process. Steam reforming is typically carried out over a catalyst bed containing oxides of copper, zinc, and aluminum ($CuO/ZnO/Al_2O_3$). With this catalyst, reformer

---

[9] Similar methods exist for other fuels. For instance, see [207] for ethanol and diesel.

operation is limited at low temperature by the formation of water condensate on the catalyst and at high temperature by sintering of the catalyst.

Since the methanol reforming reaction (6.87) is endothermic, external heat must be supplied to the reformer. Usually heat is transferred directly to the reformer reactants though, to avoid excess temperatures, certain solutions have been proposed with an intermediate heat transfer fluid (oil) heated in a separate heat exchanger. Heat is mostly produced in a burner by combustion or catalytic oxidation of excess hydrogen leaving the fuel cell anode, or by combustion of methanol extracted from the main feedstock. Of course the former solution is preferable, since the combustion of methanol is not pollution-free. Besides the recuperation of heat from the anode outlet ($\overset{*}{m}_{h,pur}$ of Sect. 6.2), other thermal integrations with the fuel-cell system are possible, including recovery of low-temperature heat from the stack cooling system [8]. Additional heat is required to preheat, vaporize, and superheat the reactant steam–methanol mixture fed to the reformer at its operating temperature. This heat can be recovered from the anode outlet as well, or from the reformer outlet, which is at a temperature normally higher than the operating temperature of a (PEM) fuel cell.

Reformed gas produced by the reformer has a small content of CO (typically, 2% by volume [8]), which, being a severe poison to the platinum catalyst used in the fuel cell, must be eliminated. Various methods to clean CO from hydrogen exist, including selective oxidation on a catalyst (platinum on alumina) bed.

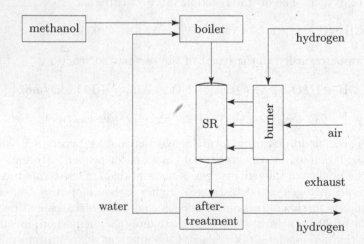

Fig. 6.14 Schematic of methanol steam reforming process

Autothermal reforming is essentially a combination of POx and steam reforming, consisting of adding air (thus oxygen) to the steam reforming reaction. Steam reforming is an endothermal reaction, while POx is an exothermal reaction. Under adiabatic conditions (insulated reformer) the equilibrium temperature is given by the relative ratios of steam and air fed into the reformer.

In autothermal reformers, the burner is only responsible for the preheating, vaporization, and superheating of methanol. The construction of the reactor itself is similar to that in steam reforming systems, but inherently less complicated. However, an additional complexity is due to the handling of reactant air, for which a compressor is required [320].

6.3.2 *Quasistatic Modeling of Fuel Reformers*

The causality representation of a fuel reformer in quasistatic simulations is sketched in Fig. 6.15. The model input variable is the hydrogen mass flow rate $\overset{*}{m}_{h,c}$. The model output variable is the methanol consumption rate $\overset{*}{m}_m$.

Fig. 6.15 Fuel reformers: causality representation for quasistatic modeling

Conversion Ratio

A simple model of a methanol steam reformer can be derived [10, 268] assuming that the reactions take place in a tube (plug-flow reactor) filled with a porous catalyst bed. Input flows to the catalyst tube are pre-vaporized methanol and steam. An isothermal process without pressure losses is assumed here. More complex, one-dimensional models accounting for detailed heat transfer, diffusion of species, and friction, are available in literature both for steam reformers [228] and for autothermal reformers [51].

For the range of conditions of interest to vehicle applications, the water-gas shift reaction (6.89) is usually neglected without a substantial loss of accuracy [10]. Experimental work has shown that the reaction rate of the reforming reaction is linear with the concentration of methanol, while the reaction rate of the gas-shift reaction is affected only slightly by the concentration of methanol or water, and thus it can be regarded as a constant [10]. Introducing the reaction rate constants k_1 for (6.87) and k_2 for (6.88), which are both functions of temperature and pressure (and k_1 also of the steam-to-methanol ratio of the feed gas σ), the rate equations for methanol and hydrogen inside the reformer (see Fig. 6.16) are

$$d\overset{*}{n}_m = - (k_1 \cdot C_m(x) + k_2) \cdot dm_c = -\overset{*}{n}_m(0) \cdot dx, \qquad (6.93)$$

$$d\overset{*}{n}_h = (3 \cdot k_1 \cdot C_m(x) + 2 \cdot k_2) \cdot dm_c, \qquad (6.94)$$

where x is the fraction of methanol converted (methanol conversion ratio), $m_c(x)$ is the mass of the catalytic bed, and $C_m(x)$ is the molar concentration of methanol. The latter is a function of x and of σ and can be calculated considering that for each mole of methanol that reacts, the total number of moles increases by two. Hence, at constant pressure and temperature the concentration of methanol is given by [268]

$$C_m(x) = (1-x) \cdot \frac{1+\sigma}{1+\sigma+2 \cdot x} \cdot C_m(0), \tag{6.95}$$

where $C_m(0)$ is the initial concentration of methanol (see below) and σ is the steam-to-methanol ratio of the feed gas (typically ranging from 0.67 to 1.5). The integration of (6.93) yields the catalyst mass that is necessary to achieve a certain conversion ratio x,

$$m_c(x) = \int_0^x \frac{\overset{*}{n}_m(0)}{k_1 \cdot C_m(\xi) + k_2} \, d\xi, \tag{6.96}$$

which, after substitution of (6.95), is solved as

$$m_c(x) = \overset{*}{n}_m(0) \cdot \left(c_1 \cdot \ln \frac{c_2}{c_2 - c_3 \cdot x} - c_4 \cdot x \right), \tag{6.97}$$

where $c_1 = (U \cdot c_3 + 2 \cdot c_2)/c_3^2$, $c_2 = U \cdot (k_1 \cdot C_m(0) + k_2)$, $c_3 = U \cdot k_1 \cdot C_m(0) - 2 \cdot k_2$, $c_4 = 2/c_3$, and $U = 1 + \sigma$. Now, (6.94) is integrated, yielding

$$\overset{*}{n}_h(x) = \overset{*}{n}_m(0) \cdot \int_0^x \frac{3 \cdot k_1 \cdot C_m(\xi) + 2 \cdot k_2}{k_1 \cdot C_m(\xi) + k_2} \, d\xi = 3 \cdot x \cdot \overset{*}{n}_m(0) - k_2 \cdot m_c(x). \tag{6.98}$$

The latter equation describes the fact that the hydrogen output molar rate is a fraction x of the theoretical value $3 \cdot \overset{*}{n}_m(0)$, diminished by the molar rate $k_2 \cdot m_c(x)$ of the CO production.[10]

Now, for a given reformer, m_c is given and $\overset{*}{n}_h(x)$ is calculated from the hydrogen mass flow rate required, as

$$\overset{*}{n}_h(x) = \frac{\overset{*}{m}_{h,c}}{M_h}, \tag{6.99}$$

where M_h is the molar mass of hydrogen. Therefore, the system of highly nonlinear coupled equations (6.97)–(6.98), which has to be solved iteratively, yields the corresponding values of x and $\overset{*}{n}_m(0)$. Notice that $C_m(0)$ is also a function of the unknown quantity $\overset{*}{n}_m(0)$,

---

[10] This model does not consider the hydrogen lost during CO removal. If, for instance, selective oxidation is used, typically about the same number of moles of hydrogen as of CO are lost.

$$C_m(0) = \frac{\overset{*}{n}_m(0)}{1+U} \cdot \frac{p_{ref}}{R \cdot \vartheta_{ref}}, \tag{6.100}$$

where R, p_{ref} and ϑ_{ref} are the universal gas constant, the pressure, and the temperature of the reformer. Typical values of p_{ref} range from 1 to 3 bar and of ϑ_{ref} from 430 to 570 K.

Finally, the methanol mass flow rate consumed is evaluated as

$$\overset{*}{m}_m = \overset{*}{n}_m(0) \cdot M_m, \tag{6.101}$$

where M_m is the molar mass of methanol ($M_m = 32\,\text{kg/kmol}$).

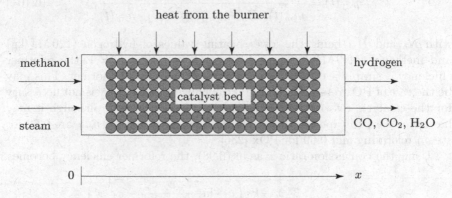

Fig. 6.16 Schematic of a catalyst bed steam reformer tube

Fuel Processing Efficiency

A clear definition of a (steam) reformer efficiency that is independent of the fuel cell operation is complicated by the critical feedback loop, in which the anode exhaust is burned to partially satisfy the heat requirements for the steam reforming reaction. If the hydrogen excess ratio (or utilization factor) λ_h is taken as a measure of the hydrogen exhaust burned, then the related losses are taken into account in the fuel cell utilization factor η_I (6.77) that can now be rewritten as

$$\eta_I(t) = \frac{\overset{*}{m}_{h,r}(t)}{\overset{*}{m}_{h,c}(t)} = \frac{1}{\lambda_h(t)}. \tag{6.102}$$

The reformer efficiency is defined as the ratio of chemical output power to total input power, which is the sum of the methanol chemical power and the fraction of the external power required[11] that is not recuperated within the

---

[11] That is the sum of the heat power required for the steam reforming process and the power required to drive the auxiliaries.

fuel-cell system. This fraction is usually rather complicated to estimate, the calculation involving various chemical, thermal, and fluid dynamic aspects. In a first approximation, it is convenient to assume that the total external power is proportional to the methanol mass flow rate,

$$P_{ref}(t) = \kappa_{ref} \cdot \overset{*}{m}_m(t). \tag{6.103}$$

Now, if a fraction μ of $P_{ref}(t)$ is recuperated from the anode outlet, from the cooling system, or within loops internal to the reformer circuits, the reformer efficiency is

$$\eta_{ref}(t) = \frac{\overset{*}{m}_{h,c}(t) \cdot H_h}{\overset{*}{m}_m(t) \cdot H_m + (1 - \mu) \cdot \kappa_{ref} \cdot \overset{*}{m}_m(t)}, \tag{6.104}$$

with H_h and H_m being the lower heating values of hydrogen (120 MJ/kg) and methanol (19.9 MJ/kg), respectively. Of course, if $\mu = 1$, the reformer efficiency is simply a ratio of the energy flows across the reformer. This may be the case of POx reformers, where the exhaust anode energy is not necessary for the catalytic reformer since the reaction is exothermal, although it may be useful, e.g., to vaporize the reactants. Typical values for η_{ref} are 0.62 for steam reforming and 0.69 for POx [236].

Using the conversion ratio x as in (6.98), the reformer efficiency becomes

$$\eta_{ref}(t) = \frac{H_h \cdot M_h}{M_m} \cdot \frac{3 \cdot x - k_2 \left(c_1 \cdot \ln \dfrac{c_2}{c_2 - c_3 \cdot x} - c_4 \cdot x \right)}{H_m + (1 - \mu) \cdot \kappa_{ref}}. \tag{6.105}$$

The overall efficiency of the fuel-cell system with reforming is finally evaluated as

$$\eta_{fcr} = \eta_{st} \cdot \eta_I \cdot \eta_{ref}. \tag{6.106}$$

6.3.3 Dynamic Modeling of Fuel Reformers

The physical causality of a fuel reformer is the reverse of that sketched in Fig. 6.15. The input side is represented by the control variables of the various circuits, while the output variables are the hydrogen mass flow rate $\overset{*}{m}_{h,c}$ and the methanol mass flow rate $\overset{*}{m}_m$. Notice that this changes the physical causality of the dynamic model of the fuel cell alone as sketched in Fig. 6.12. The fuel cell submodel now is not controlled on the hydrogen side.[12] Thus it receives the hydrogen mass flow rate as an input variable.

A simple dynamic model of a fuel processor system uses the conversion ratio defined in the previous section and describes the reformer dynamics,

---

[12] The recirculation loop is usually deactivated and the control valve is moved to the methanol side.

including the vaporizer/superheater and the gas clean-up stage, by assuming a second-order behavior [134],

$$\tau^2 \cdot \frac{d^2}{dt^2}\overset{*}{n}_h(t) + 2 \cdot \tau \cdot \frac{d}{dt}\overset{*}{n}_h(t) + \overset{*}{n}_h(t) = 3 \cdot x \cdot \frac{\overset{*}{m}_m(t)}{M_m} - k_2 \cdot m_c(x), \quad (6.107)$$

where τ is the time constant of the process and x is a static function of $\overset{*}{n}_m$ and m_c, as described in the previous section. This model describes the response to an input positive step of methanol flow as a critically damped (essentially, exponential) increase over time. The response to a decrease in methanol flow is also modeled with a second-order differential equation. However, due to the fact that the decrease of hydrogen flow requires no heat, these dynamics are much faster than the positive step. Typical values of τ are 2 s for step-up transients and 0.4 s for step-down transients [134].

6.4 Problems

Fuel Cells

Problem 6.1. For high pressures, the thermodynamic properties of gas have to be calculated using the Redlich–Kwong equation of state instead of the ideal gas law. The Redlich–Kwong equation reads

$$p = \frac{\tilde{R} \cdot \vartheta}{\tilde{V} - b} - \frac{a}{\sqrt{\vartheta} \cdot \tilde{V} \cdot (\tilde{V} + b)}, \quad (6.108)$$

where p is pressure, $\tilde{R}$ is the universal gas constant, ϑ is temperature, $\tilde{V}$ is the molar volume. The constants a and b are defined as

$$a = \frac{0.4275 \cdot \tilde{R}^2 \cdot \vartheta_c^{2/5}}{p_c}, \quad b = \frac{0.08664 \cdot \tilde{R} \cdot \vartheta_c}{p_c}, \quad (6.109)$$

where ϑ_c is the temperature at the critical point, and p_c is the pressure at the critical point. Using this equation of state, evaluate the gaseous density of hydrogen at 350 bar, 700 bar, and 300 K. *(Solution: For 700 bar, $\rho_h = 37.8\,kg/m^3$. For 350 bar, $\rho_h = 22.8\,kg/m^3$).*

Problem 6.2. A good approximation of the compressibility factor of hydrogen between pressures p and p_0 is

$$Z = 1 + 0.00063 \cdot \left(\frac{p}{p_0}\right) \quad (6.110)$$

(verify it with the results of Problem 6.1). With this assumption evaluate the energy required to compress 1 kg of hydrogen (from 1 bar) to 350 bar and

700 bar, respectively, at 300 K, under the further assumptions of (i) isothermal compression, (ii) adiabatic compression. Evaluate the result as a percentage of the energy content of hydrogen. *(Solution: For 700 bar, (i) $W_c = 8.7\,MJ/kg$ and (ii) $W_c = 21.2\,MJ/kg$. For 350 bar, (i) $W_c = 7.6\,MJ/kg$ and (ii) $W_c = 15.0\,MJ/kg$).*

Problem 6.3. Typical characteristics of various metal-hydride materials (1–4) for hydrogen storage are listed in the following table [350]. Evaluate the energy density for these storage systems. *(Solution: $\{1.86, 1.15, 1.36, 1.26\}\,kWh/kg$, respectively).*

	(1)	(2)	(3)	(4)
Material density (g/cm^3)	6.2	1.25	1.26	0.66
Porosity (%)	50	50	50	50
Mass storage capacity (%)	1.8	5.55	6.5	11.5

Problem 6.4. Evaluate the increase of energy density obtained with the cryo-compressed tank (CcH2) concept operated at 77 K with respect to conventional, ambient-temperature pressurized tanks. *(Solution: $\{104, 168\}\,\%$ for $p = \{700, 350\}$ bar, respectively).*

Problem 6.5. Explain the values of γ_{ht} in Table 6.2, for storage tanks pressurized at 350 bar. *(Hint: Use equation (5.33)).*

Problem 6.6. Evaluate the storage pressure and the specific strength (ratio of tensile strength to density) of the tank material that would be necessary to meet the 2015 DOE targets of Table 6.2 with gaseous hydrogen. *(Solution: $p = 385\,MPa$, $\frac{\sigma}{\rho} = 635\,kNm/kg$).*

Problem 6.7. Explain the explicitness of the number of cells N in (6.72). *(Hint: Review equations (6.41), (6.53), (6.66), and (6.70)).*

Problem 6.8. For the fuel cell stack of Fig. 6.11, find (i) the maximum output power $P_{fcs,max}$, (ii) the current $I_{fc,P}$ at which this power is yielded, and (iii) the current $I_{fc,\eta}$ that maximizes the overall efficiency. Compare the result with the curves shown in the figure. *(Solution: $P_{fcs,max} = 15.34\,kW$, $I_{fc,P} = 160.4\,A$, $I_{fc,\eta} = 12.9\,A$).*

Problem 6.9. Calculate the same quantities as in Problem 6.8 for a small fuel cell stack powering a racing FCHEV (see Chapt. A.6). Use the quadratic expression (6.71) for P_{aux} and the following data: $N \cdot U_{oc} = 16.8$, $N \cdot R_{fc} = 0.137$, $P_0 = 19.89$, $\kappa_1 = 6.6$, $\kappa_2 = -0.024$. *(Solution: $P_{fcs,max} = 210\,W$, $I_{fc,P} = 45\,A$, $I_{fc,\eta} = 13.25\,A$).*

Reformers

Problem 6.10. Derive (6.95).

7

Supervisory Control Algorithms

In all types of hybrid vehicles, a supervisory controller must determine how the powertrain components should operate, in order to satisfy the power demand of the drive line in the most convenient way. The main objective of that optimization is the reduction of the overall energy use, typically in the presence of various constraints due to driveability requirements and the characteristics of the components.

Based on the review article [290], this chapter describes the theoretical concepts of various types of control strategies for hybrid-electric vehicles. Appendix I contains examples of applications of these ideas.

7.1 Powertrain Control

The power output of modern powertrains is electronically controlled in order to match the request of the driver.[1] In conventional ICE-based powertrains, when the driver acts on the accelerator pedal, the pedal position is converted into a torque request. This torque is ultimately provided by the engine, which is regulated accordingly by its control unit. When the brake pedal is depressed by the driver, either the brake circuit is mechanically activated (coupled braking), or an intermediate controller converts the pedal position into a braking torque request that is split between the front and rear braking circuits (decoupled braking, ABS or similar). Similar considerations apply for purely electric powertrains (BEVs and FCEVs), with the traction motor playing the role of the engine.

Hybrid powertrains also are controlled in terms of torque. The torque control structure of an HEV can be schematically represented as in Fig. 7.1.

---

[1] In some systems, a vehicle stability controller is designed to override the driver's request when these commands lead to potentially dangerous maneuvers.

When the accelerator pedal is depressed, the first step in the torque control structure is a *driver's interpretation function* that yields the total torque required at the wheels, $T_t(t)$, as a torque request. The power demand is thus $P_t(t) = T_t(t) \cdot \omega_w(t)$. The torque request is typically evaluated on the basis of a look-up table as a function of the pedal position and the wheel or vehicle speed. The guiding principle is that 100% pedal depression should correspond to the maximum torque provided by the powertrain at the current speed, while 0% pedal depression (coasting) should provide the driver with a similar feeling as "engine braking" in conventional vehicles. Intermediate values are interpolated linearly or, more often, nonlinearly.

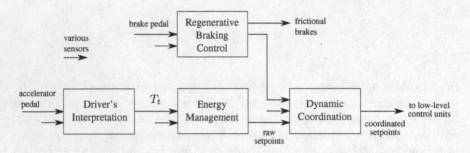

Fig. 7.1 Flowchart of an HEV supervisory controller

In contrast to ICE-based or electric powertrains, in HEVs the traction power demand can be provided in several different ways. Following the arguments of Chapter 4, parallel hybrids have one degree of freedom consisting of the torque-split ratio u at the torque coupler (4.142).[2] Typical modes of operation are conveniently described in such terms, including electric mode ($u = 1$), ICE mode ($u = 0$), power assist ($0 < u < 1$), and battery recharge using the engine ($u < 0$). In series hybrids, the degrees of freedom are the power-split ratio u_1 at the electric power link (4.132) and the APU speed command $\hat{\omega}$ that is typically given by (4.48). Consequently, typical modes of operation such as electric mode ($u_1 = 1$), APU mode ($u_1 = 0$), and battery recharge mode ($u_1 < 0$) are described purely in terms of u_1. In combined hybrids, both the speed and the torque of the generator might be considered as degrees of freedom. All other torque and speed levels depend on these through the kinetic relationships imposed by the PSD, see Sect. 4.9.2. Summarizing, the key function of a torque control structure for HEVs is encapsulated in the block that fills in the missing degrees of freedom and generates the setpoints for the low-level controllers. Such a function is often called *energy management* and will be the subject of the next sections.

---

[2] Usually gear changes are automated in parallel hybrids as well, such that the gear ratio γ becomes a further degree of freedom.

As mentioned above, the output variables of an energy-management controller are the setpoints for the low-level control units of the various components (e.g., engine torque, motor torque, etc.). These setpoints are intended to be attained under slowly varying conditions without any changes of the operating mode. When transient maneuvers implying mode changes (such as gear changes, clutch opening and closing, engine start and stop) are prescribed by the energy-management controller, the component setpoints are further coordinated before being fed to the low-level control units. In such a *dynamic coordination* block, additional setpoints are also generated for the clutch actuator, the stop-and-start motor, the automated gearbox, etc., to physically accomplish the mode change. Moreover, high-priority functions that are not based on energy considerations can be implemented in this block.[3]

When the driver acts on the brake pedal, *regenerative braking control* is activated. To benefit from regenerative braking, the activation of the friction brakes must be decoupled from the pedal depression. The brake pedal position is first converted into a braking torque request, which is subsequently split between regenerative and frictional braking requests. The former is treated as a negative torque demand $T_t(t)$ that is ultimately attributed to the electric machine.[4] When splitting the brake torque between the friction brakes and the electric machine, the guiding principle is that the regenerated energy should be maximized. Of course, the amount of energy that can be recuperated is limited by the maximum torque and power of the electric machine and by the maximum power and SoC of the battery. Certain vehicle configurations would allow splitting the brake torque between front and rear axles. Implementing such a functionality requires careful attention since it might have unpredictable effects on the vehicle dynamics and stability.[5]

In the rest of this chapter, only the energy-management control problem will be addressed. Energy-management strategies may be classified according to their dependency on the knowledge of future situations. Non-causal controllers require the detailed knowledge of the future driving conditions. This knowledge is available when the vehicle is operated along regulatory drive cycles, or for public transportation vehicles that have prescribed driving profiles. In all other cases, driving profiles are not predictable, at least not in the sense that the exact speed and altitude profiles as a function of time would be known a priori. In these cases, causal controllers must be used.

A second classification can be made regarding whether optimal control theory was used to derive the controller or not. The latter class of controllers ("heuristic" energy-management strategies) represents the state of the art in most prototypes and mass-production hybrids. Strategies derived from optimal control theory ("optimal" strategies) are the subject of research and are gradually being introduced in the industry.

---

[3] Some examples are given in Problems 7.9–7.12.

[4] For regenerative braking systems other than electric, see Sect. 5.2.

[5] See Problems 7.5–7.8.

7.2 Heuristic Energy Management Strategies

Heuristic energy management is based on intuitive rules and correlations involving various vehicular variables. Since details on heuristic strategies are seldom communicated by car manufacturers and other vehicle developers, a general description is not possible. However, some recurrent features can be identified as follows.

One guiding principle of heuristic strategies is that in hybrid vehicles the engine should be used only when its efficiency can be relatively high, while in less favorable conditions the electric mode should be given preference and the engine should be turned off. Moreover, when the engine is on, it should be operated in the highest possible efficiency regions (usually at high loads). To achieve that, the engine load is increased with respect to the driver's power demand and the extra power is directed to the battery (battery recharge mode). In parallel hybrids, such a shift of the engine operating point must take place at a given speed, since the latter is fixed by the vehicle speed and the gear. In series and combined hybrids, the shift can also involve the engine speed, which is actually a degree of freedom that can be used by the energy management.

A second guiding principle is that the battery discharge and recharge phases should be regulated such that the SoC stays within predefined limits. Therefore, when the SoC drops below a certain level, the recharge mode should be favored, while the electric mode should be more appropriate when the SoC exceeds a threshold. Further requirements might concern temperature levels, particularly the temperature of the engine catalyst during engine warm-ups. Therefore, the engine operation mode might be favored, regardless of its efficiency, to speed up the catalyst activation. Additionally, the necessity to operate engine-driven vehicle accessories, such as cabin heating or air conditioning, might require an engine-on mode.

Two common approaches to implement these intuitive principles are the map-based and the rule-based approach. In the map-based approach, the output setpoints are stored in multi-dimensional maps whose entries are measured quantities describing the state of the powertrain. To illustrate this concept for parallel HEVs [239, 355, 45, 194, 40, 22], consider the torque-split ratio u as a tabulated function of vehicle speed v and driver's torque request T_t, with the battery SoC ξ as an additional parameter. Several regions are recognizable, as shown in Fig. 7.2:

- below a certain vehicle speed and a certain wheel power the motor is used alone ($u = 1$);
- for intermediate power levels, the engine is forced to deliver excess torque to recharge the battery ($u < 0$), and that operating region becomes wider as the SoC decreases;[6]

---

[6] The operating region $u < 0$ can also be made to depend on the catalyst temperature.

- above the battery-recharge region and below the maximum engine power curve, only the engine is used ($u = 0$);
- above the maximum power of the engine, the motor is used to assist the engine ($0 < u < 1$).

In a slightly different implementation, the boundaries separating different modes are not rigid and the corresponding maps are rather created using "fuzzy logic" rules [176, 288, 110, 17, 188].

With this approach, the number of input variables must be sufficiently low to allow for a practical implementation of the corresponding maps. However, for a complete and robust energy management, the input variables should include engine speed and temperature, battery temperature, vehicle acceleration, etc., besides the ones shown in Fig. 7.2. Therefore, the map-based approach is limited to simple systems or prototypes, where several interactions or special situations can be neglected.

In the rule-based approach, the energy-management strategy is typically implemented as a finite state machine, as in the example shown in Fig. 7.3 [251, 341]. The states correspond to the modes of the HEV, including a compound state with the engine off (purely electric mode) and a compound state with the engine on (including all the hybrid modes). The transitions between the states are executed when certain conditions are fulfilled. These conditions constitute the particular strategy adopted, and they are expressed in terms of the influence variables described above. For example, stopping the engine requires *all* the following conditions to be met simultaneously:

- the vehicle speed and the power demand P_t is sufficiently low;
- the engine or catalyst temperature is sufficiently high (to avoid interrupting the engine warmup phase);
- the battery SoC is sufficiently high;
- the engine-driven accessories are not activated;

Vice versa, the transition from the engine-off state to the engine-on state is activated when at least one of the conditions above are not fulfilled (power demand is high or temperature is low, etc.). The choice of the particular hybrid mode (ICE drive, battery recharge, or power assist) is typically inspired by the second guiding principle above and other considerations:

- if the battery SoC is relatively low, then the recharge mode is activated ($u < 0$);
- if the SoC is too high, then the power assist mode can be chosen to avoid overcharging the battery ($0 < u < 1$);
- of course, the same mode is activated when the torque demand exceeds the capabilities of the engine alone.

As for the torque-split ratio in the first two cases, a common approach is to define a battery's power demand $P_{t,b}$ that is added to the driver's power demand P_t. The power $P_{t,b}$ is tabulated as a function of the SoC and possibly

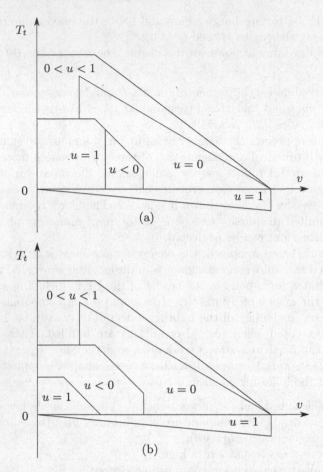

Fig. 7.2 Typical map-based, heuristic energy-management strategy for a parallel HEV. The torque-split ratio is mapped as a function of driver's torque request and vehicle speed, for (a) a relatively high battery SoC and (b) a lower SoC.

other quantities such as battery temperature. When the SoC is high, $P_{t,b}$ is negative (battery discharge), while it is positive (battery recharge) when the SoC is low. Battery recharge during engine warm-ups can be also enforced by properly setting the term $P_{t,b}$ as a function of the catalyst or engine temperature. Of course, in all practical implementations, transition rules are much more complex than in the simple example above and they can comprise several sub-rules that are combined using Boolean algebra.

The main advantage of heuristic controllers is that they are intuitive to conceive and made to directly translate control specifications. If properly tuned, they can provide good results in terms of fuel consumption reduction and charge sustainability. Unfortunately, the behavior of heuristic controllers

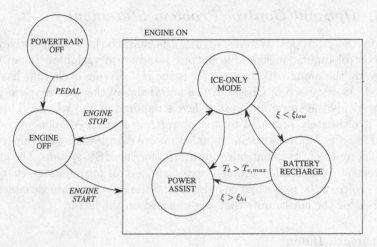

Fig. 7.3 A finite-state machine illustrating a rule-based, heuristic energy-management strategy

strongly depends upon the choice of the thresholds and the maps that define the mode switches, which actually can vary substantially with the driving conditions [293]. Besides the substantial tuning effort required, another drawback of the heuristic approach is that when several Boolean conditions are introduced, their combination and priorization becomes difficult to manage and implement. Moreover, it should be clear that expert-defined rules do not necessarily exploit the maximum performance that the system might attain. All these reasons serve to motivate the development of model-based controllers aimed at optimizing a well-defined quantitative criterion.

7.3 Optimal Energy Management Strategies

Optimizing the energy management of a hybrid powertrain consists of a two-level analysis. At a more abstract level, the energy-management strategy can be designed for a given test drive cycle, through optimization of the power flows between the system components. Such a procedure is called an *offline optimization*, since the drive cycle must be known a priori. That is obviously not possible during the real operation of the system, except for some particular scenarios. Therefore, for a practical implementation, an *online* controller is necessary. Nevertheless, an offline optimization is a very useful tool as it can be used to provide an optimal performance benchmark, which can then be used to assess the quality of any causal but suboptimal online controller. Moreover, the theoretic optimal solution might provide insights into how a realizable control system should be designed.

7.3.1 Optimal Control Problem Statement

As many other physical systems, an HEV is characterized by disturbance inputs, control inputs, measurable outputs, and state variables. In an offline framework, the main disturbance $w(t)$ is the test driving cycle that has to be followed. As discussed in Chapt. 2, the knowledge of the vehicle speed profile directly yields the speed and torque levels required at the wheels. This total power is ultimately provided by the powertrain system components. As described in Sect. 7.1, the setpoint signals that are calculated by the supervisory controller represent the controlled input $u(t)$ for the HEV system. The energy-management optimization problem can have several formulations that differ for the choice of the performance index J and the relevant state variables $x(t)$, as well as for the constraints that are imposed on the latter.

Performance Index

As illustrated in Figure 7.4, the simplest performance index $J = m_f(t_f)$ is the fuel mass m_f consumed over a mission of duration t_f. Hence, J can be written as

$$J = \int_0^{t_f} \overset{*}{m}_f(w(t), u(t))\, dt. \tag{7.1}$$

Pollutant emissions can also be included in the performance index J by considering the more general expression

$$J = \int_0^{t_f} L(w(t), u(t))\, dt, \tag{7.2}$$

where L combines the fuel consumption rate and the emission rates of the regulated pollutants. Combining these rates into a single cost function necessarily requires the introduction of a user-defined weighting factor for each pollutant species [154, 196, 15, 167, 122]. This approach is particularly relevant for systems equipped with Diesel engines, for which the urea consumption of the NO_x aftertreatment system can be additionally introduced in the performance index [63]. If the ICE is a spark-ignited engine operated with stoichiometric air/fuel ratios, its pollutant emissions can usually be reduced to negligible levels using a three-way catalytic converter. Accordingly, the pollutant emission is not considered as part of the optimization problem, although in practice "duty-cycle" (on/off operation) or engine shutoff at idle can cause problems due to excessive pollutant emissions caused by engine or catalyst cooling.

Drivability issues are sometimes included in the performance index. For example, the cost function might include an anti-jerk term, which consists of the engine acceleration squared [364], a term related to gear shifting [195], etc., multiplied by an arbitrary weighting factor. Additionally, smoothness and driver-acceptance considerations are usually included among the local constraints discussed below. Another contribution to the performance index that has been investigated is the battery life [298, 90].

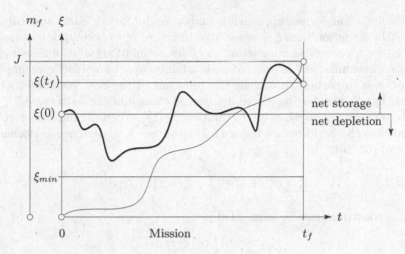

Fig. 7.4 Typical trajectories of the battery SoC $\xi(t)$ and consumed fuel mass $m_f(t)$ along a mission

State Variables

The nature of the state variables is obviously related to the dynamics of the systems, which generally include mechanical, thermal, electrical, and electro-chemical subsystems. Usually, for the purpose of energy management, HEVs can be described using quasistatic models. Thus the number of state variables strongly decreases, and the state vector can be reduced to only include integral quantities such as the battery SoC, $x(t) \equiv \{\xi(t)\}$.

The optimal control problem must respect the dynamics of the state variable,

$$\dot{x}(t) = f(w(t), u(t), x(t)). \tag{7.3}$$

If $x(t) \equiv \{\xi(t)\}$, as assumed in the rest of this Chapter, the latter equation is obviously represented by the system of (4.60)–(4.62).

Additionally, it is necessary to force the battery SoC to stay within certain admissible boundaries during the whole trajectory (state constraints). More-over, the terminal SoC (at the end of the cycle) must be close to a value that depends on the type of HEV considered. For *charge-sustaining* HEVs, only small deviations from the initial value of the SoC are permitted at the end of tests to assess vehicle energy consumption. For a plug-in HEV, a charge-depleting operation is allowed. Therefore, the optimization problem can be formalized as the problem to reach a target SoC which, at the end of the test cycle, is lower than the initial value [330, 308, 120].

In both cases, an integral constraint is imposed on $\xi(t_f)$ with respect to a target value ξ_t. In principle, this constraint can be enforced in two different ways, namely, as a soft constraint, that is, by penalizing any deviations of $\xi(t_f)$ from ξ_t, or as a hard constraint, by requiring that $\xi(t_f)$ exactly matches the value ξ_t. The former approach is mainly useful for real-time applications, where the deviation from ξ_t is associated to future fuel consumption or saving. The latter approach is only relevant in offline simulations when different solutions are benchmarked. In order to make a fair comparison of fuel-consumption values, it is important to exclude the influence of the SoC variations, which is best handled by forcing a zero net energy change in the battery.

To represent constraints on the final state $\xi(t_f)$, a penalty function $\phi(\xi(t_f))$ is added to the performance index (7.2) to obtain a constrained performance index of the form

$$J = \phi(\xi(t_f)) + \int_0^{t_f} L(w(t), u(t)) \, dt. \tag{7.4}$$

A hard constraint can be formalized as

$$\phi(\xi(t_f)) = \begin{cases} 0, & \xi(t_f) = \xi_t, \\ \infty, & \xi(t_f) \neq \xi_t. \end{cases} \tag{7.5}$$

Soft constraints can be added as functions of the difference $\xi(t_f) - \xi_t$. In [195] the quadratic penalty function $\phi(\xi(t_f)) = \alpha \cdot (\xi(t_f) - \xi_t)^2$ is used, where α is a positive weighting factor.

A quadratic penalty function tends to penalize deviations from the target SoC, regardless of the sign of the deviation. In contrast, a linear penalty function of the type

$$\phi(\xi(t_f)) = \mu \cdot (\xi_t - \xi(t_f)), \tag{7.6}$$

where μ is a positive constant, penalizes battery use while favoring the energy stored in the battery as a means for saving fuel in the future. In the regulatory standard SAE J1711 [304], μ is set to $38\,\text{kWh}$ per gallon of gasoline, which approximately corresponds to the lower heating value of the fuel. Conversely, physically meaningful definitions of μ are discussed below.

The piecewise-linear penalty function

$$\phi(\xi(t_f)) = \begin{cases} \mu_{dis} \cdot (\xi_t - \xi(t_f)), & \xi(t_f) < \xi_t, \\ \mu_{chg} \cdot (\xi_t - \xi(t_f)), & \xi(t_f) > \xi_t, \end{cases} \tag{7.7}$$

is at the core of an online energy-management strategy called T-ECMS that is presented in the next section.

Other state variables that are introduced in more general formulations of the optimal control problem are related to battery behavior [184] or thermal levels [191, 63]. The latter might prove very important, particularly in combination with a pollutant-based criterion. A thermal state variable might lead to the finding of an optimal compromise between turning off the engine to avoid pollutant emissions locally and keeping it on to avoid decreasing the catalyst temperature and reduce pollutant emissions later. Moreover, since the pollutant emission behaviour of an engine is not purely quasistatic, additional state variables representing pollutant dynamics could be considered.

Local Constraints

Local constraints are also imposed on the state and control variables. These constraints mostly concern physical operation limits, notably the maximum engine torque and speed, the motor power, or the SoC operating window. Constraints on the control variables can also be imposed to enhance smoothness and driver acceptance [348].

7.3.2 Noncausal Control Methods (Offline Optimization)

This section presents various approaches to evaluating optimal energy management laws offline. These approaches are grouped into three subclasses, namely, static optimization methods, numerical dynamic optimization methods, and closed-form dynamic optimization methods.

Static Optimization

Since a mission usually lasts hundreds to thousands of seconds, while, at each time t, multiple values of $u(t)$ must be evaluated, finding the optimal control law by inspecting all possible solutions requires excessive computational and memory resources. Simplified approaches [27] do not require detailed knowledge of $P_t(t)$ but only its average and root mean square values. However, these techniques can be applied only under certain circumstances (e.g., duty-cycle operation of series HEVs). Dynamic optimization techniques, as presented in the next sections, avoid this drawback.

Dynamic Programming

One very common technique for solving the optimal control problem stated in the previous section is dynamic programming (DP) [41, 33]. The main benefit of DP is that the solution found is guaranteed to be the global optimal solution. The drawbacks are that the computational burden increases exponentially with the number of state variables of the dynamic system. Thus, only very simple systems can be treated with this method. A useful property of the dynamic programming algorithm is that the computational burden increases only linearly with the final time t_f.

Dynamic programming requires gridding of the state and time variables. Consequently, the integral (7.4) and the state dynamics (7.3) have to be replaced by their discrete counterparts. The found solution is optimal only up to the numeric errors introduced by discretization and interpolation.

DP uses the definition (7.4) of the performance index, but it extends this definition to any point of the timestate space by defining the cost-to-go function $\Gamma(t, \xi)$ as the performance index of the optimal trajectory from the point

(t, ξ) to the point (t_f, ξ_t), see Fig. 7.5. By definition, the value $\Gamma(0, \xi(0))$ corresponds to the optimal value of J that is sought.

In a basic implementation of DP, the function $\Gamma(t, \xi)$ is calculated for the grid points $t_k = k \cdot \Delta t$, $k = 0, \ldots, N$, and $\xi_i = \xi_{min} + i \cdot \Delta \xi$, $i = 0, \ldots, p$ ($p = (\xi_{max} - \xi_{min})/\Delta \xi$). The computation starts with Γ being set as

$$\Gamma(t_f, \xi_i) = \phi(\xi_i). \tag{7.8}$$

The computation proceeds backward in time to solve the recursive algorithm

$$\Gamma(t_k, \xi_i) = \min_{u \in V} \left\{ \Gamma(t_{k+1}, \xi_i + f(w(t_k), u, \xi_i) \cdot \Delta t) + L(w(t_k), u) \cdot \Delta t \right\}. \tag{7.9}$$

The control inputs u are limited to a feasible subset V. In practice, this subset also must be discretized to restrict the search to q values u_j, $j = 1, \ldots, q$. These values can vary with time and as a function of the state.

The arguments of minimization are stored in a feedback control function

$$U(t_k, \xi_i) = \arg\{\Gamma(t_k, \xi_i)\}, \tag{7.10}$$

which is then used to reconstruct the optimal trajectories $\xi^\circ(t)$, $u^\circ(t)$, and consequently $L^\circ(t)$. Starting from $t = 0$ and proceeding forward in time,

$$u^\circ(t_k) = U(t_k, \xi^\circ(t_k))$$

$$\tag{7.11}$$

$$\xi^\circ(t_{k+1}) = \xi^\circ(t_k) + f(w(t_k), u^\circ(t_k), \xi^\circ(t_k)) \cdot \Delta t.$$

Solving equations (7.9) and (7.11) requires particular care since generally neither the state value $\xi_i + f(w(t_k), u, \xi_i) \cdot \Delta \xi$ nor the state value $\xi^\circ(t_k)$ matches any of the possible points of the grid. Therefore, the corresponding values of Γ and U must be interpolated from the values calculated for the closest points of the grid. Several interpolation methods can be used, each having their specific benefits and drawbacks. See Appendix III for further details.

Another problem arises when, as in (7.5), an infinite cost is assigned to handle unfeasible states or control inputs. If an infinite cost is used together with an interpolation scheme, the infinity value propagates backward in the grid and the number of infeasible states increases artificially. A number of techniques have been proposed to handle these interpolations correctly, using large but finite instead of infinite values or calculating exactly the boundaries between feasible and infeasible states (see Appendix III and [310]).

As mentioned above, the computational burden of DP scales linearly with the problem time N, the number of discretized state values p, and the number of discretized control input values q. Improved algorithms with reduced computational time are available. For example, the iterative dynamic programming algorithm in [21] is based on the adaptation of the state space. At each iteration, the state space is selected as a small fraction of the entire space, centered around the estimation of the optimal trajectory that was evaluated

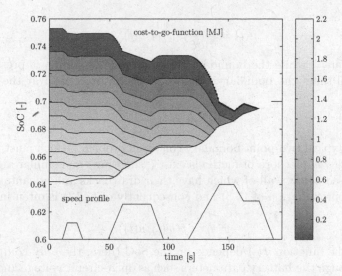

Fig. 7.5 Illustrative cost-to-go function Γ given by (7.9). The data are calculated for the operation of an A-class HEV in the ECE cycle, for a terminal time of 196 s, a target terminal SoC of 0.7, a time step of 1 s, and an SoC discretization of 0.01% of full charge. The target terminal SoC can be reached starting from an initial SoC lower than 0.753 and greater than 0.648.

in the previous iteration. Another approach, used in [362], reduces the computing time by splitting the mission into a series of time sections and solving an optimization problem for each of those sections. This approach generally does not guarantee a global optimal solution. Other approaches are discussed in Appendix III.

Connection to the Minimum Principle

Direct numerical optimization methods require substantial amounts of computational time. One approach that often permits a reduction of the computational effort is based on the minimum principle [77, 200, 54, 299, 171]. This method introduces a Hamiltonian function to be minimized at each time, that is,

$$u^\circ(t) = \arg \min_v \{H(t, \xi(t), v, \mu(t))\}, \qquad (7.12)$$

where

$$H(t, \xi, u, \mu) = L(w(t), u) + \mu \cdot f(w(t), u, \xi). \qquad (7.13)$$

In this formulation, t is a continuous variable, and the dynamics of the SoC are given by (7.3). The parameter $\mu(t)$, which corresponds to the adjoint state in classical optimal control theory, is described by the Euler-Lagrange equation

$$\dot{\mu}(t) = -\frac{\partial}{\partial \xi} f(w(t), u(t), \xi(t)). \tag{7.14}$$

Unfortunately, while the boundary condition for the state ξ is prescribed at the initial time, the boundary condition for μ is prescribed at the terminal time,

$$\mu(t_f) = \frac{\partial \phi}{\partial \xi(t_f)}. \tag{7.15}$$

This is a typical two-point boundary condition problem, which must be solved numerically. A plethora of methods exist ("shooting" and similar algorithms, as discussed below), all of which have their drawbacks and advantages.

An approximation of (7.3) and consequently of (7.13) is often introduced as

$$\dot{x}(t) \approx \widetilde{f}(w(t), u(t)). \tag{7.16}$$

In fact, the function $f(\cdot)$ depends on the SoC (here, the only component of $x(t)$) through the battery parameters such as open-circuit voltage and internal resistance. However, in many cases, this dependency can be neglected (especially for the internal resistance). This assumption may be not valid, e.g., in hydraulic hybrids [100] and in HEVs using supercapacitors, but it is reasonable for battery HEVs, where only large deviations of the SoC can cause substantial variations of the internal battery parameters. Consequently, (7.14) becomes $\dot{\mu} \approx 0$ and the adjoint state is approximately constant along the optimal trajectory. The optimization problem is thus reduced to searching for a constant parameter μ_0 that approximates $\mu(t)$ for a given mission.

The value of the adjoint state $\mu(t_f)$, or μ_0 if the approximation (7.16) is used, depends primarily on the choice of $\phi(\xi(t_f))$. In the case of a linear soft constraint such as (7.6), the value of μ_0 is

$$\mu_0 = \frac{\partial \phi}{\partial \xi(t_f)} = -\mu. \tag{7.17}$$

If the soft constraint on the final SoC is of the piecewise-linear type (7.7), then the value of the constant adjoint state must be consistent with the final sign of the SoC deviation [296], in particular,

$$\mu_0 = \frac{\partial \phi}{\partial \xi(t_f)} = \begin{cases} -\mu_{dis}, & \xi(t_f) < \xi_t, \\ -\mu_{chg}, & \xi(t_f) > \xi_t. \end{cases} \tag{7.18}$$

In the case of a hard constraint, μ_0 is the value that ensures the fulfillment of the constraint. This value must be determined iteratively, using numerical methods.

Equivalent-Consumption Minimization Strategies

In most practical cases, the relationship between μ_0 and $\xi(t_f)$ is monotonous and there is only one value μ_0 that fulfills $\xi(t_f) = \xi_t$. In particular, if μ_0 is too low, $\xi(t_f)$ will be higher than ξ_t. On the other hand, if μ_0 is too high, $\xi(t_f)$ will be lower than ξ_t.

Based on this property, μ_0 can be iteratively determined by correcting the previous estimation in accordance with the sign of $\xi(t_f)-\xi_t$ after each iteration. For example, the bisection method can be used. Of course, an estimation of μ_0 is needed to start the first iteration. Then, iterations are repeated until $\xi(t_f)$ is sufficiently close to ξ_t. Figure 7.6 shows the flowchart of this approach. At each iteration, the driving cycle is discretized into N time steps as in the previous section, and the state is initialized at the value $\xi(0)$. At each time step, the power demand is calculated from the driving cycle. Then (7.12) is applied.

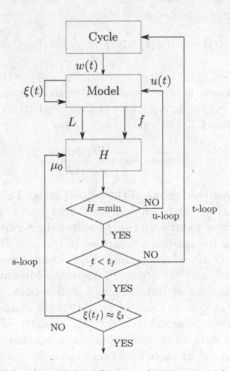

Fig. 7.6 Flowchart of the offline implementation of the ECMS

To find the control input that minimizes the Hamiltonian, the most common method discretizes the control input, as in the previous section, into q admissible values. For each control input, the corresponding Hamiltonian is calculated. The minimum among the Hamiltonian values is found and the corresponding control input is chosen as the value at the current time step. Consequently, the state is updated.

The solution of (7.12) is particularly straightforward if the Hamiltonian function can be expressed as an explicit function of the control variable. This approach requires an explicit description of $L(w(t), u(t))$ and $f(w(t), u(t))$, which can be obtained using simple, but accurate modeling tools [61, 6].[7] In some special applications, the Hamiltonian turns out to be an affine function of the control variable [297]. In this case, the minimum principle states that the optimal control variable is found at its extreme values, depending on the sign of the switching function $\partial H / \partial u$. When the switching function is zero ("singular arc" situation), additional information is required to determine the optimal control input. Typically this is the second derivative of the Hamiltonian with respect to the variable u.

In the case in which μ can be treated as a constant, the Hamiltonian function acquires a new meaning. Since the battery open-circuit voltage is constant under this assumption, both terms on the right-hand side of (7.13) can be reduced to power terms, namely,

$$P_H(t, u(t), s(t)) = P_f(w(t), u(t)) + s_0 \cdot P_{ech}(w(t), u(t)). \tag{7.19}$$

In this equation, $P_f(t, u(t)) = H_{LHV} \cdot \overset{*}{m}_f(w(t), u(t))$ is the fuel power (with H_{LHV} being the lower heating value of the fuel) and $P_{ech}(w(t), u(t)) = -\dot{\xi}(w(t), u(t)) \cdot U_{oc} \cdot Q_0 = I_b(w(t), u(t) \cdot U_{oc}$ is the battery "electrochemical" power defined in Sect. 4.5. The *equivalence factor*

$$s_0 = -\mu_0 \cdot \frac{H_{LHV}}{U_{oc} \cdot Q_0} \tag{7.20}$$

represents a dimensionless scaling of the adjoint state. The equivalence factor thus converts battery power to an equivalent fuel power that must be added to the actual fuel power to attain a charge-sustaining control strategy. It plays an important role in the approach described in the next section.

The formulation (7.12)–(7.19) is called Equivalent Consumption Minimization Strategy (ECMS) or PMP (from Pontryagin's Minimum Principle).[8]

Alternative definitions of Hamiltonian-like functions, often derived on a purely heuristic basis, can be reduced to (7.19). For example, the minimization of the overall power losses $P_f(t, u(t)) + P_{ech}(t, u(t)) - P_t(t)$ [173] is clearly equivalent to (7.19) with $s_0 \equiv 1$. Similarly, the maximization of an overall powertrain efficiency or the minimization of consumption-like performance measures [175, 154] can be interpreted as a constant-s_0 approximation of the general approach. Conversely, some heuristic definitions of a cost function might be reconducted to a value of s_0 that is heuristically tuned as a function of SoC [42]. Generally speaking, these methods are not intrinsically optimal, nor do they ensure a convergence of $\xi(t)$ toward ξ_t, and therefore they usually need heuristic correction terms. For instance, in [173] a weighted correction $\alpha(t) \cdot P_{ech}(t, u(t))$ is appended to the cost function to penalize the SoC deviations.

---

[7] See Problems 7.23–7.29

[8] Nowadays, the term ECMS is more often used for its online counterpart, see next section, while PMP is reserved for the offline application.

7.3.3   Causal Control Methods (Online Sub-Optimal Controllers)

All of the optimization techniques discussed above require knowledge about future driving conditions. This fact makes their implementation in real-time controllers a challenging task. This section lists the level of information required by various control strategies and discusses how such information can be achieved during real-time operation.

Predictive Control

The highest level of information is available when the complete mission is known at the outset. When this is the case, an offline optimization procedure, such as the dynamic programming algorithm described in the previous section, can be applied. For public transportation vehicles along fixed routes, where the mission is known in advance, such an approach may be viable. The resulting feedback function $U(t, \xi)$ given by (7.10) can then be implemented in the powertrain control systems. Instead of using an explicit dependency on time, the feedback function or the optimal cost-to-go function can be calculated, based on a previously measured velocity profile on the route, as a function of the position [157]. Obviously, this procedure delivers close to optimal results only if the driven velocity profile closely matches the measured one.

For passenger cars, the mission is usually unknown at the outset, and the estimation of future driving conditions must be made online. The combination of such an estimation with the application of dynamic programming follows the model-predictive control (MPC) paradigm [177, 21], which requires an estimation of the power demand $P_t(t)$ on a prediction horizon of duration t_f. Dynamic programming is then used to calculate the optimal control law, which is applied for a shorter control horizon $t_c < t_f$. In [21] the power demand is estimated using speed limit, curve radius, and road slope, as a function of the distance along the route. These data are obtained by combining onboard and GPS navigation. The vehicle speed is calculated using a dynamic model of the vehicle as a function of the target speed, which is constrained by the speed limit, maximum safe speed, especially in curves, and maximum speed allowed by traffic conditions (a parameter that may be available in future applications using radar sensors).

Time-Invariant Feedback Controllers

While the direct application of DP described above requires global or local estimates, a simpler approach consists of building a feedback map where the control variable u is tabulated as a function of some input variables [364, 348, 195]. The optimal solution found with dynamic programming is statistically analyzed, and implementable rules are extracted to construct the feedback map. To limit the complexity of the feedback map, only two input variables are usually allowed. Examples include torque demand and SoC [195], power demand and SoC [364], and wheel speed and power demand [348]. Although this approach performs well in real hybrid vehicles, it is based on optimization with respect to a specific drive cycle and, in general, it is neither optimal nor charge-sustaining for other cycles. Moreover, the feedback solution obtained using dynamic programming cannot be implemented directly, and the rule-extraction process is not straightforward.

To overcome these drawbacks, the procedure described in [195] is extended in [196] using stochastic dynamic programming. To obtain a time-invariant control strategy, an infinite-horizon optimization problem is solved. The feedback control law derived with stochastic dynamic programming is applicable to general driving conditions. It has been shown in simulation [196, 156] that this procedure delivers close-to-optimal results for HEVs over regulatory as well as random drive cycles. Furthermore, it has been shown that stochastic dynamic programming also can be extended easily to trade off fuel economy against emissions [317], battery health [224], or drivability [238].

ECMS-Type Controllers

In strategies derived from Pontryagin's Minimum Principle, such as ECMS, the uncertainty about future driving conditions is transferred to an uncertainty on the correct (optimal) value of the (eventually constant) adjoint state approximation μ_0, or of the equivalence factor s_0. The advantage with respect to predictive control is that only one parameter must be determined instead of a power demand $P_t(t)$ as a function of time. Various methods are available for the online estimation of s_0, usually leading to a variable estimation $s(t)$. These methods, which are described below, can be classified into three approaches depending on the information used, namely, past driving conditions, past and present driving conditions, and past, present and future driving conditions. A general flow chart of such algorithms is shown in Fig. 7.11.

- Past Driving Conditions

One technique for estimating $s(t)$ is based on ideas borrowed from pattern recognition [154, 124]. Optimal values of s_0 are pre-calculated offline for a set of representative driving patterns, which are composed of urban, expressway, and suburban driving patterns. Up to 24 characteristic parameters, such as average velocity, standstill time, and total time, can be chosen to characterize driving patterns. During real-time operation, a neural network periodically decides which representative driving pattern is closest to the current driving pattern. Then, the energy-management controller switches to the corresponding value of the parameter s_0.

- Past and Present Driving Conditions

Pattern recognition methods use information only about past driving conditions. Alternative controllers evaluate $s(t)$ continuously by adapting it to the current driving conditions or simply to the current value of the SoC.

In some implementations [158, 153], the equivalence factor $s(t)$ is calculated as the partial derivative of the present fuel power with respect to the battery power, that is,

$$s(t) = -\frac{\partial P_f}{\partial P_{ech}}(t). \tag{7.21}$$

Such an approach assumes that similar operating conditions will exist in the future, that is, the replacement energy will "cost" the same amount of fuel energy as it does in the current driving conditions. In general, this assumption leads to trajectories that are neither fuel optimal nor charge sustaining.

Another, more common strategy [166, 4, 53], mainly emphasizes the SoC control factor. The basic idea is that the estimation $s(t)$ should be adapted according to the instantaneous deviations of the SoC from its target value. A simple adaptation algorithm is given by

$$s(t) = s_t - k_p \cdot (\xi(t) - \xi_t), \tag{7.22}$$

where s_t is a first guess (possibly inspired by an offline calculation), and k_p a tunable coefficient. The rule (7.22) corrects any positive deviations of $\xi(t)$ from ξ_t by decreasing $s(t)$, that is, by favoring the use of electrochemical energy to discharge the battery. On the contrary, when the SoC is lower than its target value, $s(t)$ is increased to penalize any further use of the battery and to favor its recharging instead. In many cases, to favor the actual convergence of $\xi(t)$ toward ξ_t, an integral term might be appended to the right-hand term of (7.22), regulated by a second tunable parameter k_i [166, 53]. Also there exist some nonlinear versions of (7.22) to allow for more freedom of the SoC in the main operation window, whereas towards the bounds of the SoC more control action is forced.

- Past, Present, and Future Driving Conditions

The combination of ECMS with equation (7.22) has been shown to deliver close-to-optimal results [54] in both simulation and real hardware. However, it has been indicated [5, 280] that a non-predictive strategy performs poorly in the presence of constraints on the battery's SoC if the amount of recuperated energy is large compared to the storage capacity of the battery. This is mainly due to the fact that there will be future recuperation phases during which, if the strategy is not predictive, it is likely that the SoC runs into its constraints. In this case the braking energy is dissipated in the friction brakes and the strategy becomes largely sub-optimal. Therefore, vehicles with a large battery can be controlled optimally without detailed information of the future power demand $P_t(t)$. In fact, for plug-in HEVs it is enough to know the length of the trip to achieve close-to-optimal results [308]. However, for vehicles with limited battery capacity it is important to incorporate some predictive features in the controller, particularly in the presence of road altitude variations.

One possible solution is to improve the online estimation of the equivalence factor by including information available from the vehicle environment.

A piece of information that is usually assumed to be available is the GPS-derived altitude profile of the route that the vehicle intends to follow. The altitude profile provides the road slope as a function of the distance covered. To transform the altitude profile into a slope function of time, the future vehicle speed profile must be estimated. Information on speed limits, which is often considered to be available, can be used for such an estimation. With regard to traffic conditions, future cars are expected to include radar and other sensors that can be used to obtain this information [296].

In [5] the information about the future altitude profile is used to generate a time-varying reference trajectory $\xi_t(t)$ which is used in (7.22). The authors of [334] adjust (7.22) to account for the vehicle's actual kinetic energy, which assures that the battery can always recuperate the complete kinetic energy of the vehicle. The derivation of the *telemetry ECMS* (T-ECMS) [296] is based on similar ideas. Moreover, it provides some valuable insights to the interpretation of the equivalence factor and will therefore be discussed as an example in the next section.

- Example: T-ECMS

The T-ECMS controller is based on (7.7) and (7.18). Assuming a piecewise-linear soft constraint for the final SoC, the value of the optimal adjoint state depends on the final sign of the SoC deviation from the target value. It is further assumed that $\xi_t = \xi(0)$.

Using the approach (7.20), equation (7.18) can be rewritten as

$$s_0 = H_{LHV} \cdot \frac{\partial \phi}{\partial E_{ech}(t_f)} = \begin{cases} s_{dis}, & E_{ech}(t_f) > 0, \\ s_{chg}, & E_{ech}(t_f) < 0, \end{cases} \tag{7.23}$$

where $E_{ech}(t_f) = \int_0^{t_f} P_{ech}(\tau)\,d\tau$ is the electrochemical energy consumption. Therefore, $\phi(E_{ech}(t_f)) = H_{LHV} \cdot \phi(\xi(t_f))$ is reinterpreted as a fuel equivalent of the battery energy consumption.

In certain special cases, it is possible to evaluate $\phi(E_{ech}(t_f))$ precisely. For example, in a plug-in HEV the fuel equivalent of a given battery energy equals the fuel necessary to recharge the battery by the same amount of energy, which can be calculated from the well-to-tank (plug) efficiency of the electric grid.

In an autonomous HEV, one case for which it is rather simple to derive expressions for the fuel equivalent is the case of constant efficiencies both of the electrical path, η_e, and of the thermal path, η_f (see Fig. 7.7a) [291, 225].

In this case, the fuel equivalent $\phi(E_{ech}(t_f))$ of a positive amount of battery energy used in the mission, i.e., energy provided by the storage system, is the fuel energy necessary to reload the same amount. Figure 7.7b shows that $\phi(E_{ech}(t_f)) = E_{ech}(t_f)/(\eta_e \cdot \eta_f)$. The fuel equivalent therefore is a linear function of $E_{ech}(t_f)$. The proportionality coefficient, an equivalence factor, for this case is calculated as

$$s_{dis} = \frac{1}{\eta_e \cdot \eta_f}. \tag{7.24}$$

The fuel equivalent $\phi(E_{ech}(t_f))$ of a negative amount of energy $E_{ech}(t_f)$, i.e., one that recharges the storage system, is the fuel energy that can be saved by using $E_{ech}(t_f)$ in the future. Figure 7.7c shows that in this case $\zeta(E_{ech}(t_f))$ equals $E_{ech}(t_f) \cdot \eta_e/\eta_f$. Again, the fuel equivalent is a linear function of $E_{ech}(t_f)$. The equivalence factor for this case is calculated as

$$s_{chg} = \frac{\eta_e}{\eta_f}. \tag{7.25}$$

In general, $s_{chg} < s_{dis}$ holds. The values of s_{dis} and s_{chg} are equal only in the case in which there are no losses in the electrical path, i.e., $\eta_e = 1$.

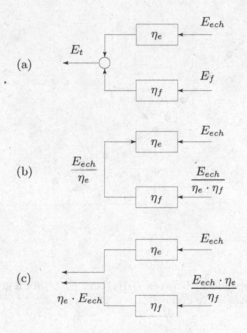

Fig. 7.7 Fuel equivalent of the electrical energy for constant efficiencies of the parallel paths (a), for positive (b), and for negative (c) electrical energy

The general case of variable path efficiencies has been analyzed in several publications. Some authors used constant, average values for the fuel and electrical efficiencies, further distinguishing between $\bar{\eta}_e^{(d)}$ in the discharge phase and $\bar{\eta}_e^{(c)}$ in the charge phase [363, 241, 242]. The equivalence factors are thus

$$s_{dis} = \frac{1}{\bar{\eta}_e^{(d)} \bar{\eta}_f} \qquad s_{chg} = \frac{\bar{\eta}_e^{(c)}}{\bar{\eta}_f}. \tag{7.26}$$

However, this approach is strongly dependent on the way the average efficiencies are defined, and it often requires heuristic corrections to avoid excessive SOC excursions. A more consistent analysis [291] has shown that s_{chg} and s_{dis}

can be evaluated purely from energy considerations, without any assumption on the path efficiencies. The procedure requires collecting data on the electrical energy use $E_{ech}(t_f)$ and the fuel energy use $E_f(t_f)$ over a mission of duration t_f, obtained with similarly structured control policies. A possible choice is to use various constant values of the control variable in the range $u \in [-u_l, u_r]$ given by the upper and lower bounds for the SOC that are admissible during the system operation. Figure 7.8 illustrates such a procedure.

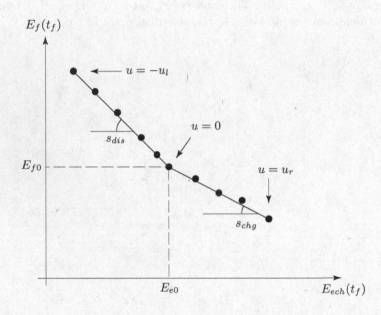

Fig. 7.8 Typical dependency between $E_f(t_f)$ and $E_e(t_f)$ for a given vehicle configuration and for a given drive cycle

In the pure thermal case $(u = 0)$ the fuel energy used, E_{f0}, is also the energy that would be used to drive the cycle if no electrical path were present. The electrical energy use in the pure thermal case, E_{e0}, is not zero due to regenerative braking power (which provides a negative contribution) and idle losses in the electrical path (i.e., the losses when the power at the output stage of the electrical path is zero, which add a positive contribution to E_{e0}). For many common scenarios and driving cycles, it has been observed that the pure thermal case separates the curve $E_f(t_f) = f(E_{ech}(t_f))$ in two branches, which are nearly linear in the range of interest. The slopes of these lines that fit the data are the two equivalence factors s_{dis} and s_{chg}.

The linear form of the curve $E_f(t_f) = f(E_{ech}(t_f))$ is observed even if the efficiencies of the parallel paths vary depending on the operating point. This may be explained by an averaging effect that is due to the large number of operating points included in a drive cycle. A straightforward but tedious analysis [295] shows that the average efficiencies of the thermal and electrical paths over the drive cycle can be computed from the equivalence factors, but not vice versa as in (7.26). The equivalence factors can be effectively used to combine the fuel consumption and the state of charge variations in an equivalent specific fuel consumption conveniently expressed, for instance, in liter/100 km.

In real-time conditions, the sign of $E_{ech}(t_f)$ is not known in advance. Therefore, the equivalence factor is continuously adapted as a quantity that varies between two limit values given by s_{chg} and s_{dis}, according to a probability factor $p(t)$, that is,

$$s(t) = p(t) \cdot s_{dis} + (1 - p(t)) \cdot s_{chg}. \tag{7.27}$$

The probability $p(t)$ in turn is calculated as

$$p(t) = \frac{E_e^+(t)}{E_e^+(t) - E_e^-(t)}, \tag{7.28}$$

as a function of the two quantities E_e^+ and E_e^- (see Fig. 7.9), which represent the maximum positive and negative values of the electric energy use that can result at the end of the mission. A mission is defined here as a trait of the vehicle's route characterized by a given value $E_h = \int_0^{t_h} P_t(\tau) \, d\tau$ of the required energy at the wheels ("energy horizon"), a quantity that is independent of the control law.

The estimation of E_e^+ and E_e^- strongly depends on the current value of the electrical energy consumption, $E_{ech}(t) = \int_0^t P_{ech}(\tau) \, d\tau$. In detail, the quantity $E_e^+(t)$ is given by the sum of three terms: (i) $E_{ech}(t)$, (ii) the electrical energy that would be used for the traction with the system driven at a constant $u = u_r$ (see Fig. 7.9) from t until the end of the mission, and (iii) a negative term due to the "available energy," i.e., the electric energy that will be stored from t until the end of the mission. The term (ii) is calculated as $u_r \cdot (E_h - E_t(t))/\bar{\eta}_e$. In fact, $E_h - E_t(t)$ is the mechanical energy that still must be delivered before the end of the mission. When it is multiplied by u_r, the mechanical energy provided at the output stage of the electrical path is obtained. To derive the energy at the input stage, the average efficiency of the electrical path (7.26) is used. The term (iii) is evaluated assuming a constant ratio λ between the available energy as a function of time and $E_t(t)$. This parameter λ is calculated for various drive cycles as the ratio E_{e0}/E_h. The final expression for $E_e^+(t)$ is

$$E_e^+(t) = E_{ech}(t) + \frac{u_r \cdot (E_h - E_t(t))}{\bar{\eta}_e} - \lambda \cdot (E_h - E_t(t)). \tag{7.29}$$

The (negative) quantity $E_e^-(t)$ is also given by three terms: (i) the current value of $E_{ech}(t)$, (ii) the electrical energy that would be recharged during the traction with the system driven at constant $u = -u_l$ (see Fig. 7.9) from t until the end of the mission, and (iii) a negative term due to the free energy, evaluated as above since it is independent of the control law. The term (ii) is calculated as $-u_l \cdot \bar{\eta}_e \cdot (E_h - E_t(t))$, applying the same considerations as above. Therefore, the final expression for $E_e^-(t)$ is

$$E_e^-(t) = E_{ech}(t) - \bar{\eta}_e \cdot u_l \cdot (E_h - E_t(t)) - \lambda \cdot (E_h - E_t(t)). \tag{7.30}$$

The resulting equation for $p(t)$ is

$$p(t) = \frac{u_r/\bar{\eta}_e - \lambda}{u_r/\bar{\eta}_e + \bar{\eta}_e \cdot u_l} + \frac{E_{ech}(t)}{(u_r/\bar{\eta}_e + \bar{\eta}_e \cdot u_l) \cdot (E_h - E_t(t))}, \tag{7.31}$$

with $p(t)$ limited between 0 and 1. For simplicity, (7.31) may be implemented with $u_r = u_l$ [291].

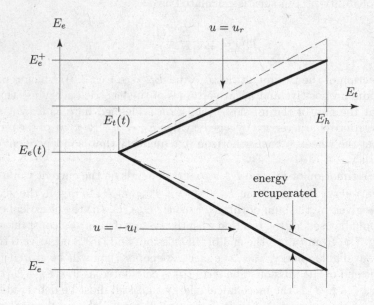

Fig. 7.9 Sketch of the quantities that lead to the evaluation of $E_e^-(t)$ and $E_e^+(t)$

It should be clear that (7.27) – (7.31) is equivalent to (7.22), except for the explicit presence of the parameter λ.[9] This parameter has a large variability and a large influence on the controller performance, thus it has to be estimated accurately. In contrast, the equivalence factors typically show a weaker influence, at least in their typical range of variability, which is rather small. As a consequence of this fact, the T-ECMS keeps s_{dis} and s_{chg} constant, i.e., a pair of average values is conveniently selected to represent the vehicle and these values are used for every driving condition. The T-ECMS includes instead a sophisticated algorithm for the on-line estimation of λ, based on

---

[9] See Problem 7.34.

the information that is provided by an on-board telemetry system during a mission (Fig. 7.10). Every mission is assumed to have a defined point that is to be reached along a defined route. All the static features of the route, including the maximum speed allowed in its various parts, the total distance to be covered, and the altitude profile are also assumed to be known.

If t_k is the time at which the k-th information I_k is available, $\hat{\lambda}_k$ is the related estimation of the parameter λ. The information I_k can be of two different types. If the presence of a moving or fixed obstacle is detected, I_k is a "stop" signal. In this case, the distance and the speed of the obstacle are assumed to be known also. When the obstacle is removed, the corresponding I_k is a "go" signal. In both cases, the estimation of λ derives from an estimation of the future velocity profile, $\hat{v}_k(t)$. The profile assumed is always the one that covers the rest of the mission in the minimum amount of time and which fulfills the constraints concerning the maximum speed and the presence of obstacles. The estimated velocity profile may only consist of: (i) trajectories at constant speed, (ii) trajectories with maximum acceleration, and (iii) trajectories with maximum deceleration. Under certain assumptions, such profiles have been proven to be fuel optimal [130].

From the estimated velocity profile $\hat{v}_k(t)$, the mechanical energy delivered at the wheels $\hat{E}_t(t)$ may be calculated. Each portion of the profile is responsible for an energy contribution, which is the sum of two terms. The former is due to altitude variations. The latter term depends on the other resistances, and it can be negative (energy being recuperated) or positive (energy being delivered). The contributions of the various portions of the estimated velocity profile to the electric energy recuperated are evaluated from the (negative) mechanical energy contributions and by using the model of the system.

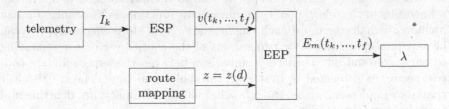

Fig. 7.10 Flowchart of the T-ECMS

Further details on the operation of the T-ECMS are illustrated in the case study A.4.

Implementation Issues

The flowchart of the online ECMS is sketched in Fig. 7.11. With respect to Fig. 7.6, the driver action replaces the driving cycle, while the s_0-finding loop is substituted by one of the estimation algorithms presented in the previous sections.

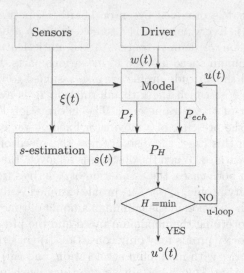

Fig. 7.11 Flowchart of the online implementation of the ECMS

In addition to the estimation of the equivalence factor, the practical implementation of ECMS requires a careful proceeding in many aspects [252, 53]. The Hamiltonian P_H given by (7.19) can be modified with additional terms to penalize situations that, though effective to improve global efficiency, could be undesirable from a drivability or component durability point of view. Such penalty terms might concern, for instance, changes of engine status, changes of engine load, changes of gear, and so on. In particular, the need to avoid too frequent engine starts or stops can lead to the introduction of additional refinements in the selection of the minimal cost function. For instance, P_H can include a penalty term on each engine start [291]. More effectively, P_H can be separately minimized for two subsets of the control vector corresponding to engine on and off. Then the comparison between the two candidate control vectors is subjected to hysteresis thresholds and time delays. It is clear that such implementations, though beneficial to drivability, are dentrimental to global powertrain efficiency.

7.4 Problems

Driver's Intepretation

Problem 7.1. An ICE-based powertrain has the following characteristics: $\gamma = \{15.0, 8.1, 5.3, 3.9, 3.1, 2.6\}$, wheel radius $r_w = 0.32\,\mathrm{m}$, rated engine power $P_{e,max} = 92\,\mathrm{kW}$ at a speed $\omega_{e,max} = 524\,\mathrm{rad/s}$, engine braking torque $T_{e,min} = 20\,\mathrm{Nm}$. Build a driver's interpretation map. Then, follow a torque control structure to generate an engine torque setpoint for a driver pedal request of 50% at a vehicle speed of 100 km/h and fourth gear. *(Solution: $T_e = 120\,Nm$).*

Problem 7.2. Add an electric machine to the powertrain of Problem 7.1, having the following characteristics: maximum torque $T_{m,max} = 140\,Nm$, base speed $\omega_b = 300\,rad/s$, maximum power $P_{m,max} = 42\,kW$. Calculate the total torque demand for the same driving situation as in Problem 7.1 if power assist is authorized at each vehicle speed. Assume coupled regenerative braking. *(Solution: $T_t = 711\,Nm$).*

Problem 7.3. Propose a driver's interpretation function for a BEV whose motor and battery have the same data as in Problem 4.9. Assume a coupled braking circuit. Calculate the torque setpoint for (i) $\omega_m = 0\,rad/s$ and 0% pedal depression, (ii) $\omega_m = 100\,rad/s$ and 0% pedal depression, (iii) $\omega_m = 250\,rad/s$ and 50% pedal depression. *(Solution: (i) $T_m = 0$, (ii) $T_m = -2.2\,Nm$, (iii) $T_m = 2.5\,Nm$).*

Problem 7.4. Derive a PI model for a human driver of an electric vehicle that tries to follow a prescribed drive cycle acting on the acceleration pedal. Derive a gain-scheduling tuning of the PI parameters. Vehicle data: mass $m_v = 1360\,kg$, $\frac{1}{2} \cdot \rho_a \cdot A_f \cdot c_d = 0.25\,N/(m/s)^2$. Evaluate the PI coefficients for a vehicle speed of 20 m/s. Assume perfect recuperation (decoupled braking). *(Solution: $w(t) = K_p \cdot z(t) + K_i \cdot \int z(t)dt$, where $w(t) = F_t(t) - (F_r + F_a(v_{des}(t)))$ and $z(t) = v(t) - v_{des}(t)$. Numerical results: $K_i = -1360$, $K_p = -1898$).*

Regenerative Braking Control

Problem 7.5. Derive an ideal law to split the braking effort between the two axles under the assumption that the adherence is the same at each wheel. *(Solution: $T_1 = T_t \cdot (1-s) - \frac{T_t^2 \cdot h}{N \cdot r \cdot (a+b)}$ and $T_2 = T_t \cdot s + \frac{T_t^2 \cdot h}{N \cdot r \cdot l}$, where T_1 and T_2 are the front and rear braking torques, respectively, T_t is the total braking torque (negative), $s = a/(a+b)$ is the static weight distribution factor, $l = a + b$ is the wheelbase, a and b are the distance from the center of gravity to the front and the rear wheels, respectively, h is the height of the center of gravity, $N = m_v \cdot g$ is the weight of the vehicle.).*

Problem 7.6. Consider a vehicle having an electric powertrain on the rear axle, with $T_{m,max} = 1540\,Nm$ (at the wheels), $P_{m,max} = 42\,kW$, and the following vehicle characteristics (see Problem 7.5): static weight distribution fraction $s = 0.40$, height of CG $h = 55\,cm$, wheelbase $l = 2.685\,m$, wheel radius $r_w = 0.32\,m$, vehicle mass $m_v = 1932\,kg$. Evaluate the regenerative braking torque and the frictional braking torque on the front and rear axles for a total braking torque $T_t = -1200\,Nm$, vehicle speed $v = 90\,km/h$, under (i) a maximum regeneration strategy, (ii) a constant braking distribution between the axles of 70%–30%, (iii) ideal braking as in the result of Problem 7.5, and (iv) a modified brake pedal that induces regenerative braking up to a deceleration of 0.05 g and then frictional braking with a constant braking distribution of 70%–30%. *(Solution: (i) $T_{rec} = -537\,Nm$, $T_1 = -663\,Nm$, $T_2 = 0\,Nm$; (ii) $T_{rec} = -360\,Nm$, $T_1 = -840\,Nm$, $T_2 = 0\,Nm$; (iii) $T_{rec} = -431\,Nm$, $T_1 = -769\,Nm$, $T_2 = 0\,Nm$; (iv) $T_{rec} = -300\,Nm$, $T_1 = -630\,Nm$, $T_2 = -270\,Nm$).*

Problem 7.7. Consider a conventional (coupled) braking system where $T_2 = k \cdot T_1$ ($T_1 < 0, T_2 < 0$). Calculate the maximum value of the adherence that can be obtained under the assumption of equal adherence between the axles (ideal distribution curve) and the corresponding total braking torque. Check what happens for higher braking torques. Then, calculate the limit value of k that can be achieved. Use the numerical values of Problem 7.6. *(Solution: For $k = 0.429$, the maximum adherence is 0.488. The maximum braking torque is $T_t = -2935\,Nm$. The limit value is $k = 0.667$).*

Problem 7.8. Derive the ideal braking distribution law as in Problem 7.5 when one axle in motoring while the other is braking (for instance, battery recharge mode in an HEV with an engine on the front axle and an electric machine on the rear axle). For simplicity, assume $a = b$. *(Solution: $T_1 = \frac{T_t}{2} - \frac{N \cdot a \cdot r_w}{2 \cdot h}$, $T_2 = \frac{T_t}{2} + \frac{N \cdot a \cdot r_w}{2 \cdot h}$).*

Dynamic Coordination

Problem 7.9. In a series HEV the supervisory control yields engine torque and speed setpoints T_e and ω_e. Derive a simple generator controller in order to achieve the desired speed of the APU. *(Solution: $T_g(t) = T_e(t) + k_p \cdot (\omega_g(t) - \omega_e(t))$ makes ω_g converge to ω_e with a time constant Θ_{apu}/k_p).*

Problem 7.10. Derive the dynamic equations to control the generator torque in a simple PSD-based system like that of the Toyota Prius. Neglect the generator inertia. *(Solution: Similar to Problem 7.9 but now with a gear ratio $z/(1+z)$ between engine and generator torque).*

Problem 7.11. In a post-transmission parallel HEV, in principle it is possible to compensate the torque gap at the wheels during a gear shift. Evaluate the time elapse after which the vehicle speed before the shift is recovered (t_r, recovery time) and the necessary electric energy for a downshift from 4th to 3rd gear occurring during a constant vehicle acceleration. Data: gear ratios including final gear $\gamma = 5,4$, motor gear ratio $= \gamma_m = 11$, transmission efficiency $\eta_t = 0.97$, motor efficiency $\eta_m = 0.85$, wheel radius $r_w = 0.29\,m$, engine shift speed $\omega_e = 4500\,rpm$, shift duration $t_s = 1\,s$; acceleration $a = 0.5\,m/s^2$, vehicle mass $m_v = 1360\,kg$, $c_r = 0.009$, $c_d \cdot A_f = 0.5\,m^2$. *(Solution: Without compensation $t_r = 1.6\,s$; with compensation $t_r = t_s$ but $\Delta E_m = 25.7\,kJ$).*

Problem 7.12. Consider a parallel HEV with an electric machine mounted on the primary shaft of the gearbox with a reduction gear ratio γ_m. During a gear shift, the inertia of the motor sums up to the inertia of the primary shaft. To reduce the synchronization lag, the motor in principle could yield a torque to compensate its own inertia. Model this situation with simple equations. Then calculate the motor energy consumption for the following data: $\gamma_m = 3.3$, downshift from 4th to 3rd gear with $\gamma_3 = 5.5$, $\gamma_4 = 3.9$, vehicle speed $v = 60\,\text{km/h}$, motor inertia $\Theta_m = 0.07\,\text{kg/m}^2$. *(Solution: $\Delta E_m = 21\,kJ$).*

Heuristic Energy Management Strategies

Problem 7.13. Consider a pre-transmission, single-shaft parallel HEV with fixed gear reduction. System data: gear ratio including final gear ratio $\gamma = 4$, engine maximum torque curve $T_{e,max}(\omega_e) = 50 + 0.7 \cdot \omega_e - 1 \cdot 10^{-3} \cdot \omega_e^2$, motor maximum torque $T_{m,max} = 150\,\text{Nm}$, motor maximum power $P_{m,max} = 25\,\text{kW}$, vehicle data $c_D = 0.33$, $A_f = 2.5\,\text{m}^2$, $c_r = 0.013$, $m_v = 1500\,\text{kg}$, $\Theta_w = 0.25\,\text{kg m}^2$, $r_w = 0.25\,\text{m}$. Consider the simple, SOC-independent heuristic energy-management strategy:

- EV mode if $\omega_e < 1000\,\text{rpm}$ or if $T_e < 40\,\text{Nm}$,
- power assist mode if $T_e > T_{e,max}$,
- else, recharge mode if $T_e > 0$
- regenerative braking if $T_e < 0$.

Evaluate the scheduled mode, the engine torque, and the motor torque for the following driving situations: (i) $v = 17\,\text{km/h}$, $a = 1.37\,\text{m/s}^2$; (ii) $v = 38.76\,\text{km/h}$, $a = 0.094\,\text{m/s}^2$; (iii) $v = 28.8\,\text{km/h}$, $a = 1.56\,\text{m/s}^2$; (iv) $v = 95\,\text{km/h}$, $a = 0.19\,\text{m/s}^2$. *(Solution: (i) $T_e = 0$, $T_m = 141\,Nm$; (ii) $T_e = 0$, $T_m = 24.2\,Nm$; (iii) $T_e = 123\,Nm$, $T_m = 37\,Nm$; (iv) $T_e = 119\,Nm$, $T_m = -59\,Nm$).*

Problem 7.14. Consider the following SOC-dependent energy-management heuristic strategy:

- engine on with $P_e = P_t - P_{t,b}(\xi)$ if $P_t < P_{e,start}(\xi)$,
- else, engine off,

with the definitions $P_{t,b} = -P_{m,max} + 2 \cdot P_{m,max}/(\xi_{hi} - \xi_{lo}) \cdot (\xi - \xi_{lo})$, $P_{e,start} = P_{m,max}/(\xi_{hi} - \xi_{lo}) \cdot (\xi - \xi_{lo})$ and the numerical values $\xi_{hi} = 80\%$, $\xi_{lo} = 40\%$. Assume a unit-efficiency motor operation. Perform again the calculations of Problem 7.13, for $\xi = \{55, 70\}\%$. *(Solution: (i) ZEV, (ii) ZEV, (iii) power assist, $P_e = \{15.95, 10.9\}\,kW$; $P_m = \{4.65, 9.7\}\,kW$, (iv) recharge and power assist, $P_m = \{-0.55, 18.2\}\,kW$, $P_e = \{21.55, 2.8\}\,kW$).*

Problem 7.15. Give an intepretation of the heuristic energy-management strategy of Problem 7.14 in terms of equivalent "cost" of the battery power with respect to the fuel power. Assume a Willans-type engine model with constant parameters e and P_0 and a unit-efficiency electric machine. *(Solution: If the nondimensional cost factor is s, the heuristic strategy is equivalent to $s \cdot e = 1 + P_0/(P_{e,start} - P_{t,b})$).*

Optimal Energy Management Strategies

Problem 7.16. Derive the exact formulation of the Euler-Lagrange equation (7.14) if the equivalent-circuit parameters of the battery are affine functions of SoC as described by (4.64) and (4.66). Consider the following system and operating point: battery capacity $Q_b = 6.5\,\text{Ah}$, nominal open-circuit voltage $U_{oc} = 250\,\text{V}$, nominal internal resistance $R_i = 0.3\,\Omega$, electric power $P_b = 15\,\text{kW}$, variation of the open-circuit voltage with respect to SOC $\kappa_2 = 20\,\text{V}$, and variation of the internal resistance $\kappa_4 = -0.1\,\Omega$. Evaluate the characteristic time constant associated with the variation of the Lagrange multiplier and assess the constant-μ approximation. *(Solution: time constant $\approx 48\,min$, which justifies the approximation of constant μ).*

Problem 7.17. Starting from the results of Problems 7.16, 4.26, find an approximated expression for the variation of the Lagrange multiplier. Evaluate the error with respect to the exact solution. *(Solution: The difference between the exact and the approximated values of $\partial I_b/\partial \xi$ is about 20%, however, the time constant is still very long).*

Problem 7.18. At low temperature operation, the variation of the internal parameters of a battery can be significant. Develop a version of the ECMS where variations of internal resistance, via the parameter κ_3 of (4.66), with temperature are accounted for. Cell data: $\kappa_1 = 3.4\,\text{V}$, $\kappa_2 = 0.5\,\text{V}$, $\kappa_4 = 0$, and

$$\kappa_3 = 0.015 - \vartheta_b \cdot \frac{0.01}{40},$$

with ϑ_b in °C. The nominal SOC is $\xi = 0.5$, temperature $\vartheta = 25\,°\text{C}$, power $P_b = 0.1 \cdot P_{b,max}$, thermal capacitance $C_{t,b} = 300$ J/K, thermal conductance $1/R_{th} = 0.5$ W/K, and capacity $Q_b = 6$ Ah. Evaluate the time constant of the adjoint states. *(Solution: Time constant of $\mu \approx 200\,h$; time constant of the temperature costate $\approx 9\,min$ but its dynamics unstable).*

Problem 7.19. Find the optimal-control formulation (Hamiltonian function and Euler-Lagrange equation) of the energy management of a hybrid powertrain with an ICE and a supercapacitor. Find under which approximation the costate is time-invariant. *(Solution: $\dot{\mu} = 0$ if $R_{sc} = 0$).*

Problem 7.20. Formulate the energy-optimal energy management in the case of a double-source electric powertrain, with a battery and a supercapacitor. *(Solution: Same Euler-Lagrange equation as in Problem 7.19).*

Problem 7.21. Formulate the optimal energy management for a parallel HEV that includes engine temperature variations. Assume that the cold-engine fuel consumption is given by an equation of the type

$$\overset{*}{m}_f(T_e, \omega_e, \vartheta_e) = \overset{*}{m}_{f,w}(T_e, \omega_e) \cdot f(T_e, \omega_e, \vartheta_e),$$

where ϑ_e is one engine relevant temperature and $\overset{*}{m}_{f,w}$ is the warm-engine fuel consumption. Moreover, assume an engine temperature dynamic of the type

$$C_{t,e} \cdot \dot{\vartheta}_e = P_{heat}(T_e, \omega_e, \vartheta_e) - \alpha \cdot (\vartheta_e - \vartheta_{amb}).$$

(Solution: The SoC costate is the same as in the usual definition of the Hamiltonian function; the temperature costate variation is $\dot{\nu} = -\overset{*}{m}_{f,w} \cdot \frac{\partial f}{\partial \vartheta_e} - \frac{\nu}{C_{t,e}} \cdot \left(\frac{\partial P_{heat}}{\partial \vartheta_e} - \alpha \right)$.)

Problem 7.22. Evaluate the optimal gear ratio profile during an ICE-based vehicle acceleration from rest to v_f on a flat road. Use (i) the acceleration time t_f or (ii) the fuel consumption m_f as the performance index. Make the following simplifying assumptions: constant engine torque $T_e = T_{e,max}$, e, P_0, continuously variable gear ratio, linearized vehicle dynamics

$$\dot{v} = \frac{F_t}{m_v} - b \cdot v,$$

where $F_t = u \cdot T_e$ and $u = \gamma/r_w$. Verify the solution given by optimal control theory by analyzing the dependency of the criterion on the gear ratio. Numerical data: $b = 10^{-2}$, $u = \gamma/r_w \in [u_{min}, u_{max}] = [2, 12]$, $m_v = 1000\,\mathrm{kg}$, $v_f = 100\,\mathrm{km/h}$, $T_e = 150\,\mathrm{Nm}$, $e = 0.4$, $P_0 = 2\,\mathrm{kW}$. *(Solution: In the case (i), $u^\circ(t) = u_{max}$ and t_f is a decreasing function of u; in the case (ii), $u^\circ(t) = u_{max}$ and m_f is a decreasing function of u).*

ECMS

Problem 7.23. Consider a parallel HEV. The engine is a Willans machine with $e = 0.3$ and $P_{e,0} = 2\,\mathrm{kW}$. The electric drivetrain has a constant efficiency $\eta_{el} = 0.8$ and a maximum/minimum power $P_{m,max/min} = \pm 20\,\mathrm{kW}$. Calculate for which values of the equivalence factor s a purely electric drive and a full recharge, respectively, are optimal for a power demand $P_d = 20$ kW. *(Solution: Purely electric mode is optimal when $s < 2.9$. Full recharge is optimal when $s > 3.4$).*

Problem 7.24. Derive a look-up table yielding the optimal engine torque T_e of a post-transmission parallel hybrid as a function of ω_w, T_t and s. Use the following engine model,

$$P_f = \begin{cases} \frac{P_{e,0} + P_e}{e}, & \text{for } T_e > 0 \\ 0, & \text{for } T_e > 0 \end{cases}$$

with the following parameters:

$$\frac{1}{e} = \begin{cases} 1.21 \cdot 10^{-5} \cdot \omega_e^2 - 0.0053 \cdot \omega_e + 2.94, & T_e < \min(T_{e,max}, T_{e,turbo}) \\ -1.63 \cdot 1^{-4} \cdot \omega_e^2 - 0.0876 \cdot \omega_e - 6.80, & T_e > T_{e,turbo} \end{cases}$$

$$\frac{P_{e,0}}{e} = \begin{cases} 0.166 \cdot \omega_e^2 + 1.174 \cdot \omega_e + 4.59 \cdot 10^3, & T_e < \min(T_{e,max}, T_{e,turbo}) \\ 5.19 \cdot \omega_e^2 - 2.83 \cdot 10^3 \cdot \omega_e + 2.27 \cdot 10^5, & T_e > T_{e,turbo} \end{cases}$$

with $T_{e,turbo} = 200\,\mathrm{Nm}$ and $T_{e,max} = -0.0038 \cdot \omega_e^2 + 2.32 \cdot \omega_e - 79\,\mathrm{Nm}$. Use the motor model

$$P_m = \omega_m \cdot T_m + (0.0012 \cdot \omega_m + 0.0179) \cdot T_m^2 + (-0.0002 \cdot \omega_m^2 + 0.789 \cdot \omega_m + 384) =$$
$$= \omega_m \cdot T_m + a(\omega_m) \cdot T_m^2 + c(\omega_m)$$

with $P_{m,max} = 42\,\mathrm{kW}$ and $T_{m,max} = 140\,\mathrm{Nm}$, and the battery model of Problem 4.26 with $R_i = 0$. Find the optimal T_e for $T_t = 1000\,\mathrm{Nm}$, $\omega_w = 39\,\mathrm{rad/s}$, $\gamma_m = 11$, $\gamma = 8.1$ (including the final gear), and $s = 2.8$. *(Solution: $T_e = 200\,Nm$).*

Problem 7.25. Find the unconstrained optimal engine torque for a post-transmission parallel hybrid with an engine model of the type

$$P_f = \frac{P_{e,0} + P_e}{e},$$

an electric machine model of the type

$$P_m = \omega_m \cdot T_m + a \cdot T_m^2 + c,$$

and the battery model of Problem 4.26,

$$P_{ech} = P_b + P_b^2 \cdot \frac{2 \cdot R_i}{U_{oc}^2},$$

where $P_b = P_m$. Neglect the SOC influence. *(Solution: $T_m = \frac{\frac{\omega_e}{e} - \frac{\gamma}{\gamma_m} \cdot s \cdot \omega_m}{2 \cdot a} \cdot \frac{1}{s} \cdot \frac{\gamma_m}{\gamma}$ and $T_e = \frac{T_t - \gamma_m \cdot T_m}{\gamma}$).*

Problem 7.26. Use the result of Problem 4.21 and a simplified battery model $P_{ech} = P_b$ to derive an analytical solution of the optimal energy management of a series hybrid. Following Problem 4.21, consider the engine Willans parameter varying as

$$\frac{1}{e} = \begin{cases} 0 & \text{for} & P_g = 0 \\ 4.01 & \text{for} & 0 < P_g \leq 14 \cdot 0.92 \cdot 10^3 \\ 3.36 & \text{for} & 14 \cdot 0.92 \cdot 10^3 < P_g \leq 62 \cdot 0.92 \cdot 10^3 \\ 3.89 & \text{for} & 62 \cdot 0.92 \cdot 10^3 < P_g \leq 68 \cdot 0.92 \cdot 10^3, \end{cases}$$

and $\eta_g = 0.92$. *(Solution: For $s < 3.76$ the engine is kept off; for $3.76 < s < 3.95$ the engine is operated at minimum power (and speed), and for $s > 3.95$ maximum engine power is demanded).*

Problem 7.27. Derive equations (7.24) – (7.25) from PMP. *(Hint: Formulate the Hamiltonian for charge and discharge. By solving for $\partial H/\partial P_e = 0$, the switching condition is obtained).*

Problem 7.28. For the simple parallel HEV model of Problem 7.23, with $e = 0.4$, $P_{e,0} = 3\,\text{kW}$, $\eta_{el} = 0.9$, find the conditions on P_t for which the ZEV mode, the ICE mode or the battery recharge with $P_b = -2 \cdot P_t$ are optimal, respectively. *(Solution: With $s_1 = \eta_{el}/e$ and $s_2 = 1/\eta_{el}/e$, the optimal modes are (i) for $s < s_1$, $P_e = 0$ (ZEV); (ii) for $s_1 < s < s_2$, ZEV for $P_t < 27/(4 \cdot s - 9)\,kW$ otherwise $P_b = 0$ (ICE mode); (iii) for $s > s_2$, ZEV for $P_t < 2700/(724 \cdot s - 1800)\,kW$ otherwise recharge).*

Problem 7.29. Use the result of Problem 7.28 to evaluate s over a drive cycle with the following characteristics: $\bar{E}_{trac} - \bar{E}_{rec} = \Delta\bar{E} = 0.183$ MJ, $\bar{E}_{trac} = 0.670$ MJ, $P_{max} = 18.9$ kW. Assume a linear relationship between cumulative energy and power demand. Then perform again the calculations for the data of Problem 7.23. *(Solution: For power demand above 15.4 kW the battery recharge mode is selected. For power demand below that threshold, the purely ICE mode is selected).*

Implementation Issues

Problem 7.30. Consider a post-transmission parallel HEV with the following simplified data: motor transmission ratio $\gamma_m = 11$, wheel radius $r_w = 0.317\,\text{m}$, engine transmission ratio $\gamma = \{15.02, 8.09, 5.33, 3.93, 3.13, 2.59\}$, transmission efficiency $\eta_t = 0.95$. Consider the following driving situation: torque demand at the wheels $T_t = 378\,\text{Nm}$, vehicle speed $v = 69.25\,\text{km/h}$, engine on, electric consumers off, 4th gear. In this situation, $T_{m,max} = 140\,\text{Nm}$, $P_{m,max} = 42\,\text{kW}$, $U_{b,min} = 300\,\text{V}$, $U_{b,max} = 420\,\text{V}$, $P_{e,start} = 3\,\text{kW}$, $P_m = 0.9345 \cdot T_m^2 + 673.97 \cdot T_m + 127.44$, $U_{oc} = 381.12\,\text{V}$, $R_i = 0.3648\,\Omega$ (discharge), $R_i = 0.3264\,\Omega$ (charge), Coulombic efficiency $\eta_c = 0.95$, $T_{e,max} = 269.7\,\text{Nm}$, $T_{e,min} = -20\,\text{Nm}$, fuel consumption

$$\overset{*}{m}_f = (T_e - T_{e,min}) \cdot (2.9 \cdot 10^{-8} \cdot T_e + 1.112 \cdot 10^{-5}) =$$
$$= 2.9 \cdot 10^{-8} \cdot T_e^2 + 1.17 \cdot 10^{-5} \cdot T_e + 2.225 \cdot 10^{-4}$$

Find the engine and motor torque calculated by the ECMS. The current estimation of the equivalence factor is $s = 3$. *(Hint: Calculate and compare H_{ev} for the ZEV mode and H_{hev} for three candidate situations, namely, (i) battery-recharge mode with $T_e = T_{e,max}$, (ii) ICE mode, and (iii) $T_e = T_{e,min}$. Solution: $T_e = 101.2\,Nm$, $T_m = 0$ (ICE mode)).*

Problem 7.31. Solve again Problem 7.30 for the situation in which the engine is turned off. *(Solution: $T_e = 0$, $T_m = 34.4\,Nm$ (ZEV mode)).*

Problem 7.32. Consider an ECMS with a stop-start strategy implementation based on hysteresis thresholds. In order to start the engine, H_{hev} (see Problem 7.30) must fulfill the condition

$$H_{hev} < x_{on} \cdot H_{ev}.$$

At the previous calculation step, the lower Hamiltonian value was H_{ev}, thus the engine is off. At the current time step, the power demand is $P_t = 13.28\,\text{kW}$. The equivalence factor estimation is $s = 3.2813$. Calculate the mode selected for a hysteresis threshold x_{on} of (i) 95% and (ii) 90%, respectively. Use the following data and models: post-transmission parallel HEV architecture, transmission efficiency, $\eta_t = 0.95$, fuel consumption $P_f = 2.5446 \cdot P_e + 9.6525 \cdot 10^3$ if $P_e > 0$, electrochemical power

$$P_{ech} = \begin{cases} 1.2707 \cdot P_m + 2.7703 \cdot 10^3 - 2.014 \cdot 10^3, & \text{if } P_m > -595 \text{ W} \\ 0.7397 \cdot P_m + 2.4544 \cdot 10^3 - 2.014 \cdot 10^3, & \text{if } P_m < -595 \text{ W}. \end{cases}$$

The cost of engine start is $P_{e,start} = 2.014$ kW (in electrochemical power units). *(Solution: In the case (i) the engine is turned on, in (ii) the engine is kept off).*

Problem 7.33. Consider an HEV under several repetitions of an elementary driving cycle. An ECMS has a PI adaptation of s as a function of SoC that yields a new estimation every cycle repetition. Assume that the overall behavior of the system on a *cycle-by-cycle* basis depends on s as follows:

$$\Delta\xi(n) = \xi(n) - \xi(n-1) = K_s \cdot (s(n) - s_0),$$

where $\xi(n)$ is the SoC at the end of the n-th repetition, s_0 is the optimal value of s, $s(n)$ is the value adopted during the n-th cycle, and $K_s > 0$ is a constant depending on the particular system. Evaluate the stability and the dynamic characteristics of the controlled system on a cycle-by-cycle basis. Evaluate the influence of the integral term in the PI controller. *(Solution: Both P and PI controllers are stable, but only with the PI control the steady-state error vanishes).*

Problem 7.34. Compare (7.22) with (7.27) – (7.31). Under which assumptions are they equivalent? *(Hint: They are equivalent if a sliding horizon is assumed, i.e., t_h is such that $E_h - E_t(t) = const$. Moreover, λ is not explicitly considered in (7.22)).*

Problem 7.35. Express s_0 as a function of u_r, u_l, s_{dis}, and s_{chg}. Evaluate s_0 for $s_{max} = 5$, $s_{min} = 2$, knowing from a cycle analysis that $u_r/\eta_e = 1.2$ and $u_l \cdot \eta_e = 1.8$. *(Solution: $s_0 = 3.2$).*

A

Appendix I – Case Studies

A.1 Case Study 1: Gear Ratio Optimization

This case study shows how the gear ratios of a manual gear box can be optimized to improve the fuel economy of a passenger car. This analysis is purely academic because it completely neglects all drivability issues and only focuses on the fuel economy of a vehicle that follows the MVEG–95 driving profile.

Nevertheless, it is instructive because it shows how a numerical parametric optimization problem can be defined and solved using a quasistatic problem formulation. The software tools used in this example are the QSS toolbox in conjunction with numerical optimization routines provided by Matlab/Simulink. This approach is quite powerful and can be used to solve nontrivial problems.

A.1.1 Introduction

The powertrain of the light-weight vehicle analyzed in Sects. 3.3.2 and 3.3.3 includes a standard five-speed manual gear box. The ratios of these five gears are chosen according to the approach introduced in Sect. 3.2.2 and satisfy the usual requirements with respect to acceleration performance, towing capability, etc.

These gear ratios do not yield the smallest possible fuel consumption when the vehicle is following the MVEG–95 test cycle and it is clear that there is a different set of gear ratios that improve the vehicle's fuel economy. However, it is not clear at the outset what gear ratios are optimal and — more importantly — what gains in fuel economy may be expected in the best case. These two questions can be answered using the approach shown below.

A.1.2 Software Structure

The vehicle model and its representation with the QSS toolbox have already been introduced in Sect. 3.3.3. The model shown in Fig. 3.11 is reused in this problem setup. As illustrated in Fig. A.1, that model is embedded into a larger software structure that has a total of four hierarchy levels.

The top level (a Matlab .m file named `optimaster.m` in this example) initializes all system parameters and defines a first guess for the optimization variables u, i.e., for the five gear ratios. After that, a numerical optimization routine is called (`fminsearch.m` in this example). This routine, which is part of Matlab's optimization toolbox, calls a user-provided .m file (named `opti_fun.m` in this example) that computes the actual value of the objective function $L(u)$, i.e., the fuel consumption of the vehicle for the chosen set of gear ratios u. For that purpose, the vehicle model that has been programmed using the QSS toolbox is used (file `sys.mdl` in this example).[1]

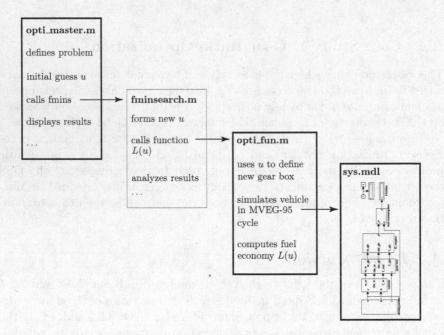

Fig. A.1 Software structure used to find the fuel-optimal gear ratios of a standard IC engine/manual gear box powertrain. The blocks with thick frames are provided by the programmer, the block framed by a thin line is a subroutine provided by Matlab's optimization toolbox.

---

[1] Note that Matlab/Simulink encapsulates all variables within the corresponding software modules. The exchange of variables across modules can be accomplished with several methods. A convenient approach is to define the necessary variables on all levels to·be global and, thus, accessible to all modules. However, this method must be used with caution in order to avoid using the same variable name for different objects.

All files necessary to solve this case study can be downloaded at the website http://www.imrt.ethz.ch/research/qss/. The programs are straightforward to understand. Some efforts have to be made in order to correctly handle situations in which infeasible gear ratios are proposed by the optimization routine.

A.1.3 Results

Obviously, each iteration started by the optimization routine requires one full MVEG–95 cycle to be simulated. Since many iterations are needed to find an optimum, short computation times become a key factor for a successful analysis. The computations whose results are shown below required approximately a total of 10 s CPU time on a 2 GHz power PC. This figure is acceptable and indicates that this method might be useful to solve more complex problems within a reasonable time.[2]

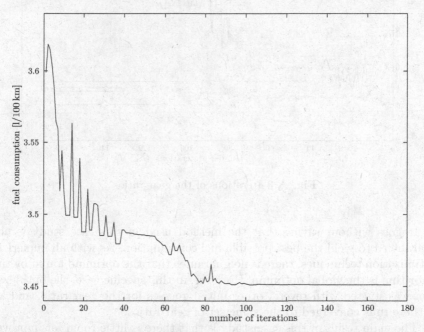

Fig. A.2 Iterations of the fuel consumption (the non-feasible solutions have been smoothed out)

Figure A.2 shows the evolution of the fuel consumption during one optimization run. The corresponding gear ratios are shown in Fig. A.3. During the optimization, several non-feasible sets of gear ratios are proposed by the

---

[2] Moreover, no attempts were made to optimize or compile the Matlab/Simulink program.

routine `fminsearch.m`. The fuel consumption of these non-feasible solutions is set to be higher than the initial fuel consumption. For clarity reasons, these outliers have been omitted in Fig. A.2 and the fuel consumption of the previous feasible solution has been used instead.

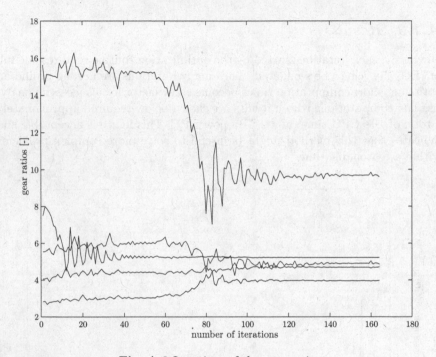

Fig. A.3 Iterations of the gear ratios

It goes without saying that the method used in this case study is not guaranteed to yield the best possible fuel consumption. As with all numerical optimization techniques, there is no guarantee that the optimum found by the algorithm is the global optimum. However, in this specific case, the optimization was started with several other initial guesses for the gear ratio, and all of these runs converged to the same set of gear ratios.

The main result of this case study is that there is little room for improvement by changing the gear ratios. As shown above, the expected gains in fuel economy (probably) are, even in the best case, less than 5%. This relatively small gain in fuel economy would not justify the poorer drivability following from the choice of fuel-optimal gear ratios.

A.2 Case Study 2: Dual-Clutch System-Gear Shifting

A.2.1 Introduction

This case study analyzes the problem of finding an optimal gear-shifting strategy for a vehicle equipped with a dual-clutch system. The vehicle model is illustrated in Fig. A.4. Which gear $x \in \{1, 2, 3, 4, 5, 6\}$ is engaged when, substantially influences the total fuel consumption. Two gear shifting strategies are compared: (i) shifting the gears as proposed by the test cycle (the MVEG–95, in this case), or (ii) shifting the gears such that the smallest possible fuel consumption is realized. Of course, at the outset this optimal gear shifting strategy is not known. Deterministic dynamic programming (see Sect. C) can be used to find it, provided the future driving profile is known. The resulting solution is not causal, but it will represent a benchmark for all possible causal control strategies.

The problem has one state variable x_k (the previous gear number) and one control input i_k (the desired future gear number), both with inherently discrete values. A DDP approach is feasible because the drive cycle, which the vehicle has to follow (the MVEG–95 cycle introduced in Fig. 2.6), is known a priori.

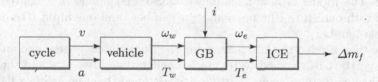

Fig. A.4 QSS powertain model, gearbox GB includes a dual clutch system

A.2.2 Model Description and Problem Formulation

The QSS model illustrated in Fig. A.4 is equivalent to the following discrete model $f(x_k, i_k)$. The chosen time step is 1 s. The vehicle model contains the air drag force

$$F_a(v) = \frac{1}{2} \cdot \rho_a \cdot c_d \cdot A_f \cdot v^2 \tag{A.1}$$

the rolling friction force

$$F_r(v) = m_v \cdot g \cdot (c_{r0} + c_{r1} \cdot v^{c_{r2}}) \tag{A.2}$$

and the inertial force

$$F_i = (m_v + m_r) \cdot a \tag{A.3}$$

The torque required at the wheel axle is

$$T_w = (F_a + F_r + F_i) \cdot r_w \tag{A.4}$$

where r_w is the wheel radius. The rotational speed and acceleration of the wheel are

$$\omega_w = \frac{v}{r_w}, \qquad \dot{\omega}_w = \frac{a}{r_w} \tag{A.5}$$

The gearbox model includes a transformation of the torque and rotational speed required at the wheel to the torque and rotational speed at the engine

$$T_e = T_w/\gamma(x), \qquad \omega_e = \omega_w \cdot \gamma(x), \qquad \dot{\omega}_e = \dot{\omega}_w \cdot \gamma(x) \tag{A.6}$$

where $\gamma(x)$ is the gear ratio of gear x. The engine model is based on the Willans approximation introduced in Chap. 3. With this approach the fuel flow can be approximated by

$$\Delta m_f = \frac{\omega_e}{e(\omega_e) \cdot H_l} \cdot \left(T_e + \frac{p_{me0}(\omega_e) \cdot V_d}{4\pi} + \Theta_e \cdot \dot{\omega}_e \right) \cdot \Delta t \tag{A.7}$$

where $p_{me0}(\omega_e)$ is the engine friction pressure, $e(\omega_e)$ is the Willans efficiency, and Θ_e is the engine inertia. In the optimization discussed below, all numerical values correspond to a two liter naturally aspirated engine and a midsize vehicle. The model is now a simplified QSS-based model of a conventional vehicle with one state (the previous gear number) and one input (the desired new gear number).

The time needed to shift gears using the dual-clutch system and an automated gearbox is assumed to be much smaller than $\Delta t = 1$ s and therefore considered as instantaneous. Further it is assumed that the gearbox has limited possibilities to change gears, i.e., that there is a constraint on the possible next gears depending on the current gear. It is this constraint that makes DDP a suitable method to solve the problem. Typically, such a gearbox contains two shafts: the first shaft carries gears one, three, and five and the second shaft gears two, four, and six. The possible instantaneous gear shifting is limited to gears from one shaft to the other. Hence, running at the gear x the possible next gear must be in the set

$$I(x) = \begin{cases} \{2,4,6\} \text{ if } x \in \{1,3,5\} \\ \\ \{1,3,5\} \text{ if } x \in \{2,4,6\} \end{cases}$$

Using the dynamic programming algorithm with the cost criteria

$$g_k(x_k, i_k) = \begin{cases} \Delta m_f & i_k \in I(x_k) \\ \infty & \text{otherwise} \end{cases} \tag{A.8}$$

it is possible to determine the optimal gear switching strategy for a given cycle that gives the minimum fuel consumption.

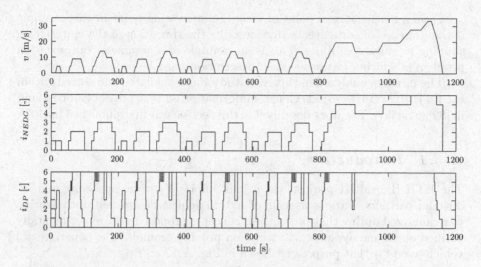

Fig. A.5 Results DDP optimization of the dual clutch gear box problem. Velocity profile of the new European drive cycle NEDC (=MVEG-95) (top); standard MVEG–95 gear switching strategy (middle); and DDP optimal gear switching strategy (bottom).

A.2.3 Results

The resulting gear switching profile for the MVEG–95 is shown in Fig. A.5. Figure A.5 also shows the standard gear switching profile for that cycle. The average CO_2 emissions for the considered vehicle using the standard MVEG–95 gear switching strategy are approximately 200 g/km. With the dual-clutch system and the optimized strategy this value is reduced to 172 g/km. It is clear that the result of the DDP problem is optimal only for the chosen test cycle. Moreover, in practice it is not possible to achieve this level of fuel economy because of the many constraints (driving comfort, energy use of the gear shifting device, etc.). Nevertheless, this result shows that there is a substantial potential to improve the fuel economy using dual-clutch systems.

A.3   Case Study 3: IC Engine and Flywheel Powertrain

This section presents an approach that can be used to improve the part-load fuel consumption of an SI engine system. The key idea is to avoid low-load conditions by operating a conventional IC engine in an on–off mode. The excess power produced by the firing engine is stored in a flywheel in the form of kinetic energy. During the engine-off phases this flywheel provides the power needed to propel the vehicle. A CVT with a wide gear ratio is necessary to kinematically decouple the engine from the vehicle.

From a mathematical point of view the interesting point in this example is the presence of state events that describe the transition of the clutch from slipping to stuck conditions. This is an example of a parameter optimization problem in which a forward system description must be used.

The problem analyzed in this case study was formulated and solved within the ETH Hybrid III project. General information on that project can be found in [82] and [357]. The work described in this section was first published in [132].

A.3.1 Introduction

The ETH Hybrid III project was a joint effort of several academic and industrial partners. Various aspects of hybrid vehicle design and optimization were analyzed during that project. All concepts proposed were experimentally verified on engine dynamometers and on proving grounds. The experimental vehicle used for that purpose is shown in Fig. A.6.

Fig. A.6 The ETH Hybrid III vehicle

This case study focusses on one particular aspect of the ETH Hybrid III design process. One of the key ideas analyzed and realized in this project was the development of a flywheel–CVT powertrain that permitted an efficient recuperation of the vehicle's kinetic energy while braking and an on–off operation of the IC engine during low-load phases. The CVT was realized in an "i^2" configuration[3] that yielded a very large gear ratio range (approximately

---

[3] With an appropriate system of external cog wheels and automatic clutches, the input and output of an "i^2" CVT can be interchanged. Therefore, the standard gear ratio range i can be used twice. Due to the fact that some overlapping is necessary at the switching point, the total gear ratio range is slightly smaller than the theoretical maximum of i^2.

1 : 20). Figure A.7 shows a schematic representation of those parts of the ETH Hybrid III powertrain[4] that are relevant for the subsequent analysis.

The flywheel has a mass of approximately 50 kg and can store sufficient energy to accelerate the vehicle from rest to approximately 60 km/h (see Sect. 5.2 for more information on flywheels). The flywheel is mounted coaxially to the engine shaft and rotates in air at ambient conditions. Its gyroscopic influence on the vehicle is noticeable but does not pose any substantial stability problems.

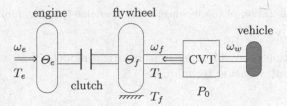

Fig. A.7 Simplified schematic representation of the ETH Hybrid III powertrain structure

For the sake of simplicity, it is assumed in this analysis that the vehicle is driving on a horizontal road with a constant velocity characterized by the constant wheel speed ω_w. Under this assumption the powertrain is operated in a periodic way as illustrated in Fig. A.8.

At $t = 0$ one cycle starts with the IC engine off and the flywheel at its maximum speed $\bar{\omega}$. At $t = \tau_o$ the command to close the clutch is issued and the engine is accelerated from rest to $\underline{\omega}$. Since the point in time $t = \tau_c$ at which the engine speed reaches the flywheel speed is not known, its detection will be an important part of the solution presented below.

In the time interval $t \subset [\tau_c, \tau_c + \vartheta)$ the engine is operated at the torque $T_{opt}(\omega_e)$ that yields the best fuel economy, i.e., almost at full load. Of course, the power produced in this phase exceeds the power consumed by the vehicle to overcome the driving resistances. Accordingly, the flywheel is accelerated until it again reaches $\bar{\omega}$. At this point in time $(t = \tau_c + \vartheta)$ the clutch is opened and the fuel is cut off such that the engine rapidly stops.

The optimization problem to be solved consists of finding for each constant vehicle speed ω_w those parameters $\underline{\omega}$ and $\bar{\omega}$ that minimize the total fuel consumption. The following two contradicting effects are the reason for the existence of an optimum: higher flywheel speeds produce smaller duty-cycles and, hence, better engine utilization, but higher flywheel speeds also produce larger friction losses. The optimal compromise that minimizes the

---

[4] In addition to the IC engine and the flywheel, the powertrain included an electric motor in a parallel configuration as well. These three power sources justified the appendix "III" in the vehicle name.

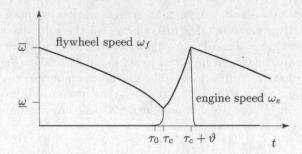

Fig. A.8 Illustration of the flywheel and the engine speed as functions of time

fuel consumption is found using mathematical models of all relevant effects and numeric optimization techniques.

A.3.2 Modeling and Experimental Validation

The system to be optimized operates in two different configurations: "clutch open" with $\omega_f \neq \omega_e$ and "clutch closed" with $\omega_f = \omega_e$. Accordingly, it is described by two different differential equations. In the case "clutch open" this equation is

$$\Theta_f \cdot \frac{d}{dt}\omega_f(t) = -T_f(t) - \frac{P_0(\omega_w)}{\omega_f(t)}, \qquad (A.9)$$

where $P_0(\omega_w)$ is the power consumed by the vehicle at the actual wheel (vehicle) speed ω_w. Following the assumptions mentioned above, the power P_0 is constant and depends on the wheel speed ω_w as follows

$$P_0(\omega_w) = p_1 \cdot \omega_w + p_3 \cdot \omega_w^3 \qquad (A.10)$$

(the numerical values of all coefficients and parameters used in this case study are listed in Tables A.1 and A.2).

The torque $T_f(t)$ stands for all friction losses in the powertrain. It includes the aerodynamic losses of the rotating flywheel. In the ETH Hybrid III project it was possible to approximate these losses by

$$T_f(t) = k_0 + k_1 \cdot \omega_f(t). \qquad (A.11)$$

In the case "clutch closed" $(\omega_e(t) = \omega_f(t))$ the powertrain dynamics are described by

$$(\Theta_f + \Theta_e) \cdot \frac{d}{dt}\omega_f(t) = T_e(\omega_f) - T_f(t) - \frac{P_0}{\omega_f(t)}, \qquad (A.12)$$

where the variable $T_e(t)$ stands for the fuel-optimal engine torque. This function can be parametrized as

$$T_e(\omega_e) = c_0 + c_1 \cdot \omega_e + c_2 \cdot \omega_e^2. \tag{A.13}$$

The last missing element of the modeling process is an approximation of the fuel mass flow during the phase when the engine fires

$$\overset{*}{m}_f(\omega_e) = f_0 + f_1 \cdot \omega_e. \tag{A.14}$$

Figure A.9 shows the comparison of the predicted and the measured flywheel speeds during on on–off cycle. The deviations at lower speed are noticeable, but do not substantially influence the final result. This can be seen by comparing the predicted and the measured fuel consumption as illustrated in Fig. A.12. Five different pairs of switching speeds are shown in that figure, including the optimal solution $\underline{\omega} = 114\,\mathrm{rad/s}$ and $\overline{\omega} = 278\,\mathrm{rad/s}$.

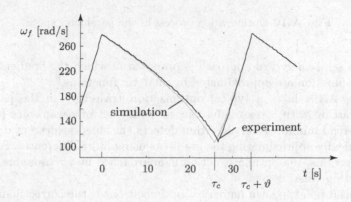

Fig. A.9 Comparison of the predicted and measured flywheel speed during one full cycle

A.3.3 Numerical Optimization

The optimization criterion that has to be minimized is the powertrain's fuel consumption per distance travelled in one cycle, i.e.,

$$J = \frac{\int_0^{\tau_c + \vartheta} \overset{*}{m}_f(\omega_e(t))\, dt}{\tau_c + \vartheta}. \tag{A.15}$$

The two variables to be optimized are $\{\underline{\omega}, \overline{\omega}\}$. The fuel consumption $\overset{*}{m}_f$ and the times τ_c, ϑ depend on these quantities. Note that since the velocity of the vehicle is assumed to be constant, the distance it travels is proportional to the duration of one cycle.

The solution to the problem analyzed in this section can be found using numerical optimization techniques. Here, a fully numeric approach is presented.

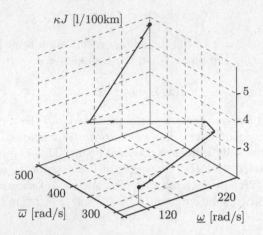

Fig. A.10 Optimization process in the parameter space

In [132] a semi-analytical approach is proposed in which the gradients of the objective function are approximated by analytic functions.[5]

Figure A.10 shows a typical optimization trajectory in the parameter space. Starting with a reasonable but arbitrary set of parameters $\{\underline{\omega}_0, \bar{\omega}_0\}$, the numerical minimization algorithm detects the steepest-descent directions by numerically approximating the gradients using finite differences. Following these directions, the optimal solution is approached in a reasonable amount of computing time.

A crucial point in such numerical optimizations is the correct handling of state events. In the problem analyzed in this case study the relevant state event is the switch between the system structure described by (A.9) and the structure described by (A.12). The time instant τ_c at which this event takes place is not known a priori but is determined by the evolution of the system state variables. This effect can cause problems in numerical optimizations.

In fact, numerical optimizations rely on approximations of the gradients of the objective functions by finite differences. In an idealized setting, the smaller the differences are, the better the gradients can be approximated. In real situations many errors limit the minimal differences that can be used. One of the most important sources of errors are too large step sizes used by the integration routines. While relatively large step sizes are acceptable whenever the system does not change its structure, close to a state event the integration steps must be adapted to localize the event within a predefined error bound that is compatible with the differences chosen to approximate the gradients. Figure A.11 illustrates the main idea of such an iteration that has to take place each time a state event has been detected.

---

[5] Compared to the fully numerical solution, a speed-up factor of the order of 20 was observed in this particular example.

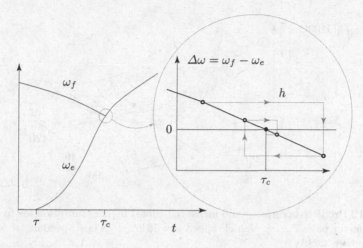

Fig. A.11 Illustration of the iterations necessary to correctly detect the state event "clutch sticks"

A.3.4 Results

The procedure outlined above was used to determine the fuel-optimal engine-on and engine-off speeds $\{\underline{\omega}, \bar{\omega}\}$ for various vehicle speeds. The results shown below are valid for the case $v = 50\,\text{km/h}$. Compared with a standard ICE-based powertrain, the fuel consumption was reduced by more than 50%. Extending the approach introduced above to the case of non-constant vehicle speeds and using the energy recuperation capabilities of the ETH Hybrid III prowertrain, the total fuel consumption in the city part of the MVEG–95 cycle could be reduced by almost 50%. A detailed description of all results of the ETH Hybrid III project can be found in [81].

Table A.1 shows the model parameters and Table A.2 lists the coefficients of the polynomials (A.10), (A.11), (A.13), and (A.14) used in the numerical optimization. The results of these calculations are shown in Fig. A.12. Also shown in that figure are five measured data points. As can be seen in that figure, the measured data matches well the predicted values and — more

Table A.1 Model parameters used in all calculations shown in this section

wheel speed	$\omega_w = 51.3\,\text{rad/s}$
engine inertia	$\Theta_e = 0.125\,\text{kg}\,\text{m}^2$
flywheel inertia	$\Theta_f = 2.8\,\text{kg}\,\text{m}^2$
dissipated power	$P_0 = 3.2\,\text{kW}$
scaling factor	$\kappa = 9.37 \cdot 10^3\,\text{l s/kg}\,100\,\text{km}$

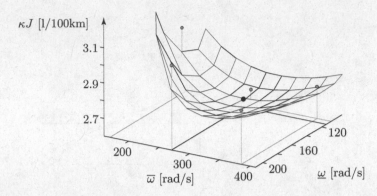

Fig. A.12 Predicted (surface) and measured (dots) fuel consumption for five different parameter pairs $\{\underline{\omega}, \bar{\omega}\}$. Vehicle speed $v = 50\,\text{km/h}$ (wheel speed $\omega_w = 51\,\text{rad/s}$, driving power $3.2\,\text{kW}$).

Table A.2 Coefficients of polynomials (A.10), (A.11), (A.13) and (A.14) (compatible units)

$p_1 = 4.0319 \cdot 10^1$	$p_3 = 8.3518 \cdot 10^{-3}$
$f_0 = -2.4796 \cdot 10^{-4}$	$f_1 = 7.3260 \cdot 10^{-6}$
$c_0 = 5.3375 \cdot 10^1$	$c_1 = 1.8222 \cdot 10^{-1}$
$c_2 = -2.4455 \cdot 10^{-4}$	
$k_0 = 1.0043$	$k_1 = 3.6707 \cdot 10^{-3}$

importantly — the best fuel economy is obtained using very similar duty cycle parameters $\{\underline{\omega}, \bar{\omega}\}$.

A.4 Case Study 4: Supervisory Control for a Parallel HEV

The problem discussed in this section is the supervisory control, i.e., the power split control of a parallel hybrid vehicle. This study is based on a quasistatic model of the system, which is validated with respect to experimental data of overall fuel consumption over regulatory drive cycles.

The availability of a validated model discloses the possibility of comparing the various control strategies treated in Chap. 7. In particular, the improvements of the sub-optimal controllers ECMS and T-ECMS with respect to a heuristic controller are assessed, with the performance calculated with the dynamic programming technique taken as a global optimum reference.

The problem was formulated and solved for the DaimlerChrysler Hyper, a prototypical parallel hybrid car based on the series-production Mercedes A-Class A 170 CDI. The work described in this section was first published in [291]. The description and the validation of the T-ECMS approach was first published in [296].

A.4.1 Introduction

The Hyper is a parallel hybrid vehicle with the two prime movers acting separately on the front and the rear axles. The thermal path consists of a front-wheel driven powertrain, with a 1700 cm^3 Diesel engine that yields 44 kW maximum power (66 kW in the series-production setup), and a 5-speed, automated manual gearbox. The electrical path includes a 6.5 Ah, 20 kW NiMH-battery pack and a motor/generator connected to the rear wheels via a second 5-speed gearbox. The two gearboxes are connected in such a way that they shift simultaneously.

The supervisory controller determines at each time how to split the mechanical power between the two parallel paths. In principle all of the ideas discussed in Chap. 7 could be applied to achieve minimum fuel consumption and a good self-sustainability of the battery during typical vehicle operation. The control results obtained with different strategies may be compared using a validated model of the system.

A.4.2 Modeling and Experimental Validation

The representation of the Hyper model with the QSS toolbox is shown in Fig. A.13. The relevant model parameters are listed in Table A.4.

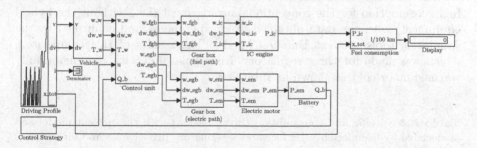

Fig. A.13 QSS model of the Hyper parallel hybrid vehicle

The rolling resistance is calculated as a fifth-order polynomial function of the vehicle speed,

$$F_r(v) = g \cdot m_v \cdot \left(a_0 + a_1 \cdot v(t) + a_2 \cdot v^2(t) + \right.$$
$$\left. + a_3 \cdot v^3(t) + a_4 \cdot v^4(t) + a_5 \cdot v^5(t)\right). \tag{A.16}$$

At the wheel axle, the basic relationship is the balance of torques

$$T_{fgb}(t) + T_{egb}(t) = T_w(t).$$

The control variable is the torque split factor u that regulates the torque distribution between the parallel paths. As described in Sect. 4.8), this factor is defined as $u(t) = T_{egb}(t)/T_w(t)$. The value $u(t) = 1$ therefore means that all the (positive) torque needed at the wheels is provided by the electrical path, or that all the (negative) torque available at the wheels from regenerative braking is entirely absorbed by the electrical path. When $u(t) = 0$, it means that all the torque needed at the wheels is provided by the fuel path. When $T_w(t) = 0$, no control is needed and u remains undefined.

The fuel consumption of the engine is calculated as a tabulated function of engine speed and torque (engine map)

$$\overset{*}{m}_f(t) = f_{ic}\left(T_{ic}(t), \omega_{ic}(t)\right). \tag{A.17}$$

The output power of the battery is calculated as a tabulated function of motor speed and torque

$$P_b(t) = f_{em}(T_{em}(t), \omega_{em}(t)). \tag{A.18}$$

For the battery, the equivalent circuit model illustrated in Fig. 4.29 has been adopted, where the inner resistance R_b is a constant and the open circuit voltage $U_{b,oc}$ is a tabulated function of the state of charge.

The reliability of the model developed is demonstrated by comparing the fuel consumption for the conventional arrangement (i.e., without the hybrid equipment) with the data obtained from experiments and confirmed by the simulation tools used at DaimlerChrysler. The results of this comparison, which was made for three regulatory drive cycles and with the engine fully warmed-up or cold, is shown in Table A.3.

Table A.3 Specific fuel consumption (liters/100 km) with the present model and as declared by DaimlerChrysler for various regulatory drive cycles and for warm as well as for cold start

Cycle	Model	Warm	Cold
ECE	5.51	5.54	6.54
EUDC	3.91	3.92	4.04
MVEG–95	4.49	4.52	4.96

Table A.4 Numerical values of the vehicle parameters

a_0	rolling resistance coefficient	$8.80 \cdot 10^{-3}$
a_1	rolling resistance coefficient	$-6.42 \cdot 10^{-5}$
a_2	rolling resistance coefficient	$9.27 \cdot 10^{-6}$
a_3	rolling resistance coefficient	$-3.30 \cdot 10^{-7}$
a_4	rolling resistance coefficient	$6.68 \cdot 10^{-9}$
a_5	rolling resistance coefficient	$-4.46 \cdot 10^{-11}$
A_f	frontal area	$2.31 \, \text{m}^2$
c_d	drag coefficient	0.32
$I_{b,max}$	battery maximum current	$\pm 200 \, \text{A}$
m_v	vehicle total mass	$1680 \, \text{kg}$
Q_0	battery charge capacity	$6.5 \, \text{Ah}$
r_w	wheel radius	$0.29 \, \text{m}$
R_b	battery inner resistance	$0.65 \, \Omega$
$\{\gamma_{egb}\}$	electric machine gear ratios	$\{12.38, \, 8.98, \, 6.45, \, 4.57, \, 3.32\}$
$\{\gamma_{fgb}\}$	engine gear ratio	$\{9.97, \, 5.86, \, 3.84, \, 2.68, \, 2.14\}$
η_{sm}	starter motor efficiency	0.6
Θ_{ic}	engine inertia	$0.195 \, \text{kg} \, \text{m}^2$
Θ_v	vehicle total inertia	$145 \, \text{kg} \, \text{m}^2$

A.4.3 Control Strategies

Three cumulative quantities are introduced for control purposes. They are the fuel energy use, $E_f(t) = \int_0^\tau H_l(\tau) \cdot \overset{*}{m}_f(\tau) \, d\tau$, the electrical energy use, $E_e(t) = \int_0^\tau I_b(\tau) \cdot U_{b,oc}(\tau) \, d\tau$, that is the variation (positive or negative) of the electrical energy stored, and the mechanical energy delivered at the wheels, $E_m(t) = \int_0^\tau T_w(\tau) \cdot \omega_w(\tau) \, d\tau$. The quantity H_l is the lower heating value of the fuel.

The control strategies whose performance will be compared are:

1. pure thermal operation, with $u(t) = 0$ except when (i) the required power at the wheels would be too high for the engine alone, and (ii) in case of regenerative braking, in which the regenerative power is collected with $u(t) = 1$;
2. the "parallel hybrid electric assist" strategy derived from the literature [158], according to which the engine is on when (i) the vehicle speed is higher than a set point, or (ii) if the vehicle is accelerating, or (iii) if the battery SoC is lower than a minimum value; when the engine is on, all the torque required at the wheels has to be provided by the engine path ($u = 0$), and an additional (positive or negative) torque is required that is proportional to the current SoC deviation; the minimum torque that the engine can provide is given by a specified fraction of the maximum torque at the current engine speed;

3. the global optimal control strategy, calculated off-line using dynamic programming techniques;
4. a pattern recognition technique, which estimates the parameter $s_0(t)$ from a comparison of the driving conditions in the near past with a set of reference patterns;
5. the ECMS as defined in Chap. 7; and
6. the T-ECMS also defined in Chap. 7.

The control performance considered in the following includes (i) the specific fuel consumption SFC (liters/100 km), (ii) the final battery SoC reached, starting from a value of 0.7, and (iii) the equivalent specific fuel consumption eSFC (liters/100 km). This latter figure is the sum of the fuel energy use and of the the electrical energy use weighted by the equivalence factor s_{dis} or s_{chg}. For the definition and the procedure to calculate these equivalence factors, see Chap. 7. For the Hyper and the MVEG–95 cycle, the values calculated are $s_{dis} = 2.7$, $s_{chg} = 1.6$. The corresponding plot $E_f(t_f) = f(E_e(t_f))$ is shown in Fig. A.14. Other parameter values for the MVEG–95 cycle are $\lambda = 0.21$ and $E_m(t_f) = 4.98\,\mathrm{MJ}$.

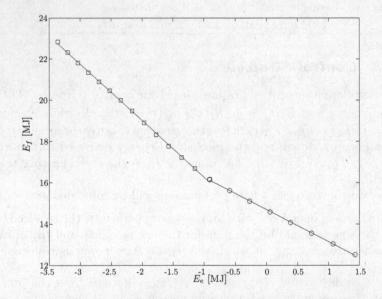

Fig. A.14 Dependency between $E_f(t_f)$ and $E_e(t_f)$ for the MVEG–95 cycle, data calculated (circles) and linearly fitted (solid lines). The slopes of the fitting curves are the two equivalence factors. Values of $E_{e0} = -0.90\,\mathrm{MJ}$, $E_{f0} = 16.16\,\mathrm{MJ}$ (see Chap. 7 for their definition.

The values of the mentioned parameters evaluated for other drive patterns are listed in Table A.5. Clearly, λ has a wider variability, being allowed to take every value from zero (no energy recuperable) to infinity (no positive energy delivered). For the regulatory drive cycles (first five rows), which all have a zero road slope, the values of λ depend only on the relative importance of deceleration trajectories. Therefore, this parameter is higher for urban cycles such as the ECE or the Japanese 10–15. The variability of λ increases for drive patterns with altitude variations, such as the proprietary AMS cycle. In contrast, the two equivalence factors show a much smaller variability. Also shown in Table A.5 is the parameter s_0 which is the constant equivalence factor that yields, when used in the policy given by (7.19), the globally optimum control law. This controller minimizes the fuel consumption and keeps the state of charge at the end of the mission equal to the initial value [295]. The values of s_0 are always between s_{chg} and s_{dis}.

Table A.5 Values of λ and of the equivalence factors for different drive patterns

Cycle	λ	s_{dis}	s_{chg}	s_0
MVEG–95	0.21	2.7	1.6	2.5
ECE	0.35	2.9	1.4	2.6
EUDC	0.12	2.9	1.6	2.2
FTP75	0.17	2.9	1.7	2.4
JP 10–15	0.35	3.2	1.4	2.2
AMS	0.09	3.1	1.8	2.3
SMD/mean	52.3%	5.6%	13.3%	

A.4.4   Results

Table A.6 summarizes the control performance evaluated for the MVEG–95 cycle with the control strategies mentioned in the previous section. The table clearly shows that, even without any particular control strategy, the hybridization process is beneficial for the fuel economy. In fact, the fuel consumption with the pure thermal mode is lower than the fuel consumption of the conventional arrangement. This reduction is due to two contributions: (i) an increase of 1.92 MJ of fuel energy due to the additional mass, and (ii) a reduction of 2.70 MJ due to the suppression of the idle consumption by turning the engine off. An additional amount of 0.91 MJ stored in the battery due to regenerative braking can be used to reduce the fuel consumption even further.

The operation of the heuristic controller ("parallel hybrid electric assist") with the parameters as in [158] yields only a limited improvement with respect to the pure thermal mode. Much more significant is the reduction of fuel consumption obtained with the causal optimal controller with pattern recognition. The reduction of the value of SFC of about 20% with respect to

Table A.6 Specific fuel consumption SFC (liters/100 km), final SoC (initial 0.7) and equivalent specific fuel consumption eSFC (liters/100 km) for the MVEG–95 cycle; [a]: the SFC in this case is lower than the global optimum because there is an associated depletion of SoC

Strategy, MVEG–95	SFC	SoC	eSFC
Conventional	4.49 (100%)	–	–
Pure thermal hybrid	4.28 (95%)	0.83	3.91 (87%)
Electric assist	4.25 (95%)	0.74	4.12 (92%)
Global optimum	3.18 (71%)	0.70	3.18 (71%)
Pattern recognition	3.41 (76%)	0.75	3.27 (73%)
ECMS[a]	3.13 (70%)	0.68	3.21 (71%)
T-ECMS	3.35 (75%)	0.74	3.25 (72%)

the pure thermal mode is due to the relatively good agreement between the "true" value of s_0 and the values estimated during the drive cycle. A further reduction of the SFC of about 30% with respect to the pure thermal mode is obtained with the ECMS. This value is very close to the global optimum, calculated with the dynamic programming technique, and it corresponds to a zero deviation of the battery SoC. The control performance of the T-ECMS is reasonably close to the global optimum as well.

The traces of the battery SoC during the MVEG–95 obtained with the ECMS and the T-ECMS are shown in Fig. A.15 together with the optimal trajectory calculated with the dynamic programming technique. The figure clearly shows that differences among the computed trajectories arise only in the second (i.e., extra-urban) part of the drive cycle.

Table A.7 extends the of the various control strategies to more drive patterns, including five regulatory drive cycles, a proprietary test drive cycle, and five patterns with strong altitude variations. The last ones (VN1 to VN5) are partial records of 100 s each of a drive in a mountainous region. The control performance is summarized by the eSFC and the final state of charge.

For the regulatory drive cycles the performance of the T-ECMS is very close to the global optimum. In some cases (e.g., ECE) the T-ECMS performs even better than the ECMS. In fact, on one hand, the ECMS profits from the a priori knowledge of λ. On the other hand, this is only an average value, while the various values of $\hat{\lambda}_k$ used by the T-ECMS are more accurate estimations since they are based on the future driving conditions. Due to the fact that for each cycle a good approximation of the actual value of s_0 is always found in the reference set, the pattern recognition technique is very effective as well.

When the road grade is considered, the pattern recognition technique seems no longer to be able to sustain the battery state of charge at reasonable levels. Indeed, in some cases the battery is completely depleted or overcharged before the end of the mission. This is due to the fact that, in the presence of grade, even patterns that show a similar set of characteristic parameters

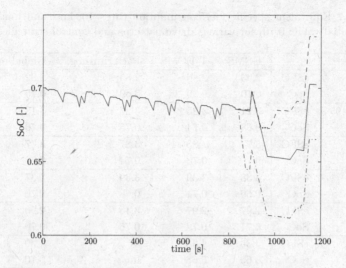

Fig. A.15 ECMS – Battery state of charge for the MVEG–95: optimal trajectory (solid), with the ECMS (dashdot) and the T-ECMS (dashed)

could have significantly different values of s_0. Thus their estimation could be incorrect. This effect is not clearly visible in the flat terrain cycles, where the variations of s_0 are less prominent. In contrast, the T-ECMS seems to be able to sustain the charge even in the presence of large altitude variations. This is accomplished without reducing the capability of optimizing the fuel consumption. In fact, the values for the eSFC that are obtained with the T-ECMS are still close to the global optimum.

The performance of the ECMS strongly depends on the choice of the parameters λ, s_{dis}, s_{chg}. The influence of λ on the equivalent fuel consumption is shown in Fig. A.16 for the ECE cycle. The cycle-averaged value $\lambda = 0.35$ actually yields a minimum value for the eSFC. For variations of λ in an interval of twice its standard mean deviation, variations of about 10% in the eSFC are calculated. For the MVEG–95 cycle, the same calculation yields a sensitivity of about 5%.

The two parameters s_{dis}, s_{chg} are responsible for weaker variations of the eSFC. In Fig. A.16 the curves at given values of eSFC in the ECE cycle are plotted as a function of the two parameters (for $s_{dis} > s_{chg}$). The cycle values $(2.91, 1.41)$ yield an eSFC close to the minimum. For variations of s_{dis} and s_{chg} in an interval of twice their standard mean deviation, variations in the eSFC of about 2% and 1%, are calculated, respectively. Similar numbers have been obtained for the MVEG–95 cycle.

As described in Chap. 7, such an analysis confirms that the parameter λ has a large variability and a large influence on the controller performance. Thus, it has to be estimated accurately. In contrast, the equivalence factors have a weaker influence, at least in their typical range of variability which

Table A.7 Equivalent specific fuel consumption (liter/100 km) and final state of charge (initial value: 0.70) for various drive patterns and control strategies

		ECMS	T-ECMS	Pattern recog.	Global optimum
ECE	eSFC	2.93	2.86	2.85	2.82
	SoC	0.70	0.69	0.69	0.70
MVEG–95	eSFC	3.21	3.25	3.27	3.18
	SoC	0.68	0.74	0.75	0.70
EUDC	eSFC	3.77	3.85	3.87	3.77
	SoC	0.70	0.73	0.74	0.70
FTP	eSFC	3.38	3.39	3.39	3.37
	SoC	0.69	0.71	0.68	0.70
10–15	eSFC	2.97	2.97	3.15	2.90
	SoC	0.71	0.72	0.76	0.70
AMS	eSFC	4.35	4.33	–	4.30
	SoC	0.69	0.68	lim	0.70
VN1	eSFC	7.45	7.45	7.50	7.42
	SoC	0.68	0.65	0.47	0.70
VN2	eSFC	3.54	3.61	–	3.50
	SoC	0.70	0.67	lim	0.70
VN3	eSFC	6.12	6.13	6.24	6.11
	SoC	0.60	0.60	0.32	0.70
VN4	eSFC	0.79	0.68	0.84	0.68
	SoC	0.60	0.71	0.52	0.70
VN5	eSFC	3.31	3.25	3.27	3.18
	SoC	0.67	0.63	0.00	0.70

is rather small. It is as a consequence of this fact that the T-ECMS keeps s_{dis} and s_{chg} constant, i.e., a pair of average values is conveniently selected to represent the vehicle and is then used for every drive condition.[6]

An example of how the T-ECMS estimates the velocity profile $\hat{v}_k(t)$ during a typical urban drive cycle is shown in Fig. A.17. The various plots refer to sequential values of t_k, $k = 1, \ldots, 8$. The instants t_k with the corresponding signals I_k are listed in Table A.8. This set of information has been derived from the velocity profile of the ECE drive cycle. All the changes in the vehicle acceleration as scheduled in the drive cycle have been translated to a corresponding value I_k to simulate the real-time output of a telemetry system. Figure A.17 clearly shows that the more information is used, the closer the estimated velocity profile is to the original drive cycle.

---

[6] For the system considered here, a suitable choice made after inspection of Table A.5 was: $s_{dis} = 2.6$, $s_{chg} = 1.7$.

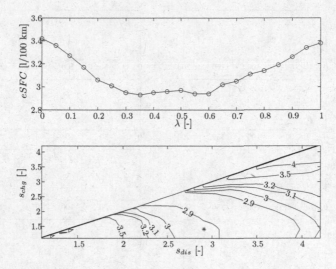

Fig. A.16 Equivalent specific fuel consumption with the ECMS as a function of λ (top) and as a function of s_{dis} and s_{chg} (bottom), for the MVEG–95 cycle. The asterisk corresponds to the values listed in Table A.5.

Table A.8 Scheduled information used to simulate the ECE cycle and estimated values of λ (EOM: end-of-mission)

k	t_k	I_k	d_o (m)	v_o (m/s)	$\hat{\lambda}_k$
1	11	GO	–	–	0.25
2	15	STOP	45.8	0	0.28
3	49	GO	–	–	0.26
4	61	STOP	266.6	0	0.41
5	117	GO	–	–	0.33
6	143	STOP	263.1	9.7	1.52
7	176	STOP	–	–	∞
8	180	EOM	–	–	–

The estimation of λ associated with the velocity profile of Fig. A.17 is presented in Fig. A.18. The successive estimates made at t_k, $k = 1, \ldots, 7$ are plotted as dots and shown with the "true" curve $\lambda(t)$. This is obtained by applying its definition at the various instants, with the energy terms evaluated from the current time to the end of the mission. The correspondence between the values of $\hat{\lambda}_k$ and $\lambda(t_k)$ is evident.

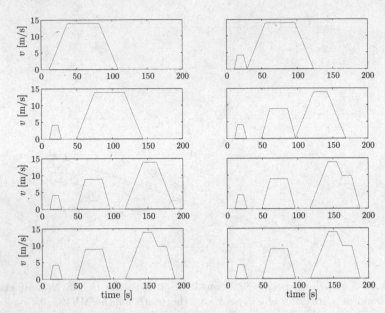

Fig. A.17 Estimated velocity profile during the ECE cycle. The various plots refer to the different instants (see Table A.8) at which a new piece of information is available.

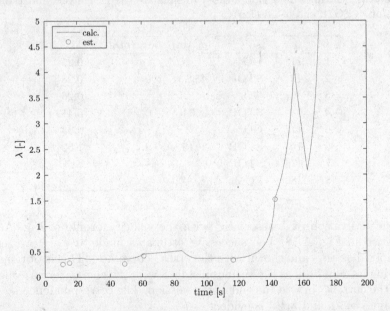

Fig. A.18 Calculated curve of λ and values estimated with the T-ECMS, for the ECE cycle. The cycle-averaged value of λ is 0.35.

A.5 Case Study 5: Optimal Rendez-Vous Maneuvers

The problem discussed in this section is the fuel-optimal control of a vehicle which approaches a leading car whose speed is known using telemetry. The speed of the leader can be zero (fixed obstacle) or greater than zero (moving obstacle) and can also vary with time. The process ends when the follower has reached the speed of the leader. The distance between the two vehicles at this point can be rigidly specified (e.g. zero), or just forced to be nonnegative. Similarly, the time needed for the process can be specified or unconstrained. In all cases the legal speed limits have to be taken into account.

Two problem settings will be discussed:

- unspecified rendez-vous time t_f and specified final distance $\delta(t_f)$; and
- specified rendez-vous time t_f and unspecified final distance $\delta(t_f)$.

For both cases, the optimal trajectory will be found with the theory of optimal control, in such a way as to minimize the specific fuel consumption. The work described in this section was first published in [294].

A.5.1 Modeling and Problem Formulation

The two vehicles are assumed to drive at time $t = 0$ with speed $w(0) = w_0$ (leader) and $v(0) = v_0$ (follower). It is assumed that the speed trajectory of the leading vehicle may be predicted to be

$$w(t) = \begin{cases} w_0 + \dfrac{v_1 - w_0}{t_1} \cdot t & t < t_1 \\ v_1 & t \geq t_1 \end{cases} . \tag{A.19}$$

Accordingly, the distance it travels will be given by

$$e(t) = \begin{cases} w_0 \cdot t + \dfrac{v_1 - w_0}{2 \cdot t_1} \cdot t^2 & t < t_1 \\ v_1 \cdot t - \dfrac{v_1 - w_0}{2} \cdot t_1 & t \geq t_1 \end{cases} . \tag{A.20}$$

The velocity v_1 is the actual top speed limitation and may be obtained using GPS-based on-board navigation systems. The time interval t_1 is the leading vehicle's acceleration time which may be estimated using measurements of its previous velocity.

The dynamics of the leader vehicle are completely described by the equations (A.19) and (A.20). The follower's dynamics are given by

$$m \frac{d}{dt} v(t) = F(t) - c_0 - c_2 v^2(t), \tag{A.21}$$

$$\frac{d}{dt} d(t) = v(t), \tag{A.22}$$

with the two initial conditions $v(0) = v_0 \geq 0$ and $d(0) = 0$. Accordingly, the distance between the two vehicles is given by

$$\delta(t) = e(t) - d(t) + \delta(0), \tag{A.23}$$

where of course the constraint $d(t) \geq 0$ has to be satisfied for all times t by all admissible solutions (the initial distance $\delta(0)$ is assumed to be known as well). In the case analyzed in Sect. A.5.2, the conditions that have to be met at the final time are

$$v(t_f) = v_1 \quad \text{and} \quad d(t_f) = d_f = e(t_f) + \delta(0). \tag{A.24}$$

Constant gear ratios are assumed here, i.e., vehicle and engine/motor speed are assumed to be strictly proportional. The extension to more than one gear ratio is straightforward and can be accomplished using the same ideas as introduced below.

An important point to clarify is the connection between the mass flow $\overset{*}{m}_f(t)$ of the fuel consumed by the engine and the mechanical power $F(t)v(t)$ delivered to the vehicle. Many approaches have been proposed for this relation, most of which are too complicated to be used for practical optimal control problems. A simple yet sufficiently precise description of this relation is given by the Willans formulation as introduced in Sect. 3.1.3. For the purposes of this case study, the following formulation is used

$$F(t) = \frac{e \cdot H_l}{v(t)} \cdot \overset{*}{m}_f(t) - F_0. \tag{A.25}$$

Here, the positive constants e, H_l, and F_0 describe the engine's internal efficiency, the fuel's lower heating value, and the engine's mechanical friction and gas exchange losses, respectively. The propulsion force is assumed to be limited by

$$F \in [-F_0, F_{max}], \tag{A.26}$$

where the upper limit corresponds to the maximum force (torque) of the engine and the lower limit to its friction force (torque). In general, these parameters are functions of the engine speed, however, since the dependency is not very strong it may be neglected in a first analysis. Notice that $F = -F_0$ implies a fuel mass flow of zero. In this condition the engine's internal friction losses are compensated by the diminishing kinetic energy of the vehicle (fuel cut-off). Several extensions of this problem formulation could be investigated, for instance the effects of irreversible or regenerative braking. However, this would exceed the scope of this case study.

The criterion which has to be minimized is the fuel consumed per distance travelled while satisfying the requirements of the rendez-vous maneuver as described above, i.e.,

$$J = \frac{\int_0^{t_f} \overset{*}{m}_f(t)\, dt}{\int_0^{t_f} v(t)\, dt} = \frac{\int_0^{t_f} (F(t) + F_0) \cdot v(t)\, dt}{e \cdot H_l \cdot d(t_f)}. \tag{A.27}$$

A.5.2 *Optimal Control for a Specified Final Distance*

The classical theory of optimal control is applied to the problem described above, with t_f treated as a constant in a first step. The optimal final time may then be found in a second step solving a one-parameter nonlinear programming problem.

Obviously, for a fixed final time t_f and distance $d(t_f)$, minimizing the criterion $J(F)$ given by (A.27) is equivalent to minimizing the criterion

$$\tilde{J} = \int_0^{t_f} F(t) \cdot v(t)\, dt, \tag{A.28}$$

because e, H_l, F_0, and $d(t_f)$ are all positive constants and because

$$\int_0^{t_f} F_0 \cdot v(t)\, dt = F_0 \cdot d_f \tag{A.29}$$

is a positive constant for all admissible solutions as well.

The Hamiltonian function associated with this problem is given by

$$H(t) = F(t) \cdot \left(v(t) + \frac{\lambda_1(t)}{m} \right) - \\ \frac{\lambda_1(t)}{m} \cdot (c_0 + c_2 \cdot v^2(t)) + \lambda_2(t) \cdot v(t), \tag{A.30}$$

and the dynamic behavior of the adjoint state variables is described by

$$\frac{d}{dt}\lambda_1(t) = 2 \cdot c_2 \cdot v(t) \frac{\lambda_1(t)}{m} - F(t) - \lambda_2(t), \tag{A.31}$$

$$\frac{d}{dt}\lambda_2(t) = 0. \tag{A.32}$$

Notice that the Hamiltonian is time invariant and affine in the control signal $F(t)$. Obviously, this Hamiltonian reaches its minimum at the extreme values of the control signal and, according to the Maximum Principle, the optimal control signal will be discontinuous ("bang-bang control") in the regular case

$$F_{opt} = \begin{cases} F_{max} & \sigma(t) > 0 \\ F_\sigma & \sigma(t) \equiv 0 \\ -F_0 & \sigma(t) < 0 \end{cases}, \tag{A.33}$$

where $\sigma(t) = v(t) + \lambda_1(t)/m$. In the singular case ($F_{opt} = F_\sigma$) the optimal force has to be chosen such that the vehicle velocity is constant (see Lemma 1 below). In the regular case, i.e., if the solution does not include singular arcs, there is at most one switching event. This can be explained by analyzing the limiting case where $c_2 \approx 0$ (negligible aerodynamic drag). In this case the time derivative of the switching function $\sigma(t)$ is given by the expression

$$\frac{d}{dt}\sigma(t) = -\frac{c_0 + \lambda_2}{m}. \tag{A.34}$$

Since $\lambda_2(t) = \lambda_2$ is constant – see (A.32) – the switching function is an affine function of the time t and it can switch its sign at most once. Since the ODEs (A.21) and (A.31) are smooth functions this assertion will be true for sufficiently small values of c_2 as well.

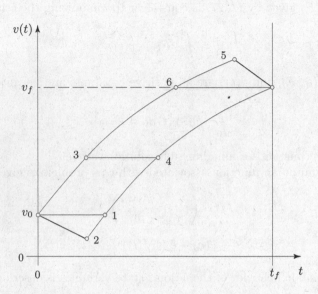

Fig. A.19 Classification of possible solution families

Lemma 1: During the singular arc phase the vehicle velocity is constant.

Proof: The singular arc is characterized by the condition $\sigma(t) \equiv 0$ for all $t \in [t_a, t_b]$ and therefore the condition $\dot{\sigma}(t) \equiv 0$ has to be satisfied in this interval as well. From the first condition the equality $v(t) = -\lambda_1(t)/m$ may be derived and from the second condition

$$\frac{c_0 + \lambda_2}{m} = \frac{c_2}{m} \cdot \left(\frac{2 \cdot v(t) \cdot \lambda_1(t)}{m} - v^2(t) \right). \tag{A.35}$$

Inserting in this expression the equality $v(t) = -\lambda_1(t)/m$ the following expression is obtained

$$\frac{c_0 + \lambda_2}{c_2} = -\frac{3 \cdot \lambda_1^2(t)}{m^2}. \tag{A.36}$$

Since the expression on the left-hand side of (A.35) is a constant – again, see (A.32) – during the time interval $t \in [t_a, t_b]$ the adjoint variable $\lambda_1(t)$ has to be constant as well. Using the relation $v(t) = -\lambda_1(t)/m$, which is true for $t \in [t_a, t_b]$, the assertion follows.

The specific value of the singular velocity may be found as

$$v_\sigma = \sqrt{\frac{-c_0 - \lambda_2}{3 \cdot c_2}}.$$

(A.37)

Going back to the general problem the following qualitative points are important to be understood:

- With the exception of two particular combinations of v_1 and d_f, the problem is over-determined. The two special cases are indicated in Fig. A.19 by the curves $v_0 \to 2 \to v_f$ and $v_0 \to 5 \to v_f$, respectively, and they correspond to the case where d_f is exactly equal to the area under the speed trajectory obtained for F equal to either F_{max} or $-F_0$.
- For values of d_f larger than $d_f(v_0 \to 5 \to v_f)$ or smaller than $d_f(v_0 \to 2 \to v_f)$, no solution is possible. For values of d_f that are intermediate between the two extreme values, only solutions that include singular arcs are possible, as shown by the example $v_0 \to 3 \to 4 \to v_f$.
- The two values of d_f corresponding to the curves $v_0 \to 1 \to v_f$ and $v_0 \to 6 \to v_f$ separate two regions in which the control signal changes its sign at both sides of the singular arc, from a region in which the control signal keeps its sign.
- In all cases, the control law is determined by one or two real parameters: either the regular switching time t_s or the singular arc initial and final time $\{t_a, t_b\}$. These one or two degrees of freedom are used to satisfy the two boundary conditions.

As mentioned above, the optimization problem analyzed in this section has no unused degrees of freedom, i.e., the two conditions $v(t_f) = w(t_f)$ and $d(t_f) = d_f$ fully define the unknown control parameters $\{t_a, t_b\}$ after optimal control theory has been used to determine the *qualitative* form of the optimal control signal. Therefore, the problem may be re-stated as a simple parameter optimization procedure.

Moreover, the differential equations (A.21)–(A.22) have to be solved for a piecewise constant input F. The resulting ODEs are separable and a closed-form solutions becomes possible. An example for the case $v_0 \to 6 \to v_f$ and $v_0 \to 5 \to v_f$ of Fig. A.19 is presented in the original paper [294]. Analogously, closed-form expressions for $v(t)$ and $d(t)$ may be obtained for all the other situations depicted in Fig. A.19.

Unfortunately, it is generally not possible to explicitly solve the equations

$$v(t_f) = w(t_f),$$

(A.38)

$$d(t_f) = d_f$$

(A.39)

using the closed-form expressions for $v(t)$ and $d(t)$. Numerical approaches will have to be used for that, for instance a minimization of the criterion

$$\epsilon = \frac{(v(t_f) - w(t_f))^2}{w^2(t_f)} + \frac{(d(t_f) - d_f)^2}{d_f^2}.$$

(A.40)

However, these computations are not very heavy because:

- No differential equations have to be solved numerically, i.e., the velocity and the distance of the follower vehicle are defined in closed form.
- The gradients of the criterion (A.40) with respect to the switching times $\{t_a, t_b\}$ (the only unknown parameters) may be computed explicitly.

Figure A.20 shows two solutions obtained with the approach described above, for the values $t_1 = 25\,\text{s}$ ($d_f = 1.15\,\text{km}$) and $t_1 = 35\,\text{s}$ ($d_f = 1.05\,\text{km}$), respectively. The other data are identical in both cases: $v_0 = 15\,\text{m/s}$, $w_0 = 10\,\text{m/s}$, $v_1 = 30\,\text{m/s}$, $t_f = 40\,\text{s}$, $\delta_0 = 200\,\text{m}$, and the vehicle parameters were chosen to be $F_{max} = 1900\,\text{N}$, $F_0 = 300\,\text{N}$, $c_0 = 150\,\text{N}$, $c_2 = 0.43\,\text{N}\,\text{s}^2/\text{m}^2$, $m = 1500\,\text{kg}$.

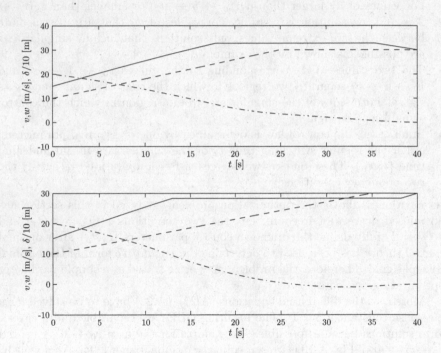

Fig. A.20 Case a) feasible fuel-optimal trajectory (top) and case b) not feasible fuel-optimal trajectory (bottom). Solid lines: $v(t)$; Dashed lines: $w(t)$; Dashdot lines: $\delta(t)$.

Figure A.20 also shows that solutions in terms of $v(t)$ and $d(t)$ are not automatically guaranteed to be feasible. The additional constraint $\delta(t) > 0$ $\forall t \in [0, t_f]$ has to be checked a posteriori.

Up to now the final time t_f has been assumed to be constant. If this parameter is varied as well, the fuel consumption decreases for $t_f > t_1$ because the acceleration part becomes less important compared to the cruising part in which the follower vehicle has reached the desired speed v_1. Assuming the leader vehicle to follow the expected trajectory (A.19) for all times, the limiting case $t_f \to \infty$ is characterized by the fuel consumption (expressed in the more familiar $1/100\,\mathrm{km}$ unit)

$$\overset{*}{m}_{f,\infty} = \frac{c_0 + c_2 \cdot v_1^2 + F_0}{e \cdot H_l \cdot \rho_f}, \tag{A.41}$$

with ρ_f being the fuel density.

A.5.3 Optimal Control for an Unspecified Final Distance

When the distance traveled is not imposed, the performance index defined by (A.27) is still valid. However, the distance is not a constant anymore and therefore the simplified formulation (A.28) is no longer correct. Using the plant dynamics (A.21)–(A.22) to express $F(t)$ as a function of $v(t)$ and $\dot{v}(t)$ and using partial integration, the performance index (A.27) can be written as

$$J = \frac{1}{e \cdot H_l} \cdot \left\{ F_0 + c_0 + m \cdot \frac{v^2(t_f) - v_0^2}{2 \cdot d(t_f)} + \frac{c_2}{d(t_f)} \int_0^{t_f} v^3(t)\, dt \right\}. \tag{A.42}$$

The main difference with respect to the case of Sect. A.5.2 is that (A.42) now includes also an integral term. To transform it into the standard Euler–Lagrange formulation, a third state variable is needed. One possible choice is the normalized mass of fuel consumed m_f, so that the new third state equation will be

$$\frac{d}{dt} m_f(t) = (F(t) + F_0) \cdot v(t). \tag{A.43}$$

The performance index is now simply

$$J = \frac{m_f(t_f)}{d(t_f)} \tag{A.44}$$

which does not contain any integral term. The resulting Euler–Lagrange equations are

$$\frac{d}{dt} \lambda_1(t) = \frac{2 \cdot c_2 \cdot v(t)}{m} \cdot \lambda_1(t) - \lambda_2(t) - (F(t) + F_0) \cdot \lambda_3(t), \tag{A.45}$$

$$v(t_f) = w(t_f), \tag{A.46}$$

$$\frac{d}{dt} \lambda_2(t) = 0, \tag{A.47}$$

$$\lambda_2(t_f) = -\frac{m_f(t_f)}{d^2(t_f)}, \tag{A.48}$$

$$\frac{d}{dt} \lambda_3(t) = 0, \tag{A.49}$$

$$\lambda_3(t_f) = \frac{1}{d(t_f)}. \tag{A.50}$$

Obviously, the optimal control law is again discontinuous with its switching time defined by the switching function $\sigma(t)$

$$F = \begin{cases} F_{max} & \sigma(t) < 0 \\ F_\sigma & \sigma(t) \equiv 0 \, , \\ -F_0 & \sigma(t) > 0 \end{cases} \tag{A.51}$$

where $\sigma(t) = \lambda_1(t)/m + v(t) \cdot \lambda_3$. A sensitivity analysis with respect to the parameter c_2 of the solutions in terms of speed (obtained using parameter optimization procedures) is shown in Fig. A.21. The following points are worth mentioning:

- From $c_2 = 0$ up to a certain limit value ($c_2 = c_{2,crit} \approx 0.13$), the optimal speed trajectory is qualitatively similar to the case $c_2 = 0$. A "bang-bang" controller is optimal, and the switching time is given by the condition that the switching function σ changes its sign. The switching time increases with an increase of c_2.
- The limit value of $c_{2,crit}$ is smaller than typical values of c_2 for series production vehicles ($c_2 = 0.5 \cdot \rho_a \cdot c_d \cdot A_f > 0.25$). Singular arcs are therefore likely to appear.
- From the limit value of c_2 onward, the switching function reaches at a time t_a the zero value with a zero derivative, i.e., $\sigma(t_a) \to 0$, $\dot{\sigma}(t_a) \to 0$, and tends to remain constant. In this case, a singular arc occurs. The corresponding control signal cannot be determined by the condition $\dot{\sigma} = 0$, because the first derivative

$$\frac{d}{dt}\sigma(t) = \frac{1}{m} \left(\frac{2 \cdot c_2 \cdot \lambda_1(t)}{m} \cdot v(t) - \lambda_2 - \right.$$
$$\left. - (F_0 + c_0) \cdot \lambda_3 - c_2 \cdot \lambda_3 \cdot v^2(t) \right) \tag{A.52}$$

is independent of the control signal. However, using the second derivative of $\sigma(t)$ the missing information may be found.
- Moreover, $\ddot{\sigma}$ shows that along the singular arcs the vehicle velocity is constant. In fact, from

$$\frac{d^2}{dt^2}\sigma(t) = \frac{2 \cdot c_2 \cdot v(t) \cdot \dot{v}(t) \cdot \lambda_3}{m} = 0 \tag{A.53}$$

it follows that $\dot{v} = 0$ (all other quantities are larger than zero).
- The velocity on the singular arc is given by

$$v_\sigma = \sqrt{-\frac{\lambda_2 + F_0 \cdot \lambda_3 + c_0 \cdot \lambda_3}{3 \cdot c_2 \cdot \lambda_3}}. \tag{A.54}$$

This expression is not directly applicable because both λ_2 and λ_3 depend on the unknown final distance $d(t_f)$. Nevertheless, (A.54) is useful because it shows the influence of c_2 on the optimal solution. In fact, for sufficiently small values of c_2, the singular velocity is larger than the top speed of the regular "bang-bang" solution and no singular arcs will appear.

- As c_2 increases, v_σ decreases until it falls below v_1 for a certain limit value $c_2 = c_{2,sw} \approx 0.25$. In this case, the singular arc separates two periods with $F(t) = F_{max}$.
- The (unrealistic) upper limit of $c_2 \approx 1.80$ corresponds to the situation in which the final speed no longer can be reached, even with constant maximum thrust.
- The optimal trajectories obtained represent the best trade-off between the losses $\left(v_f^2 - v_0^2\right)/(2 \cdot m \cdot d_f)$, which should require a maximum d_f and correspondingly higher speeds, and the term $c_2/d_f \cdot \int v^3(t)\,dt$ which, for sufficiently large values of c_2, requires lower speed values.

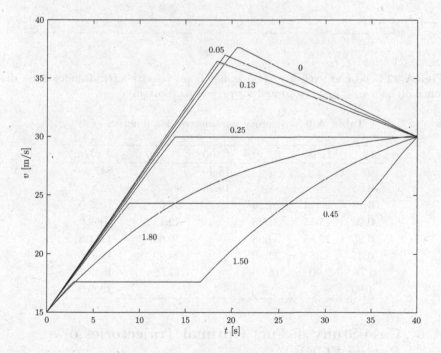

Fig. A.21 Parameter sweep ($c_2 = 0.0 \to 1.80$)

The solutions for the two cases $c_2 = 0.20 < c_{2,sw}$ and $c_2 = 0.45 > c_{2,sw}$ are shown for illustration purposes in Fig. A.22 (the vehicle parameters are the same as those used in Sect. A.5.2). Table A.9 shows the parameters of the optimal trajectories as a function of c_2.

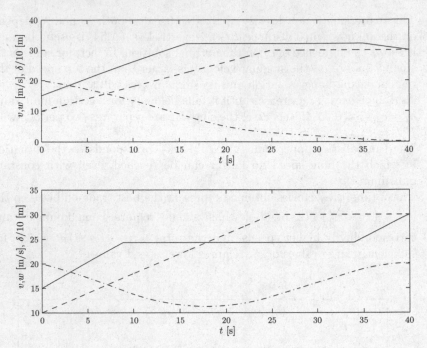

Fig. A.22 Solid lines: velocity $v(t)$; dashed lines: velocity $w(t)$; dashdot lines: distance $\delta(t)$, for $c_2 < c_{2,sw}$ (top) and for $c_2 > c_{2,sw}$ (bottom)

Table A.9 Trajectory parameters as a function of c_2

c_2	t_a (s)	t_b (s)	t_s (s)	v_σ (m/s)	J (J/m)
0	–	–	18.40	–	847.2
0.05	–	–	19.26	–	922.8
0.13	20.54	20.86	–	37.62	1002.0
0.20	15.76	34.96	–	32.13	1069.6
0.25	13.88	39.98	–	30.00	1111.8
0.45	8.88	33.96	–	24.29	1246.6
1.50	2.90	16.66	–	17.59	1843.1
1.80	–	–	–	–	2200.0

A.6   Case Study 6: Fuel Optimal Trajectories of a Racing FCEV

The problem discussed in this section consists of determining the trajectories of an ultra-light, three-wheeler FCEV that minimize the hydrogen consumption over a given race circuit. The latter is characterized by the total length and road slope versus distance covered. Moreover, the particular case analyzed here imposes a lower limit to the average speed in such a way that the time to cover the circuit is also a problem constraint.

The problem is solved using optimal control theory. This requires a mathematical model of the vehicle powertrain that is developed and validated in the first section. Then the optimal control law is analytically determined. Finally, the results presented are compared with those yielded by a conventional PID controller.

The work described in this section was first published in [297].

A.6.1 Modeling

The system considered here consists of: (i) a PEM fuel cell system, (ii) a DC traction motor equipped with a DC–DC converter, (iii) a fixed gear reduction, (iv) a lightweight, three-wheeler car body.

The model derived belongs to the class of dynamic models. Each power converter is represented by a submodel, whose input and output variables are the power factors at the input and output stages of the component. The choice of the input variables and of the output variables is made according to the physical direct causality. Figure A.23 illustrates the variables exchanged by the various submodels.

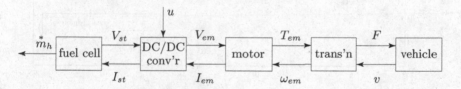

Fig. A.23 Block diagram of the FCEV

The fuel-cell submodel describes a fuel-cell stack with all the auxiliary devices that are necessary to supply hydrogen, air, and the coolant flows. A fraction of the current delivered by the fuel cell stack $I_{fc}(t)$ is drawn off to drive such auxiliaries. The power balance of (6.28) is written in terms of currents as

$$I_{fc}(t) = I_{st}(t) + I_{aux}(t). \tag{A.55}$$

The stack current is related to the fuel cell voltage by means of the static polarization curve. Measurements taken at the fuel cell terminals are shown in Fig. A.24. The experimental data have been fitted with a nonlinear polarization model (M1)

$$U_{st}(t) = c_0 + c_1 \cdot \exp\left(-c_2 \cdot I_{fc}(t)\right) - c_3 \cdot I_{fc}(t) \tag{A.56}$$

derived from (6.21)–(6.22) and with a linear model (M2)

$$U_{st}(t) = c_4 - c_5 \cdot I_{fc}(t), \tag{A.57}$$

which is derived from (6.25). In both cases, the effect of the concentration overvoltage has been neglected, as it occurs only at very high current densities.

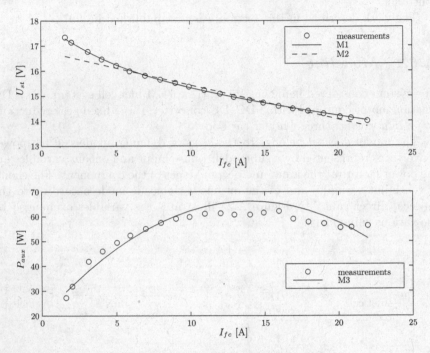

Fig. A.24 Fuel cell voltage (top) and auxiliary power (bottom) as a function of the stack current. Measured data (circles), fitted with (A.56) (solid), and with (A.57) (dashed).

The power drained by the auxiliaries depends on the mass flow rates of air and hydrogen and on the heat generated by the stack that is removed through the coolant flow. Ultimately, the auxiliary power is a function of the fuel cell current, as shown in Fig. A.24. The data measured are fitted with (6.71). Since (6.32) shows that the hydrogen mass flow rate is proportional to the fuel cell current, $\overset{*}{m}_h(t) = c_9 \cdot I_{fc}(t)$, (6.71) is rewritten as

$$P_{aux}(t) = c_6 + c_7 \cdot I_{fc}(t) + c_8 \cdot I_{fc}^2(t). \tag{A.58}$$

The motor controller senses the actual motor speed and the accelerator signal $u(t)$ (control input to the system) and determines the motor armature voltage according to the following rule

$$U_{em}(t) = \kappa_{em} \cdot \omega_{em}(t) + K \cdot R_{em} \cdot u(t), \tag{A.59}$$

where κ_{em} is the motor torque constant, R_{em} the armature resistance, and K a static accelerator gain ($K = 30$).

The current required at the input stage of the converter is given by a balance of power

$$I_{st}(t) = \frac{U_{em}(t) \cdot I_{em}(t)}{\eta_c \cdot U_{st}(t)}, \tag{A.60}$$

where η_c is the constant efficiency of the converter.

The motor submodel follows the standard armature representation of DC motors, (4.23)

$$I_{em}(t) = \frac{U_{em}(t) - \kappa_{em} \cdot \omega_{em}(t)}{R_{em}}, \tag{A.61}$$

$$T_{em}(t) = \kappa_{em} \cdot I_{em}(t). \tag{A.62}$$

The transmission consists of a fixed gear placed between the motor and the drive wheel axle. The tractive force and input speed are calculated as

$$F(t) = \eta_t^{\pm 1} \cdot \frac{\gamma \cdot T_{em}(t)}{r_w}, \tag{A.63}$$

$$\omega_{em}(t) = \frac{\gamma \cdot v(t)}{r_w}, \tag{A.64}$$

where η_t is the constant transmission efficiency, γ is the constant transmission ratio, and r_w is the wheel radius. The positive sign in (A.63) is valid for $T_{em} > 0$ (traction), the negative sign for $T_{em}(t) < 0$ (braking). The numerical values of the model parameters are listed in Table A.10.

Table A.10 Numerical values for the fuel cell vehicle model

c_0	15.93	γ	20.4
c_1	2.06	r_w	0.25 m
c_2	0.200	m_v	115 kg
c_3	0.0876	m_r	1.1
c_4	16.80	ρ_a	1.2 kg/m^3
c_5	0.137	A_f	0.3 m^2
c_6	19.89	c_d	0.3
c_7	6.60	c_r	0.002
c_8	−0.236	g	9.81 m/s^2
c_9	0.208	κ_{em}	0.0168
η_c	0.95	R_{em}	0.08 Ω

Combining (A.55)–(A.64) an overall equation is derived, relating the stack current to the acceleration signal and the vehicle speed,

$$I_{fc}(t) = \frac{P_{aux}(t)}{U_{st}(I_{fc}(t))} + \frac{K \cdot u(t)}{\eta_c \cdot U_{st}(I_{fc}(t))} \left(K \cdot R_{em} \cdot u(t) + \kappa_{em} \cdot \frac{\gamma}{r_w} \cdot v(t) \right).$$

(A.65)

This equation is implicit since $P_{aux}(t)$ and $U_{st}(t)$ are functions of $I_{fc}(t)$ through (A.58) and (A.56) or (A.57). The dependency of the tractive force on the acceleration signal is given by

$$F(t) = \frac{\eta_t \cdot \gamma}{r_w} \cdot \kappa_{em} \cdot K \cdot u(t).$$

(A.66)

Using (A.66), the vehicle dynamics may be described by[7]

$$\frac{d}{dt}v(t) = h_1 \cdot u(t) - h_2 \cdot v^2(t) - g_0 - g_1 \cdot \alpha(x(t)),$$

(A.67)

where the new parameters h_1, h_2, g_0, and g_1 are defined as functions of the physical quantities c_r, c_d, A_f, and m_v, and $x(t)$ is the distance travelled

$$\frac{d}{dt}x(t) = v(t).$$

(A.68)

The vehicle model (A.65)–(A.67) is not suitable for use in optimal control theory due to its highly nonlinear structure. Therefore, a simpler model was sought to relate the hydrogen mass flow rate to the vehicle speed and the acceleration signal. On the basis of (A.65) and (6.32), the following three-parameter structure (M4) was selected

$$\overset{*}{m}_h(t) = b_0 + b_1 \cdot v(t) \cdot u(t) + b_2 \cdot u^2(t).$$

(A.69)

Notice that the model (A.69) is strictly equivalent to (A.65) if (i) the auxiliary current is approximated by an affine function of the hydrogen mass flow rate, and (ii) the fuel cell voltage is approximated by a constant value.

A comparison of the validated models M1 and M3 and its three-parameter counterpart M4 is shown in Fig. A.25 in terms of hydrogen mass flow rate as a function of the vehicle speed, with $u(t)$ as a parameter. The comparison shows a good agreement at low acceleration and speed, while higher differences are observed at higher power levels (both higher $u(t)$ and $v(t)$). Therefore, the three-parameter model should be considered to be a valid approximation for low-power operating conditions. The constants b_0, b_1, b_2 are listed in Table A.11.

A.6.2 Optimal Control

The optimization problem can be stated in mathematical terms as follows: find the control law $u(t)$, $t \in [0, t_f]$ that minimizes the vehicle fuel consumption

---

[7] An approximation valid for small grade has been introduced, $\sin \alpha \approx \alpha$, $\cos \alpha \approx 1$. Moreover, the vehicle mass m_v is increased by the quantity m_r representing the inertia of the rotational masses.

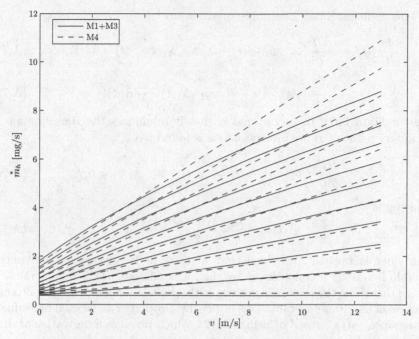

Fig. A.25 Hydrogen consumption and mass flow rate with the physical model (solid) and the three-parameter model (dashed) as a function of the vehicle speed, for values of u from 0 to 1 regularly spaced with step 0.1

Table A.11 Fitted values for the three-parameter model

b_0	0.475 mg/s
b_1	0.707 mg/m
b_2	1.24 mg/s

$\int_0^{t_f} \overset{*}{m}_h(t)\, dt$, subject to the constraint of a given average speed v_m. The latter condition can be written as $x(t_f) = x_f$ where x_f is the given length of the drive cycle and $t_f = x_f/v_m$. The problem is defined by the choice of v_m and by the particular grade profile $\alpha(x)$.

Using the results of the previous sections, the problem is formally described in the framework of optimal control theory. The system state variables are $v(t)$ and $x(t)$, the system state equations being (A.67) and (A.68). The incremental cost $\overset{*}{m}_h(t)$ is expressed by the three-parameter model of (A.69) as a function of the state and control variables. The Hamiltonian function is thus constructed as

$$H(u, v, x, \lambda_1, \lambda_2) = b_0 + b_1 \cdot v(t) \cdot u(t) + b_2 \cdot u^2(t) + \lambda_1(t) \cdot \{h_1 \cdot u(t) - $$
$$- h_2 \cdot v^2(t) - g_0 - g_1 \cdot \alpha(x(t))\} + \lambda_2(t) \cdot v(t),$$

$$(A.70)$$

where $\lambda_1(t)$ and $\lambda_2(t)$ are two Lagrange multipliers.

The resulting Euler–Lagrange equations are

$$\frac{d}{dt}\lambda_1(t) = -\frac{\partial H}{\partial v} = -b_1 \cdot u(t) + 2 \cdot h_2 \cdot \lambda_1(t) \cdot v(t) - \lambda_2(t), \qquad (A.71)$$

$$\frac{d}{dt}\lambda_2(t) = -\frac{\partial H}{\partial x} = g_1 \cdot \lambda_1(t) \cdot \frac{d}{dx}\alpha(x). \qquad (A.72)$$

The condition for u to be optimal is that it minimizes the Hamiltonian. A stationary point of H with respect to u is found as

$$\frac{\partial H}{\partial u} = b_1 \cdot v(t) + 2 \cdot b_2 \cdot u(t) + h_1 \cdot \lambda_1(t) = 0, \qquad (A.73)$$

resulting in

$$u^o(t) = -\frac{h_1 \cdot \lambda_1(t) + b_1 \cdot v(t)}{2 \cdot b_2}. \qquad (A.74)$$

This value is optimal, provided that $u^o(t) \in [0,1]$. Otherwise the control variable lies along a constrained arc, i.e., it is either $u^o(t) = 0$ or $u^o(t) = 1$. Notice that since the system is not explicitly time dependent, the Hamiltonian function of (A.70) is constant. This noticeable property can be used to evaluate for instance $\lambda_2(t)$ instead of using (A.72), which involves a derivative of $\alpha(x)$ with respect to x.

The boundary conditions of the optimization problem are split between the initial time $t = 0$ and the terminal time t_f. Two cases will be discussed. The former is representative of a vehicle start scenario, the latter of steady repetitions of the same periodic route. For a start cycle, the boundary conditions are

$$v(0) = 0, \quad x(0) = 0, \qquad (A.75)$$

$$\lambda_1(t_f) = 0, \quad x(t_f) = x_f = v_m \cdot t_f, \qquad (A.76)$$

with the second of (A.76) replacing any condition of the Lagrange multiplier $\lambda_2(t_f)$, which is therefore unconstrained.

For a periodic route the boundary conditions are

$$x(0) = 0, \qquad (A.77)$$

$$\lambda_1(t_f) = \lambda_1(0), \quad x(t_f) = x_f = v_m \cdot t_f, \quad v(t_f) = v(0), \qquad (A.78)$$

with the initial value of the speed not assigned but a new constraint imposed at the terminal time.

The system of (A.67), (A.68), (A.71), (A.72), and (A.74) is highly nonlinear and subject to inequality constraints on the control variable. Thus a fully analytical solution cannot be found.

A way of numerically solving the optimal control problem consists of searching the initial values for the Lagrange multipliers and the initial speed, $\lambda_1(0)$, $\lambda_2(0)$, and $v(0)$ which, when applied to the aforementioned system of differential equations, lead to a fulfillment of the terminal boundary conditions. These particular values will be referred to as $\lambda_1^o(0)$, $\lambda_2^o(0)$, and $v^o(0)$ since they generate the optimal control law $u^o(t)$. Notice that in the start-cycle case the initial speed assigned is $v^o(0) = 0$.

The optimization results will be compared with a conventional driving strategy, in which a PID regulator tries to achieve the desired average speed. Such a strategy emulates the behavior of a human driver who tries to follow a target speed. Its mathematical definition for the start-cycle case is

$$u(t) = K_p \cdot (f \cdot v_m - v(t)) + K_i \cdot \int_0^t (f \cdot v_m - v(t))\, dt, \qquad (A.79)$$

where f is a factor, usually very close to unity, which is tuned to achieve an exact match between v_m and the actual average speed. For the periodic route case the reference control law is

$$u(t) = K_1 \cdot \int_0^t (v(0) - v(t))\, dt + K_2 \cdot \int_0^t (v_m \cdot t_f - x(t))\, dt, \qquad (A.80)$$

where the two integral terms try to keep the terminal speed close to the initial value and the average speed close to the target value, respectively.

A.6.3 Results

As previously stated, the dynamic optimization system is analytically solved through the Euler–Lagrange equations, and thus it is converted to a static parametric optimization with two or three degrees of freedom, namely $\lambda_1^o(0)$, $\lambda_2^o(0)$, and possibly $v^o(0)$.

The optimal controller was developed for a drive cycle having a required average speed of 30 km/h and a known grade profile. The total length of the cycle is 3636 m.

Figure A.26 shows the optimization results obtained for the start-cycle case. The optimal speed trajectory and the optimal control law are compared with the corresponding traces calculated with the reference PID strategy. The figure also shows the hydrogen mass flow rate and the hydrogen mass consumption as a function of time, as well as the traces of $\lambda_1(t)$ and $\lambda_2(t)$ corresponding to the optimal trajectory.

The PID controller uses the values $K_i = 0$, $K_p = 0.5$, $f = 1.054$. After some trial-and-error tests, these values seemed to provide the best result in terms of fuel consumption, and they guarantee the fulfillment of the constraint over the average speed.

The values found using the initial Lagrange multipliers are $\lambda_1^o(0) = -9.92$, and $\lambda_2^o(0) = -0.305$. The mass of fuel consumed with the PID strategy is 758 mg, with the optimal controller 700 mg. The gain in fuel economy is thus around 7.5%. In both cases the desired average speed is achieved within a

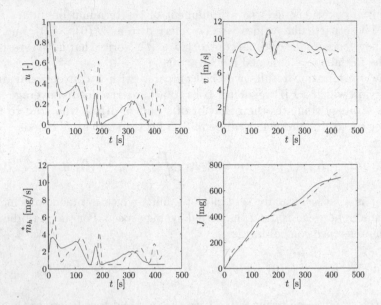

Fig. A.26 Optimization results: control law, speed trajectory, hydrogen mass flow rate, hydrogen mass consumption as a function of time for the start-cycle case. Solid lines: optimal control; dashed lines: PID control.

tolerance of 0.1%. The optimal trajectory $\lambda_1^o(t)$ converges to zero at the end of the cycle with an even narrower tolerance.

Figure A.26 clearly shows how both strategies start with the maximum acceleration ($u = 1$) to reach a speed close to the required average value. However, this phase is shorter in the optimal trajectory, thus the optimal speed increases less rapidly. The PID controller is characterized by two long constrained arcs, i.e., time intervals in which the control signal stays along its lower bound $u = 0$. This behavior seems to be avoided by the optimal controller, except for a very short time at the end of the two main downhills (approximately, $t = 150$ s and $t = 190$ s). In the final part of the route, the optimal controller completely releases the accelerator, with the vehicle free to roll. Another difference between the two strategies is that the average value of the optimal control law (0.19) is lower than for the PID controller (0.23).

Figure A.27 shows the optimization results obtained for the periodic route case. The PID controller uses the values $K_1 = 0.111$, $K_2 = 5.92 \cdot 10^{-6}$. The optimized parameters found are $\lambda_1^o(0) = -20.00$, $\lambda_2^o(0) = -0.25$, $v^o(0) = 8.20$. The mass consumption with the PID strategy is 672 mg, with the optimal controller 613 mg. The gain in fuel economy is thus around 9%.

These traces also show that the optimal control law is much smoother than the PID controller output. The latter consists of several acceleration spikes

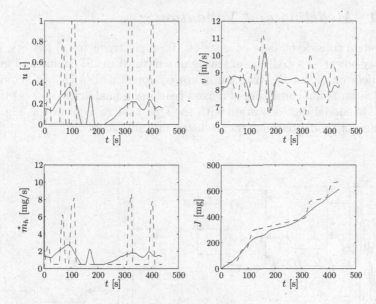

Fig. A.27 Optimization results: control law, speed trajectory, hydrogen mass flow rate, hydrogen mass consumption as a function of time, periodic route case. Solid lines: optimal control, dashed lines: PID control.

which try to boost the vehicle speed when it becomes too low. Consequently, the fuel mass flow rate shows a largely irregular variation as well. However, the average values of the optimal and the PID control laws are almost the same (0.15).

Similarly to vehicle speed, the optimal Lagrange multipliers are both periodic also. The periodic nature of $\lambda_1(t)$ is imposed through the constraint of (A.78), while for $\lambda_2(t)$ it is a consequence of the invariance of the Hamiltonian.

A.7 Case Study 7: Optimal Control of a Series Hybrid Bus

The problem illustrated in this section consists of determining the control law that minimizes the fuel consumption of a series hybrid bus over a specified route, which is assigned in terms of length and altitude profile.

The problem is solved using optimal control theory. This requires a mathematical model of the vehicle powertrain that is developed and validated in the first section. Then the optimal control law is analytically determined. Finally, the results presented are compared with those yielded by a conventional, thermostat-type controller.

The work described in this section was first published in [14].

A.7.1 Modeling and Validation

The system considered here consists of: (i) an electrochemical battery, (ii) an auxiliary power unit (APU) consisting of a natural gas SI engine and an electric generator, (iii) a DC traction motor equipped with a DC–DC converter, (iv) a transmission consisting of a fixed gear and a final drive, and (v) the bus body. The model representation with the QSS toolbox is shown in Fig. A.28. The relevant model parameters are listed in Table A.12.

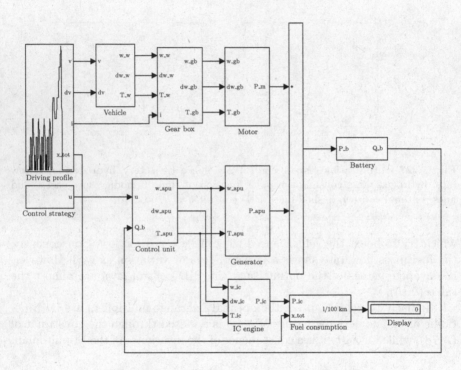

Fig. A.28 QSS model of the series hybrid bus

Table A.12 Relevant data of the system studied in this section

m_v	4436 kg
c_r	0.015
c_d	0.40
A_f	4.7 m^2
r_w/γ_{gb}	0.0259 m
η_{gb}	0.9

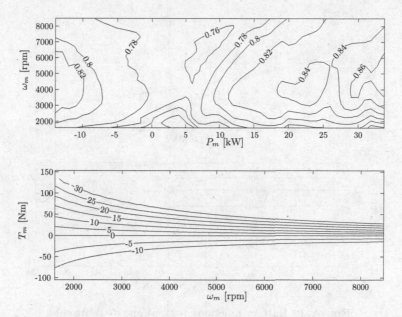

Fig. A.29 Traction motor efficiency as a function of speed and electric power (top). Motor electric power (kW) as a function of speed and torque (bottom).

The traction motor is described here by its quasistatic efficiency map. The motor efficiency is provided by the manufacturer as a function of input (electric) power and shaft speed, as shown in Fig. A.29. From these data, the input power is calculated as a function of speed and torque

$$P_m = f_m\left(\omega_m, T_m\right). \tag{A.81}$$

The battery current is provided by the manufacturer as a function of the power required P_b and the battery depth of discharge $\xi = 1 - q_b$,

$$I_b = f_b\left(P_b, \xi\right). \tag{A.82}$$

The variation of the battery depth of discharge is tabulated as a function of the battery current and power,

$$\frac{d}{dt}\xi(t) = \begin{cases} \dfrac{P_b(t)}{f_\xi(I_b(t))}, & P_b(t) > 0 \\[2ex] \kappa \cdot P_b(t), & P_b(t) < 0 \end{cases}. \tag{A.83}$$

The form of the two functions f_b and f_ξ are shown in Fig. A.30.

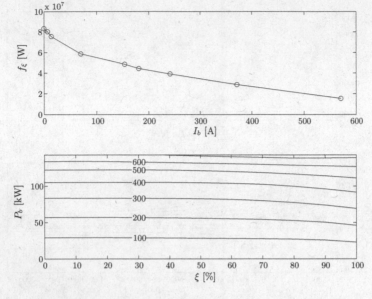

Fig. A.30 Battery functions f_ξ (top) and f_b (bottom)

The combination of (A.82) and (A.83) yields the variation of $\xi(t)$ as a function of the battery power. This dependency is fitted with the control-oriented model

$$\frac{d}{dt}\xi(t) = c_{b,0}(\xi) + c_{b,1}(\xi) \cdot P_b(t) + c_{b,2}(\xi) \cdot P_b^2(t), \qquad (A.84)$$

$$c_{b,i}(\xi) = c_{b,i0} + c_{b,i1} \cdot \xi(t), \quad i = 0, 1, 2, \qquad (A.85)$$

whose validation is presented in [14].

The input variable of the APU submodel is the electrical power P_{apu} required at the APU output stage. The operating point of the engine is related only to the power rather than to its single factors, i.e., voltage and current. Thus the operating point is selected in such a way as to maximize the APU efficiency at every power request. The efficiency is a tabulated function of the engine torque and speed,

$$\eta_{apu} = f_\eta \left(\omega_{apu}, T_{apu} \right). \qquad (A.86)$$

The map f_η provided by the manufacturer is illustrated in Fig. A.31. The engine speed is controlled as a function of the APU power in such a way as to maximize the efficiency of the APU. A lower limit of 2000 rpm has been introduced to enhance drivability. Combining these two pieces of information the input power, i.e., the chemical power associated with the fuel, is evaluated as a function of the output power. This dependency is shown in Fig. A.31, together with an affine approximation of the same curve. The agreement between the two curves clearly leads to a control-oriented model of the Willans type [272] for this component,

$$P_f(t) = \frac{\dot{P}_{apu}(t) + P_{f,0}}{e_{apu}},$$ (A.87)

where $P_{f,0} = 22.5\,\mathrm{kW}$ is the external power loss and $e_{apu} = 0.329$ the internal efficiency. The fuel consumption mass flow rate is $\overset{*}{m}_f(t) = P_f(t)/H_l$, where H_l is the lower heating value of the fuel.

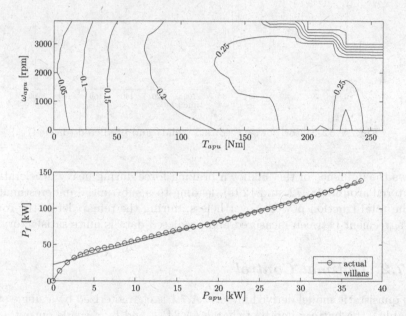

Fig. A.31 Model of the APU: efficiency as a function of engine speed and torque (top) and fuel power as a function of the APU power (bottom)

The motor speed $\omega_m(t)$ and the motor power $P_m(t)$ were recorded in a test drive of the bus. The vehicle acceleration is calculated from the speed and filtered. The braking force was not recorded, thus it represents a major source of uncertainty. A relatively flat route was chosen for the test with the aim of isolating the strong influence of grade on the traction power. Although always smaller than 2–3%, the road slope has been estimated accurately from altitude data.

The motor power predicted with the system model previously presented is compared with the recorded trace shown in Fig. A.32. The figure clearly

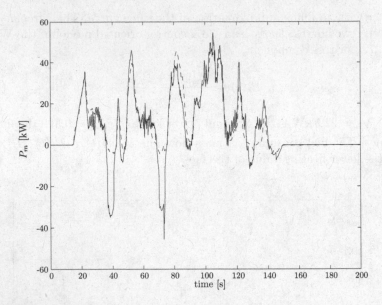

Fig. A.32 Motor power, recorded (solid) and predicted (dashed)

shows the influence of the unknown braking force during heavy decelerations (centered around 38, 72, and 129 s), leading to a substantial underestimation of the total traction power. Nevertheless, during the remainder of the route, the agreement between measured and predicted data is quite satisfactory.

A.7.2 Optimal Control

The quasistatic model derived in Sect. A.7.1 is characterized by a single state variable – the battery depth of charge (A.84) – and by a single output – the fuel consumption mass flow rate (A.87). The power flows from the two paths are balanced at the DC link level,

$$P_b(t) + P_{apu}(t) = P_m(t). \tag{A.88}$$

The control variable $u(t)$ is defined here as the ratio between the power flowing in the battery path and the total power

$$P_b(t) = u(t) \cdot P_m(t), \quad P_{apu}(t) = (1 - u(t)) \cdot P_m(t). \tag{A.89}$$

The optimization problem can be stated in precise mathematical terms as follows: find the control law $u(t)$, $t \in [0, t_f]$ that minimizes the fuel consumption with a constraint over the battery depth of discharge $\xi(t_f) = \xi(0)$.

Using the results of the previous sections, the problem is formally described in the framework of optimal control theory introduced in Chap. 7. The Hamiltonian function is constructed as

$$H(\xi, u, P_m) = \overset{*}{m}_f(t) + \lambda \cdot \dot{\xi}(t), \tag{A.90}$$

where $\lambda(t)$ is a Lagrange multiplier. The corresponding Euler–Lagrange equation is

$$\frac{d}{dt}\lambda(t) = -\frac{\partial H}{\partial \xi} = -\lambda(t) \cdot \frac{\partial \dot{\xi}}{\partial \xi}. \tag{A.91}$$

The condition for $u(t)$ to be optimal is that it must minimize the Hamiltonian. Notice that since the system is not explicitly time-dependent, the Hamiltonian function is constant with time.

The derivative $\partial \dot{\xi}/\partial \xi$ in (A.91) may be calculated using the parameterization of (A.85). A further simplification consists of neglecting the dependency of $\dot{\xi}$ on ξ which is justified if ξ varies only slightly around its initial level. Since this is actually one of the goals of the supervisory controller, such a simplification will be used in the following, leading to

$$\dot{\lambda}(t) = 0, \quad \lambda(t) = \text{constant} = \lambda^o. \tag{A.92}$$

The value λ^o must be chosen such that the condition $\xi(t_f) = \xi(0)$ is satisfied. Under this assumption, the following sub-optimal control law is found as

$$u^o(t) = \arg \min_u \left\{ \overset{*}{m}_f(u, P_m(t)) + \lambda^o \cdot \dot{\xi}(u, P_m(t)) \right\}. \tag{A.93}$$

This control law corresponds to the minimization of a cost function that only depends on the current driving conditions (via $P_m(t)$) and on λ^o. Therefore, it may be interpreted as a real-time controller, provided that the optimal value of λ^o is known. This in turn depends on the whole drive cycle, which can be reasonably estimated using, for instance, the techniques proposed in [291, 296] for parallel hybrid vehicles.

An alternative way to apply (A.93) is in terms of power flows. The first part of the right-hand term $\overset{*}{m}_f(t)$ can be easily converted into fuel chemical power after multiplication by the lower heating value of the fuel. The term $\dot{\xi}(t)$ can be also converted into electrochemical power after multiplication by the battery open-circuit voltage. This is slightly dependent on the state of charge but, if the latter is kept reasonably constant, this dependency can be neglected as it has been done for $\dot{\xi}(t)$. This means that (A.93) can be rearranged as

$$u^o(t) = \arg \min_u \left\{ P_f(u, P_m(t)) + \lambda^o \cdot P_b(u, P_m(t)) \right\}. \tag{A.94}$$

The search for the optimal values $u^o(t)$ may be done with a simple inspection procedure, i.e., several values of u are tested at each time and the respective values of the cost function are compared to find the minimum. The accuracy of such a method increases with the dimension of the test set and thus with the complexity.

An alternative approach may use the control-oriented model derived in the previous section to calculate the Hamiltonian as an analytical function. After minimization with respect to u the optimal control law then follows. The operating limits of u are given as

$$1 - \frac{P_{apu,max}}{P_m(t)} \le u(t) \le 1, \quad P_m(t) > 0, \tag{A.95}$$

$$1 \le u(t) \le 1 - \frac{P_{apu,max}}{P_m(t)}, \quad P_m(t) < 0, \tag{A.96}$$

since the motor power is always lower than the maximum power available from the battery.

To analytically calculate the optimal control law, various cases should be considered (in the following, $u^-(t) = 1 - P_{apu,max}/P_m(t)$).

For $P_m(t) < 0$ the limits of $u(t)$ given by (A.95)–(A.96) are both positive. Thus the battery power is negative and the Hamiltonian, see (A.83), is calculated as

$$H(u, P_m) = \frac{P_{f,0}}{H_l} + \frac{(1 - u(t)) \cdot P_m(t)}{e_{apu} \cdot H_l} + \lambda^o \cdot \kappa \cdot u(t) \cdot P_m(t). \tag{A.97}$$

The Hamiltonian varies linearly with the control signal u and it exhibits a minimum for $u = 1$ or $u = u^-(t)$, according to the sign of the derivative of (A.97) with respect to u

$$u^o(t) = \begin{cases} 1, & \lambda^o \cdot \kappa < \dfrac{1}{e_{apu} \cdot H_l} \\ u^-(t), & \lambda^o \cdot \kappa > \dfrac{1}{e_{apu} \cdot H_l} \end{cases}. \tag{A.98}$$

For $0 < P_m(t) < P_{apu,max}$ the limits of $u(t)$ have opposite signs. In the range $u^-(t) \le u < 0$ the Hamiltonian varies linearly with u. In $u = 0$ there is a discontinuity, since $H(0^+) > H(0^-)$. For $0 < u < 1$ the Hamiltonian may have a local minimum. The latter is found by setting to zero the derivative of H with respect to u

$$\frac{\partial H}{\partial u} = -\frac{P_m(t)}{e_{apu} \cdot H_l} + \lambda^o \cdot \big(c_{b,1}(\xi(t)) \cdot P_m(t) + 2 \cdot c_{b,2}(\xi(t)) \cdot u^* \cdot P_m^2(t)\big) = 0, \tag{A.99}$$

which yields

$$u^* = \frac{\dfrac{1}{e_{apu} \cdot H_l \cdot \lambda^o} - c_{b,1}}{2 \cdot P_m(t) \cdot c_{b,2}} \tag{A.100}$$

if $0 < u^* < 1$. Summarizing, the optimal control variable in this case is defined by

$$
u^o(t) = \begin{cases} u^-(t), & \lambda^o \cdot \kappa > \dfrac{1}{e_{apu} \cdot H_l} \\[3ex] \arg\min_u\{H(0^-), H(u^*)\}, & \lambda^o \cdot \kappa < \dfrac{1}{e_{apu} \cdot H_l} \end{cases} .
\tag{A.101}
$$

Again, for $P_m(t) > P_{apu,max}$ the limits of u are both positive, thus the battery power is also positive. A possible stationary point for the Hamiltonian is found using (A.100). It is optimal if it respects the constraints on u, otherwise the optimal control trajectory lies on a constrained arc

$$
u^o(t) = \arg\min_u\{H(u^-(t)), H(u^*), H(1)\}.
\tag{A.102}
$$

A.7.3 Results

The sub-optimal control law given by (A.93) has been tested along a given route from the city of L'Aquila, Italy to the Faculty of Engineering located on a hill about 950 m high. The road slope $\alpha(x)$ has been calculated from topographic data, while the speed profile $v(t)$ represents a typical mission of a public bus connecting the two localities.

The sub-optimal controller adopted minimizes the Hamiltonian function with a search procedure. The test set of the control variable includes the values $u = -1, -0.5, \ldots, 1$ plus the lower and upper values dictated by the APU operating limits. Moreover, the optimal control law depends on the value of λ. There is only one value of that parameter that fulfills the condition $\xi(t_f) = \xi(0)$ (continuous curve in Fig. A.33). Higher values of λ tend to excessively penalize the use of the battery as a prime mover, thus the final state of charge is too high (dash-dotted curve in Fig. A.33). Lower values of λ tend to overfavor the use of the battery, thus the final state of charge is too low (dashed curve in Fig. A.33).

The control results are compared with those calculated using: (i) the dynamic programming technique, (ii) a thermostatic-type controller (as specified below), and (iii) the procedure described in Sect. A.7.2 to find the minimum of the Hamiltonian analytically. The results of dynamic programming are practically coincident with those of the sub-optimal controller [14] and thus they do not appear in Fig. A.34 that shows a comparison of state of charge trajectories and fuel mass consumed. Bellman's dynamic programming algorithm [295] was implemented with a time step of 1 s and a state of charge step size of $5 \cdot 10^{-5}$. The test set of u is the same as for the sub-optimal controller. The thermostatic-type controller uses two limits for the depth of discharge, ξ_{hi} and ξ_{lo}. When $\xi > \xi_{lo}$ the APU output is set to 30 kW. When $\xi < \xi_{hi}$

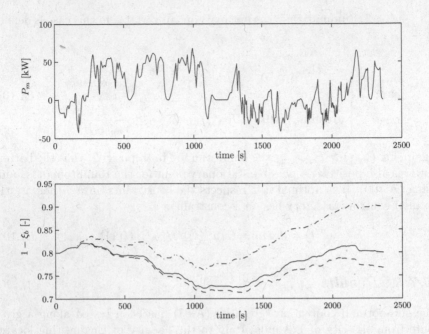

Fig. A.33 Route L'Aquila–faculty of engineering. Top: electric power required (positive) and provided (negative) by the traction motor. Bottom: trajectory of the battery state of charge (1-ξ) obtained with $\lambda = 7.5 \cdot 10^8$ (dashdot), $\lambda = 6.7 \cdot 10^8$ (solid), and $\lambda = 6 \cdot 10^8$ (dashed).

the APU is turned off. For intermediate levels of SoC the APU status is kept equal to that at the previous time step.

The improvement in fuel economy obtained with optimal control with respect to a conventional controller is evident from the figures. Due to its inefficient on/off nature, the thermostatic controller exhibits a fuel mass consumption which is 29% higher than the global optimum calculated using the dynamic programming approach.

The figures also confirm that the control laws calculated with the minimization of the Hamiltonian are nearly optimal. Only a difference of less than 1% arises between the fuel mass consumption obtained with dynamic programming and the minimization of the Hamiltonian. The difference slightly increases to 5% if the analytical minimization is considered, mainly because of the approximation errors inherent to the control-oriented model, and despite the fact that the analytical minimization may find optimal values of u which are not included in the test set of the search procedure. However, the analytical minimization has the advantage of a smaller computation time required.

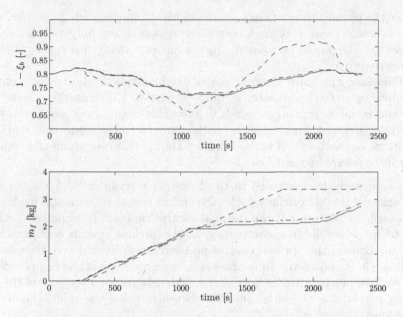

Fig. A.34 Battery state of charge trajectory (top) and mass of fuel consumed (bottom): sub-optimal controller (solid), sub-optimal controller with analytical minimization (dashdot), on/off thermostat-type controller (dashed line).

A.8 Case Study 8: Concept Evaluation for a Hybrid Pneumatic Engine

This case study aims at evaluating two different concepts of a hybrid pneumatic engine (HPE). Three engine concepts are evaluated with respect to their fuel economy and drivability using quasi-static models and deterministic dynamic programming:

- As a baseline benchmark, a conventional naturally aspirated internal combustion engine with a rated power of 99 kW is evaluated. Arguments that speak for such a system are low complexity and cost. However, the downsizing potentials of such an engine are limited since in order to reach the same maximum power a large turbo charger is needed, which greatly reduces drivability due to the "turbo-lag".
- In order to estimate the full potential of a HPE, the second system under consideration is an HPE where all valves are actuated by an electro-hydraulic valve system (EHVS) and are therefore fully variable. Such an engine has the potential to reach maximum fuel economy: the engine can be drastically downsized and turbocharged without sacrificing drivability, since air from the pressure tank can be used to overcome the turbo lag. Furthermore, the fully variable valves allow for a large degree of freedom

for the optimization of each engine mode. However, such a complex system would incur very high cost since at least three fully variable valves per cylinder would be needed. During this case study this concept will be referred to as FV.

- Therefore, the third concept under consideration is the realistic option where only the charge valve is equipped with an EHVS and the intake and exhaust valves remain camshaft driven. This engine combines low system complexity and a moderate cost with improved fuel economy that is mainly due to downsizing and turbocharging. During this case study this concept will be referred to as FCS.

This case study is structured in the following way: in Sect. A.8.1 the control oriented models of the vehicle, the engine and the pneumatic modes are presented. In Sect. A.8.2 the optimal control problem is formulated which has to be solved for the comparison of the various system configurations. The solutions of the optimal control problem are presented and discussed in Sect. A.8.3. Based on the simulation results a prototype of the engine system was realized in hardware. In Sect. A.8.4 the hardware is presented and the fuel saving potential is validated on a test bench utilizing the vehicle emulation technique.

A.8.1 Control Oriented Models

The hybrid pneumatic vehicle (HPV) is modeled in a quasi-static way. All vehicle variables are calculated based on a given velocity profile that the driver intends to follow. The information is propagated through the vehicle in a noncausal sense. The discretized model is described as

$$x_{k+1} = f(x_k, u_k, v_k, a_k, i_k), \qquad (A.103)$$

where x_k represents the internal energy U_t of the pressure tank, the control input u_k is the chosen engine mode, v_k is the vehicle speed, a_k is the vehicle acceleration, and i_k stands for the gear, each at the time instant k. Since the tank volume V_t does not change, the internal energy is a function of the tank pressure only:

$$U_t = \frac{p_t \cdot V_t}{\kappa - 1}, \qquad (A.104)$$

where κ is the ratio of specific heats of air. The elements of the HPV model are described in the following subsections.

Vehicle

For this investigation, a mid-size vehicle is chosen. The parameters are listed in Table A.13. The total vehicle mass m_v is

$$m_v = m_{v0} + m_{ice} + m_t,$$

Table A.13 Vehicle parameters

parameter	symbol	value	note
vehicle base mass	m_{v0}	1450 kg	
ICE mass	m_{ice}	$V_d \cdot 67\,\text{kg/l}$	
air tank mass	m_t	25 kg	hybrid only
air tank volume	V_t	30 l	hybrid only
gearbox efficiency	η_{gb}	0.93	
auxiliary power	P_{aux}	400 W	
rolling friction	c_{r0}	0.0065	
	c_{r1}	$1.22 \cdot 10^{-5}\,\text{s/m}$	
	c_{r2}	$2.34 \cdot 10^{-6}\,\text{s}^2/\text{m}^2$	
air resistance	$A_f \cdot c_d$	$0.83\,\text{m}^2$	
wheel radius	r_w	0.31 m	
air density	ρ_a	$1.293\,\text{kg/m}^3$	

where m_{v0} is the vehicle base mass, m_{ice} is the mass of the ICE which depends on the displacement V_d of the engine, and m_t is the mass of the air tank.

The signals relevant for the HPE are calculated based on the drive cycle (vehicle speed v and acceleration a). The force on the vehicle F is determined by the vehicle acceleration force, the air resistance, and the rolling friction force

$$F = m_v \cdot a + \frac{1}{2} \cdot c_d \cdot A_f \cdot \rho_a \cdot v^2 + m_v \cdot g \cdot (c_{r0} + c_{r1} \cdot v + c_{r2} \cdot v^2).$$

The rolling friction is approximated by a second-order polynomial, found by fitting actual measurement data. Assuming a 5-speed gearbox with a constant efficiency η_{gb}, F and v translate to an engine rotational speed ω_e and a demanded engine torque T_e as follows,

$$\omega_e = \frac{v}{r_w} \cdot \gamma(i), \tag{A.105}$$

$$T_e = \frac{F \cdot r_w}{\gamma(i)} \cdot \eta_{gb}^{-sign(F)} + T_{EHVS} + \frac{P_{aux}}{\omega_e}, \tag{A.106}$$

where T_{EHVS} is the torque demand of the hydraulic pump used for the actuation of the EHVS, P_{aux} is the electric power needed for the auxiliaries, r_w is the wheel radius, and $\gamma(i)$ is the gear ratio of the gear chosen. The gear shifting is realized using a rule-based strategy for all configurations.

Internal Combustion Engine (ICE)

To capture the behavior of a throttled and a dethrottled engine, the extended Willans approximation presented in [128] is used

$$p_{me} = \underbrace{\left(\hat{e}(\omega_e) + \frac{p_{mgas0}}{p_{mf,NA}(\omega_e)} \right)}_{e(\omega_e)} \cdot p_{mf} - \underbrace{(p_{mfric}(\omega_e) + p_{mgas0} + p_{mheat})}_{p_{me0}(\omega_e)},$$

where p_{mfric} is the mechanical friction mean pressure, p_{mgas0} denotes the maximum pumping losses and p_{mheat} captures other effects contributing to the drag torque such as heat losses. The Willans parameter $e(\omega_e)$ consists of the thermodynamic efficiency $\hat{e}$, increased by the reduction of the pumping work for higher loads. The variable $p_{mf,NA}(\omega_e)$ is the fuel mean pressure for NA conditions.

The mechanical friction mean pressure is modeled by a second-order polynomial,

$$p_{mfric}(\omega_e) = r_0 + r_1 \cdot \omega_e + r_2 \cdot \omega_e^2. \tag{A.107}$$

The coefficients are listed in Table A.14.

Table A.14 Base engine parameters

parameter	symbol	value
max. power	P_{max}	99 kW
compression ratio	ε	10.5
displacement	V_d	2.0 l
no. cylinders	z	4
no. intake valves	z_{IV}	2
no. exhaust valves	z_{EV}	2
mechanical friction	r_0	$3.7 \cdot 10^4$ Pa
	r_1	40.58 Pa $\cdot$ s/rad
	r_2	0.137 Pa $\cdot s^2/\text{rad}^2$
idle speed	$\omega_{e,idle}$	720 rpm
min. speed for fuel cut-off	$\omega_{e,cutoff}$	1100 rpm

The maximal pumping losses p_{mgas0} for a throttled SI engine are assumed to be independent of the engine speed. A value of $p_{mgas0} = 0.8$ bar is taken. For a dethrottled engine it can be assumed that there are no pumping losses in the whole operating range, i.e., $p_{mgas0} = 0$ bar. The resulting relations between the mean fuel pressure and the mean effective pressure for both engine types are visualized in Fig. A.35.

Replacing the camshaft by an EHVS reduces the engine's mechanical friction. Consequently, in the model of the dethrottled engine, the mechanical friction is reduced by 25%. In turn, the energy necessary for the hydraulic power has to be taken into account for all fully variable valves, which is done in the next subsection. These adjustments are not shown in Fig. A.35.

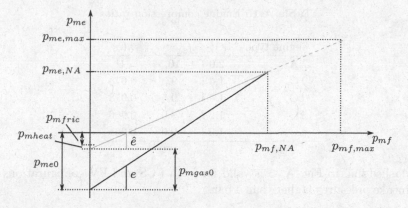

Fig. A.35 Willans approximations of a downsized and turbocharged throttled engine (black solid and gray dashed lines) and of a downsized and turbocharged dethrottled engine (gray solid and gray dashed lines). The variable $p_{me,max}$ indicates the engine's full load.

Downsizing and Turbocharging

The advantage of using a Willans approximation is that it can be scaled to other engine sizes. The baseline engine configuration is chosen to be a 2.0 l NA gasoline engine with standard fixed camshafts. The Willans fit is also adapted for all turbocharged (TC) engines. The following modeling assumptions regarding engine scaling are made:

- The mean pressures p_{mfric}, p_{mheat} and p_{mgas0} remain constant for all engine sizes.
- The identified efficiency $e(\omega_e)$ depends on the compression ratio ε of the engine as follows:

$$e(\omega_e) = k(\varepsilon) \cdot e_{\varepsilon=10.5}(\omega_e), \tag{A.108}$$

where $k(\varepsilon)$ was determined using an engine process simulation. Its values are listed for all considered engine sizes in Table A.15 (see also [128]). The more the SI engine is downsized the lower its compression ratio has to be chosen to avoid knocking.
- All engines are operated in a strictly stoichiometric manner.
- The rated power of the downsized engines is at least equal to the power of the baseline engine due to the use of turbochargers.
- All turbochargers are designed for maximum efficiency, since high instantaneous torque can be provided using the boost mode. This means that the intake pressure level is assumed to be equal to the exhaust pressure level for intake pressures higher than 1 bar. This assumption directly indicates that for intake pressures higher than 1 bar no pumping work is performed, i.e., $p_{mgas0} = 0$ bar, independent of the valve-train. Thus, the

Table A.15 Engine compression ratios

engine type	V_d	ε	$k(\varepsilon)$
NA	2.0 l	10.5	1
TC	1.6 l	9.5	0.978
TC	1.4 l	9.5	0.978
TC	1.2 l	9.0	0.966
TC	1.0 l	9.0	0.966

dashed line in Fig. A.35 is valid for both FCS and FV configurations for intake pressures higher than 1 bar.

Electro-Hydraulic Valve System

The additional torque at the crankshaft needed for the actuation of the hydraulic pump of the EHVS [220] is calculated as follows,

$$T_{EHVS} = \frac{V_{hyd,tot} \cdot p_{hyd}}{4 \cdot \pi} \cdot \frac{1}{\eta_{hyd}}, \tag{A.109}$$

where $V_{hyd,tot}$ is the total hydraulic volume and η_{hyd} denotes the mean efficiency of the hydraulic pump which is assumed to be constant at $\eta_{hyd} = 0.6$. The hydraulic pressure p_{hyd} is chosen such that it always ensures that the charge valve resists the gas force of the pressurized air in the tank. This is assumed to be guaranteed if the hydraulic pressure is eight times higher than the tank pressure, i.e., $p_{hyd} = 8 \cdot p_t$. The hydraulic pressure also has to respect the operating boundaries for the EHVS system, which are $p_{hyd,min} = 50$ bar and $p_{hyd,max} = 200$ bar.

The total hydraulic volume $V_{hyd,tot}$ to be displaced by the hydraulic pump during two revolutions yields

$$V_{hyd,tot} = \sum_i n_i \cdot V_{hyd,i} \qquad i = IV, EV, CV \quad , \tag{A.110}$$

where n_i is the number of the EVHS actuations of the intake, exhaust and charge valves during two engine revolutions for a given engine. In Table A.16, the assumed displaced hydraulic volume V_{hyd} per one valve actuation is listed.

Pneumatic Modes

According to Sect. 5.6 the enthalpy transferred from and to the tank W_t is a function of the tank pressure p_t and the demanded engine torque T_e. The optimization of the two- and four-stroke pneumatic modes can be found in [86].

Table A.16 Displaced hydraulic volumes for one actuation

Valve	$V_{hyd,i}$
Intake valves	$407 \, \text{mm}^3$
Exhaust valves	$385 \, \text{mm}^3$
Charge valve	$414 \, \text{mm}^3$

The assumptions for the pneumatic modes are the following:

- Air is an ideal and calorically perfect gas.
- The valves are idealized by neglecting opening and closing times as well as flow restrictions.
- The mechanical friction and heat losses (p_{mfric} and p_{mheat}) are assumed to be equal to the ones that occur when dragging the engine. The pumping losses, however, depend on the pneumatic operating mode and point.
- The fuel injection is cut off for all pneumatic modes.
- The air tank is modeled adiabatically. This assumption seems to be quite optimistic. However, since the tank could be heated using exhaust gas, it is assumed here that the heat losses are equal to the heat gains. Thus, the change in internal energy is equal to the enthalpy transferred

$$\Delta U_t = W_t(p_t, T_e). \tag{A.111}$$

The two-stroke pneumatic modes require variable valve actuation for all valves, while the four-stroke pneumatic modes require only a fully variable charge valve. In the following each pneumatic mode is briefly discussed. Further details on the two-stroke modes can be found in Sect. 5.6.

The *two-stroke pneumatic motor mode* is characterized by one air power stroke and one exhaust stroke per revolution per cylinder. The air power stroke takes place during the downward motion of the piston, while pressurized air from the tank is injected into the cylinder. During the upward stroke, the air is exhausted either via the intake valves or via the exhaust valves.

The *two-stroke pneumatic pump mode* consists of an upward compression stroke with an air transfer to the pressure tank once the cylinder pressure has reached the tank pressure. The subsequent downstroke is an expansion stroke with an air intake phase once the cylinder pressure has reached the ambient pressure. This happens at every engine revolution for each cylinder.

The *four-stroke pneumatic motor mode* works like a conventional combustion engine cycle with fuel cut-off. However, during the expansion stroke, pressurized air is injected to generate a positive torque.

The *four-stroke pneumatic pump mode* works like a conventional combustion engine cycle with fuel cut-off as well. In this case, however, air is transferred during the compression stroke to the pressurized tank once the cylinder pressure has reached the tank pressure.

The efficiencies of the pneumatic engine modes are an important measure for the overall fuel saving potential of the hybrid propulsion system. As mentioned in Sect. 5.6, the efficiency of each mode can be represented as a function of the mean indicated pressure p_{mi} and the tank pressure p_t. The mean indicated pressure and the engine torque are related in the following way

$$p_{mi} = \frac{T_e \cdot N \cdot \pi}{V_d} + p_{mfric}, \qquad (A.112)$$

where N indicates the number of strokes, e.g., $N = 4$ for a four-stroke engine.

In Fig. A.36, the indicated efficiency η and COP levels are shown for all pneumatic modes. In the area of very low values of $|p_{mi}|$ the influence of the gas exchange strokes of the four-stroke mode is clearly visible. The operating range in terms of p_{mi} and p_t of the two- and four-stroke mode coincide. However, the two-stroke pneumatic motor mode offers double the torque of the four-stroke motor mode for the same p_{mi}, see (A.112).

A.8.2 Optimal Control Strategy

Compared to a conventional drive train, the HPE has an additional degree of freedom in choosing its mode of operation. Thus, the system requires a designated supervisory controller that determines the operating mode at each time instant. For a fuel-efficient operation, it is important to choose the right mode at the right moment. In order to exclude the influence of a possibly suboptimal controller, the theoretically minimal fuel consumption of the system is evaluated using the DP-algorithm described in Sect. C. In order to be able to compare the performance of several systems based only on the fuel consumption, a charge sustaining strategy is required, i.e., the tank pressure at the end of the drive cycle is required to be equal or higher than the one in the beginning.

To use the deterministic dynamic programming technique, the HPV model has to be reduced and reorganized. First, the control options have to be examined. Each mode choice is equal to a control input u. This control input u substitutes low-level control inputs such as throttle or charge valve actuation. The system configurations listed in Table A.17 are investigated for various engine sizes. This allows to determine the contribution of the various effects on the fuel consumption reduction, e.g., downsizing (DS), stop/start (SS), hybrid modes.

Since the drive cycle is assumed to be known in advance for every time step k, the operating point variables ω_e and T_e can be included into each discrete state change function f_k,

$$x_{k+1} = f_k(x_k, u_k), \qquad k = 0, 1, 2, ..., N - 1. \qquad (A.113)$$

Furthermore, the same can be done with the cost function $g_k(x_k, u_k)$ for every time step k. The cost is the fuel energy consumed,

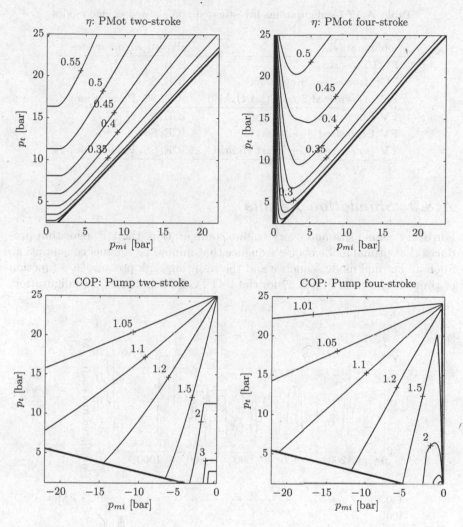

Fig. A.36 Efficiency and COP for two- and four-stroke pneumatic modes ($\varepsilon = 10$)

$$\sum_{k=0}^{N-1} g_k(x_k, u_k) = H_l \sum_{k=0}^{N-1} \Delta m_f(x_k, u_k), \qquad (A.114)$$

where H_l is the lower heating value of gasoline and Δm_f is the fuel mass used.

The initial state is chosen to be $x_0 = p_{t,0} = 10\,\text{bar}$. The final cost is set to ∞ for $x_N < x_0$, and it is set to zero otherwise. This way, charge sustenance is enforced.

Table A.17 Configurations investigated and allowed engine modes

configuration	allowed engine modes
FCS Downsized	ICE
FCS Downsized Stop/Start	ICE, SS
FCS Downsized Stop/Start Hybrid	ICE, SS, Pmot, Pump
FV Downsized	ICE
FV Downsized Stop/Start	ICE, SS
FV Downsized Stop/Start Hybrid	ICE, SS, Pmot, Pump

A.8.3 Simulation Results

For every engine size and every engine configuration, the DP algorithm produces the optimal mode choice sequence that minimizes the fuel consumption. Such an optimal mode sequence and the resulting tank pressure as a function of time are shown in Fig. A.37 for the 1.4 l TC HPE in the FCS configuration.

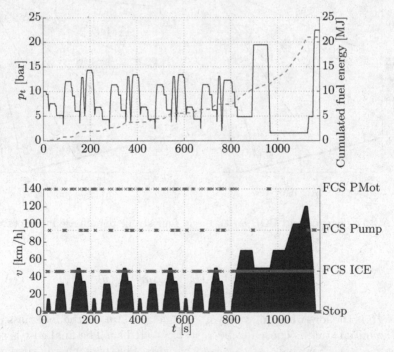

Fig. A.37 Simulated tank pressure, fuel consumption, vehicle speed and mode for the 1.4 l TC HPE with camshaft driven intake and exhaust valves (FCS configuration)

Here, the algorithm mainly decides when to use the recuperated energy for pneumatic propulsion and when to use the conventional combustion mode instead. The engine is always started with the pneumatic motor mode. Additionally, it is also used to propel the vehicle. The pneumatic pump mode is always used for recuperation, with the exception of a part of the last braking phase since the tank pressure is already higher than in the beginning. The algorithm chooses the conventional mode since it does not yield any cost because of the fuel cut-off.

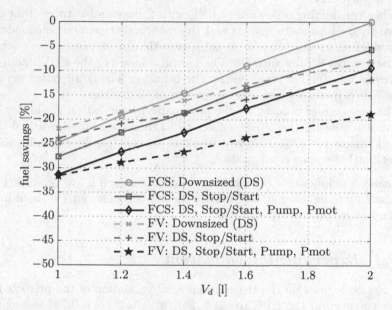

Fig. A.38 Simulated fuel savings relative to the 2.0 l NA engine for various engine configurations on the MVEG-95 drive cycle

Figure A.38 shows the fuel saving results on the MVEG–95 for the system configurations listed in Table A.17. It is found that:

- Downsizing a conventional engine (FCS) by a factor of 2 can save as much fuel as 25%. It is the most important contributing factor to the overall fuel gain. Note that without the HPE concept, the turbo lag would be an issue.
- The relative downsizing effect for dethrottled engines is smaller than for throttled engines.
- Adding a stop/start (SS) capability to the engine results in 6% fuel savings for the baseline FCS engine. This effect is lower for downsized engines and dethrottled engines, since the engine idling losses are highest for large FCS engines.

- Dethrottling the baseline engine leads to fuel savings of 8%. For strongly downsized engines, the fuel burned due to the higher torque resulting from the hydraulic system energy need exceeds the advantage of having a dethrottled engine. For an engine smaller than 1.25 l, the fuel energy demand of the hydraulic system is larger than the fuel savings due to dethrottling.
- The overall fuel saving potential of a downsized and turbocharged HPE system based on the FCS configuration amounts to 31%. The hybridization itself contributes 3% points. This result shows that the hybridization is not so important.
- When considering a downsized HPE with fully variable valves that allow two-stroke pneumatic modes and the dethrottled combustion mode, the overall fuel saving potential is not higher than in the case where only the charge valve is fully variable. The parasitic losses of the EHVS actuation prevent any further fuel savings. Furthermore, having all valves actuated by EHVS increases cost and complexity of the system.
- The fuel savings induced by the pneumatic hybridization are higher for a non-downsized dethrottled engine ($= 7\%$) than for a non-downsized throttled engine ($= 4\%$). This is a result of the two-stroke capability of the dethrottled engine configuration.

The study concludes that pneumatic hybridization itself is less promising than its capability to enable a strong downsizing and turbocharging without having to compromise on the driveability.

A.8.4 Experimental Validation

This section focuses on the the experimential validation of the predicted fuel saving potential of the HPE concept. For this purpose, a 0.75 l two-cylinder port-fuel injection (PFI) TC gasoline engine, whose data are listed in Table A.18, is modified by replacing one exhaust valve per cylinder by an EHVS actuated charge valve. The exhaust valve and the two intake valves per cylinder remain camshaft-driven. Figure A.39 shows a schematic of the resulting system while Fig. A.40 shows a picture of the engine test bench.

Table A.18 Engine data of the modified engine

parameter	symbol	value
engine type		PFI TC gasoline
max. power	P_{max}	61 kW
compression ratio	ε	9
displacement	V_d	0.75 l
no. cylinders	z	2

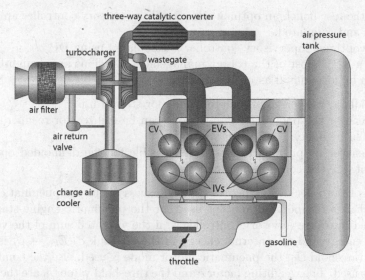

Fig. A.39 Schematic of the downsized and turbocharged two-cylinder PFI SI engine

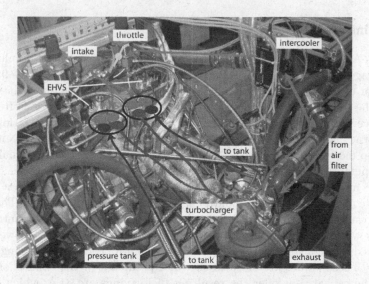

Fig. A.40 Picture of engine test bench

The engine is mounted on a dynamometer test bench that emulates the forces acting on the crankshaft. This method allows to determine the fuel consumption of a vehicle equipped with such an engine, without actually having to integrate the engine into a vehicle. Using this experimental setup, the consumed fuel of the engine in realistic drive cycle conditions can be measured very accurately and reproducably using a fuel scale.

On the test bench an optimal and causal supervisory controller are implemented and compared.

The optimal supervisory controller is calculated using DP.

For a HPV most of the energy management decisions are straightforward and can be formulated as intuitive rules:

- No idling, use pneumatic motor mode for stop/start.
- Whenever the pneumatic pump mode can be utilized for recuperation, it should be done.
- Whenever no pneumatic modes are feasible for the demanded operating point, use the conventional mode.

What is left to decide is when to use the excess air in the pneumatic motor mode that is recuperated and not used for the pneumatic engine start. The proposed causal supervisory controller is: if the weighted sum of the vehicle's kinetic energy and the internal energy in the air tank $c \cdot E_{kin} + U_t$ is larger than a threshold U_0, the pneumatic motor mode is used, else the combustion mode is used. The weighting factor c and the threshold value U_0 are the tuning parameters. More details on the experimential validation can be found in [87].

Experimental Results

Figures A.41 and A.42 show the vehicle emulation measurement results for a Volkswagen Polo (2005 model) equipped with the HPE on the MVEG–95 drive cycle for the optimal and the causal controllers. The figures show the modes used, the tank pressure trajectories, and the normalized deviations from the reference vehicle velocities. In both measurements the velocity deviations stay within the mandated bounds that are indicated by the dashed lines. Note that the stop mode is also activated when switching from the conventional mode to the pump mode, to eliminate the fuel wall film of the PFI engine.

The measured fuel consumption as well as the fuel consumption of the vehicle equipped with the standard engine are listed in Table A.19. The measured results show that the evaluated engine concept yields substantial fuel savings when compared with the NA engine of the same rated power. The measured fuel consumption of the HPE is 35% lower for a downsizing factor of 1.86. These values are close to the ones predicted in the simulation part. With the causal controller the resulting fuel savings are 31%, which is only slightly lower than with the optimal controller.

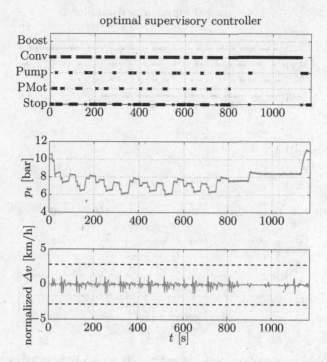

Fig. A.41 Vehicle emulation measurement results for a VW Polo '05 on the MVEG–95 drive cycle using the optimal supervisory controller. The dashed lines indicate the boundaries.

Table A.19 Vehicle emulation measurement results for a VW Polo '05 on MVEG–95 drive cycle

engine	conv. engine	HPE	savings	HPE	savings
controller	-	optimal		causal	
displacement	1.39 l	0.75 l		0.75 l	
max. power	63 kW	61 kW		61 kW	
ECE	8.30 l	4.20 l	−50%		
EUDC	5.20 l	3.99 l	−23%		
MVEG–95	6.30 l	4.07 l	−35%	4.36 l	−31%

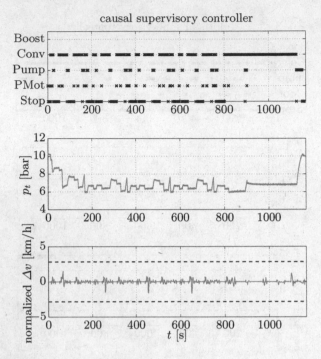

Fig. A.42 Vehicle emulation measurement results for a VW Polo '05 on the MVEG-95 drive cycle using the causal supervisory controller. The dashed lines indicate the boundaries.

B

Appendix II – Optimal Control Theory

This appendix briefly summarizes the most useful results of optimal control theory. In the first section static problems are analyzed, i.e., the objective function only includes time-independent control variables. This formulation yields a parameter optimization or nonlinear programming problem for which several closed-form and numerical solution algorithms are known. Several excellent textbooks are available on that subject, for instance [28] and [33].

The second section analyzes dynamic optimal control problems. Starting with a brief repetition of the classical variational calculus theory, the concepts of adjoint (Lagrange) states and Hamiltonian formulations are introduced. To be able to deal with the case of constrained input variables, Pontryagin's minimum principle is briefly introduced. As with the first section, the main objective here is to collect the main facts without any proofs and to introduce the notation. Readers interested to learn more about this field are referred to one of the several available textbooks, for instance [41].

B.1 Parameter Optimization Problems

B.1.1 Problems without Constraints

The vector $u = [u_1, \ldots, u_m]^T \in I\!\!R^m$ consists of arbitrary parameters and the mapping $L : I\!\!R^m \to I\!\!R$ is a sufficiently differentiable function (the *performance index*) that has to be minimized.[1] Sufficient conditions for a point u^o to be a local minimum are

$$\left. \frac{\partial L(u)}{\partial u} \right|_{u=u^o} = 0, \quad \text{and} \quad \left. \frac{\partial^2 L(u)}{\partial u^2} \right|_{u=u^o} > 0, \tag{B.1}$$

i.e., the gradient of the performance index must be zero at the minimum (u^o is a stationary point), and the Hessian matrix of the performance index must

---

[1] A maximization problem can be obtained from a minimization problem by simply multiplying the performance index by -1.

be positive definite (in the neighborhood of u^o, $L(\cdot)$ increases throughout). The condition (B.1) is globally sufficient only for specific cases, for instance if it is known that the function $L(\cdot)$ is globally convex.

Necessary conditions for a local minimum are:

$$\left.\frac{\partial L(u)}{\partial u}\right|_{u=u^o} = 0, \quad \text{and} \quad \left.\frac{\partial^2 L(u)}{\partial u^2}\right|_{u=u^o} \geq 0. \qquad (B.2)$$

To establish whether a minimum exists, additional information is needed.

Example 1. Performance index:

$$L(u) = \frac{1}{2} \cdot u^T \cdot M \cdot u, \quad u = \begin{bmatrix} u_1 \\ u_2 \end{bmatrix}, \quad M = \begin{bmatrix} 1 & 1 \\ 1 & \mu \end{bmatrix}, \quad \mu \in \mathbb{R}. \qquad (B.3)$$

Minimization:

$$\frac{\partial L}{\partial u} = M \cdot u, \quad \frac{\partial^2 L}{\partial u^2} = M, \qquad (B.4)$$

that is, if M is non-singular ($\mu \neq 1$) then only one minimum $u^o = [0,0]^T$ exists. For $\mu = 1$ M is semidefinite positive and all the points on the line $\lambda \cdot [1,-1]^T$ are minima. For $\mu > 1$, M is positive definite and u^o is a global minimum. For $\mu < 1$, M is indefinite and u^o is not a minimum, but a saddle point, see Fig. B.1.

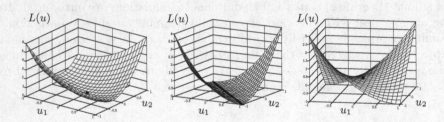

Fig. B.1 Performance index of Example 1, left $\mu > 1$, center $\mu = 1$, right $\mu < 1$

Example 2. Performance index:

$$L(u) = (u_1 - u_2^2) \cdot (u_1 - 3 \cdot u_2^2). \qquad (B.5)$$

Minimization:

$$\frac{\partial L}{\partial u} = \begin{bmatrix} 2 \cdot u_1 - 4 \cdot u_2^2 \\ -8 \cdot u_1 \cdot u_2 + 12 \cdot u_2^3 \end{bmatrix} = \begin{bmatrix} 0 \\ 0 \end{bmatrix} \Rightarrow u^0 = \begin{bmatrix} 0 \\ 0 \end{bmatrix}, \quad \left.\frac{\partial^2 L}{\partial u^2}\right|_{u=u^o} = \begin{bmatrix} 2 & 0 \\ 0 & 0 \end{bmatrix}, \tag{B.6}$$

that is,[2] $L_u(u) = 0$ has a triple but isolated solution, and this solution is not a local minimum, as shown in Fig. B.2.

---

[2] L_u denotes the partial derivative of L with respect to u.

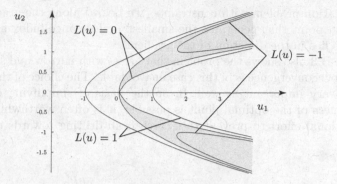

Fig. B.2 Performance index of Example 2

B.1.2 Numerical Solution

Only in very rare cases a closed-form solution may be obtained. In the majority of all cases numerical methods are needed. Several different approaches are possible. One class of algorithms utilizes information on the gradients of $L(u)$. These algorithms can be subdivided into:

- semi-analytical methods where the performance index $L(u)$ and its gradient $\partial L/\partial u$ are available in a closed form; and
- fully numerical methods in which only the performance index is known and whose gradient must be approximated through finite differences.

The semi-analytical methods in general converge much faster than the fully numerical methods and they are less sensitive to rounding effects.

The numerical search can be either of first order or of second order. The first order numerical methods use the idea of the steepest descent such that all first-order algorithms have approximately this structure:

1. guess an initial value for $u(1)$;[3]
2. evaluate the gradient $\Gamma_L(i) = \frac{\partial L}{\partial u}\big|_{u=u(i)}$ either from a known relationship for the gradient or through numerical approximation using finite differences;
3. determine the new iteration point according to the rule $u(i+1) = u(i) - h(i) \cdot \Gamma_L(i)$;
4. check if the variation of the performance index $|L(u(i+1)) - L(u(i))|$ is smaller than a predetermined threshold. In this case, the algorithm ends, otherwise it is repeated starting at point 2.

Crucial for the convergence is the choice of the relaxation factor $h(i)$. If it is chosen too large (see Fig. B.3), the algorithm may overshoot and even become unstable. One possibility to choose $h(i)$ in an optimal way consists of solving

---

[3] The argument denotes the iteration index.

a minimization problem with constraints (see below) along the gradient, i.e., $u(2)$ is the point that generates the smallest performance index and at the same time lies in the direction of the gradient.

Figure B.3 also shows the problem that arises with narrow and steep "valleys": a slow convergence is in this case unavoidable. The choice of the starting values is very important as well. If, on the basis of simplifying considerations, a guess of the optimal point is possible, it is often worthwhile making this additional effort to prevent the search from drifting towards unrealistic solutions.

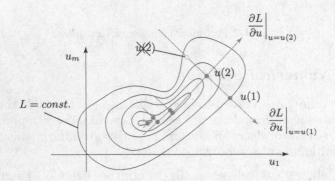

Fig. B.3 Schematic of the steepest descent algorithm

Besides first-order methods (which use only the first-order derivatives), second-order methods are known (Newton-Raphson approaches). They converge faster in the neighborhood of the optimum, but are known to be inefficient far away from it. The basic idea is to approximate the performance index through a quadratic form

$$L(u) \approx L(u(i)) + \Gamma_L(i) \cdot (u - u(i)) + \frac{1}{2} \cdot (u - u(i))^T \cdot \Gamma_L'(i) \cdot (u - u(i)) \tag{B.7}$$

where $\Gamma_L'(i) = \left. \frac{\partial^2 L}{\partial u^2} \right|_{u=u(i)}$, and then to use the solution of the approximate problem (which is solvable in closed form) as a new iteration value

$$u(i+1) = u(i) - \left[\Gamma_L'(i) \right]^{-1} \cdot \left[\Gamma_L(i) \right]^T . \tag{B.8}$$

All the algorithms discussed above are available in program libraries or for instance in Matlab^{TM} (optimization toolbox). In most cases the use of this software is recommended over any attempts to re-invent the wheel.

B.1.3 Minimization with Equality Constraints

Let $u = [u_1, \ldots, u_m]^T \in \mathbb{R}^m$ be a vector of arbitrary control variables, $x = [x_1, \ldots, x_n]^T \in \mathbb{R}^n$ a vector of state variables[4] and $f : \mathbb{R}^{m+n} \to \mathbb{R}^n$, $L : \mathbb{R}^{m+n} \to \mathbb{R}$ sufficiently differentiable functions of these quantities. The optimization problem consists of finding the values u^o, x^o that minimize the performance index L and, at the same time, fulfill the constraint $f(u^o, x^o) = 0$. With the adoption of a new variable $z = [u, x]^T$, the following expressions may be written in a more compact way.

In the case $n = 1$, the solution of this problem can be immediately given with the help of Fig. B.4. An optimal point z^o is characterized by the fact that both the gradient on an iso-level curve of the performance index and the gradient on the subset defined by the constraint are co-linear,

$$\left.\frac{\partial L}{\partial z}\right|_{z=z^o} + \lambda \cdot \left.\frac{\partial f}{\partial z}\right|_{z=z^o} = 0, \tag{B.9}$$

where λ is a new arbitrary parameter. These $m + 1$ equations, together with the constraint $f(z^o) = 0$, define the $m + 2$ unknown quantities $z^o = [u^o, x^o]$ and λ^o.

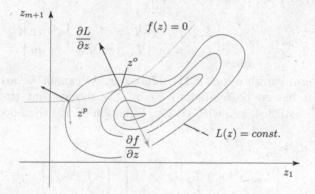

Fig. B.4 Constrained optimization problem with one constraint

The solution of the general case can be derived from the fact that in the optimal point any *arbitrary* variation dz that satisfies the constraints may not cause any variation of the performance index, i.e., for

$$df(u, x) = f_u \cdot du + f_x \cdot dx = 0, \tag{B.10}$$

---

[4] The distinction between control and state variables is only a matter of convenience.

it has to be true that

$$dL(u, x) = L_u \cdot du + L_x \cdot dx = 0. \tag{B.11}$$

From (B.10), it follows that the variation dx depends on the variation du,

$$dx = -f_x^{-1} \cdot f_u \cdot du. \tag{B.12}$$

The $n \times n$ matrix f_x has to be nonsingular in the optimal point, otherwise the problem is not well posed. Inserting (B.12) into (B.11),

$$dL(u, x) = [L_u - L_x \cdot f_x^{-1} \cdot f_u] \cdot du = 0. \tag{B.13}$$

Since this has to be valid for any arbitrary variation du, the following necessary condition results

$$L_u - L_x \cdot f_x^{-1} \cdot f_u = 0. \tag{B.14}$$

Sufficient conditions for a local minimum have to take into account the second-order variations. Expanding in a Taylor series the performance index around the optimal point x^o, u^o up to the second-order variations,

$$dL \approx [L_x(x^o, u^o), L_u(x^o, u^o)] \cdot \begin{bmatrix} dx \\ du \end{bmatrix} +$$

$$+ \frac{1}{2} \cdot [dx^T, du^T] \cdot \begin{bmatrix} L_{xx}(x^o, u^o) & L_{xu}(x^o, u^o) \\ L_{ux}(x^o, u^o) & L_{uu}(x^o, u^o) \end{bmatrix} \cdot \begin{bmatrix} dx \\ du \end{bmatrix}. \tag{B.15}$$

The linear term vanishes for (B.11). Moreover, dx cannot be arbitrarily selected, but it has to be related to du in such a way that the constraint $f(x, u) = 0$ is satisfied, i.e., with (B.12). Equation (B.15) then becomes

$$dL \approx \frac{1}{2} \cdot du^T \cdot [-f_u^T \cdot (f_x^{-1})^T, I] \cdot$$

$$\cdot \begin{bmatrix} L_{xx}(x^o, u^o) & L_{xu}(x^o, u^o) \\ L_{ux}(x^o, u^o) & L_{uu}(x^o, u^o) \end{bmatrix} \cdot \begin{bmatrix} -f_x^{-1} \cdot f_u \\ I \end{bmatrix} \cdot du, \tag{B.16}$$

which has to be fulfilled for arbitrary values of du. The sufficient condition for the point x^o, u^o being optimal is therefore

$$\left. \frac{\partial^2 L}{\partial u^2} \right|_{opt} = (L_{uu} - f_u^T \cdot (f_x^{-1})^T \cdot L_{xu} -$$

$$- L_{ux} \cdot f_x^{-1} \cdot f_u + f_u^T \cdot (f_x^{-1}) \cdot L_{xx} \cdot f_x^{-1} \cdot f_u)|_{x^o, u^o} > 0, \tag{B.17}$$

where $opt = \{x^o, u^o, f(x, u) = 0\}$. In other words, the Hessian matrix of the performance index in the point x^o, u^o and for variations that satisfy the constraints has to be positive definite. If (B.17) is only semidefinite, x^o, u^o could not be a minimum.

An analogous formulation of the optimization problem, which will be easily extended later, uses a formalism that is based on the approach (B.8). Instead of the original problem, the function

$$H(x, u, \lambda) = L(x, u) + \lambda^T \cdot f(x, u) \tag{B.18}$$

is minimized. The function $H(\cdot)$ is the *Hamiltonian* function of the optimization problem, while $\lambda = [\lambda_1, \ldots, \lambda_n]^T$ are new arbitrary variables that will be referred to as Lagrange multipliers.

In the optimal point $f(x^o, u^o) = 0$ must be zero and therefore in that point $H(x^o, u^o, \lambda^o) = L(x^o, u^o)$. For the formulation (B.18) to be equivalent to the original optimization problem, the variation

$$dH(x, u, \lambda) = H_x(x, u, \lambda) \cdot dx + H_u(x, u, \lambda) \cdot du + f(x, u) \cdot d\lambda \tag{B.19}$$

must vanish if, and only if, dL vanishes. The last addend in (B.19) is in any case identically null (constraint). The new degrees of freedom λ can be defined in such a way that $H_x(x, u, \lambda) = 0$, i.e.,

$$\begin{aligned} H_x(x, u, \lambda) &= L_x(x, u) + \lambda^T \cdot f_x(x, u) = 0 \quad \Rightarrow \\ &\Rightarrow \lambda^T(x, u) = -L_x(x, u) \cdot f_x^{-1}(x, u). \end{aligned} \tag{B.20}$$

Inserting λ defined in this way in H_u, the necessary condition (B.14) is obtained. The original optimization problem and the Hamiltonian formulation are therefore equivalent. The necessary condition of the optimization problem can be written as

$$dH(x, u, \lambda) = 0 \tag{B.21}$$

or

$$i) \quad \left. \frac{\partial H}{\partial u} \right|_{u=u^o, x=x^o, \lambda=\lambda^o} = 0 \Rightarrow \text{Condition for optimality}$$

$$ii) \quad \left. \frac{\partial H}{\partial x} \right|_{u=u^o, x=x^o, \lambda=\lambda^o} = 0 \Rightarrow \text{Fulfilled by the choice (B.19)}. \tag{B.22}$$

$$iii) \quad \left. \frac{\partial H}{\partial \lambda} \right|_{u=u^o, x=x^o, \lambda=\lambda^o} = 0 \Rightarrow \text{Constraint}$$

These $m + 2n$ equations are necessary conditions for u^o, x^o, and λ^o representing the solution of the optimization problem. Similarly, also the sufficient condition (B.17) can be derived in the Hamiltonian formulation.

The Lagrange multipliers allow an interesting interpretation, which will be useful in the following,

$$\left. \frac{\partial L}{\partial f} \right|_{u=u^o} = -\lambda^T, \tag{B.23}$$

i.e., considering instead of the constraint $f(x, u) = 0$ the slightly modified constraint $f(x, u) - df = 0$ (df being a constant vector), the value of the performance index in the new optimum is approximated as

$$L|_{u=u^o(df)} \approx L|_{u=u^o(0)} - \lambda^T \cdot df. \qquad (B.24)$$

A further application of the Hamiltonian formalism will be shown in the next section.

B.1.4 Minimization with Inequality Constraints

Let $u = [u_1, \ldots, u_m]^T \in I\!R^m$ be a vector of arbitrary variables and the mappings $f : I\!R^m \to I\!R^n$, $L : I\!R^m \to I\!R$ sufficiently differentiable functions of these quantities. The optimization problem consists of finding that value of u^o which minimizes the performance index L and at the same time satisfies the condition $f(u^o) \leq 0$. Such a problem is often denoted as *nonlinear programming*, in analogy to the concept of *linear programming*, in which both L and f are linear. Often, the number of inequalities n is larger than the number of variables m. Indeed, some of the inequalities in the optimum are not active at all, but it is not known a priori which inequality does not have to be considered. The distinction between control and state variables thus loses its meaning.

If no inequality is active at all, the problem is reduced to the original optimization problem without constraints with the sufficient conditions (B.1). If only one inequality is active, the situation shown in Fig. B.5 arises.

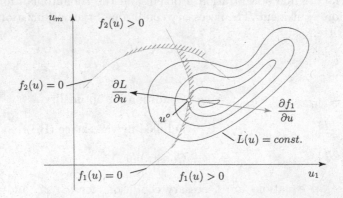

Fig. B.5 Optimization with inequality constraints: two constraints, with only one active

This problem is analogous to the situation shown in Fig. B.4. In particular, the boundary of the inequality constraint becomes a regular constraint, such as the one that was discussed in the last section. Nevertheless, there is a difference: the sign of the Lagrange multiplier is now given, since only positive multipliers are admissible. The sign of the differential dL has to be positive or zero

$$dL|_{u=u^o} = \left.\frac{\partial L}{\partial u}\right|_{u=u^o} \cdot du \geq 0 \qquad (B.25)$$

for admissible variations du, i.e. for

$$df|_{u=u^o} = \left.\frac{\partial f}{\partial u}\right|_{u=u^o} \cdot du \leq 0. \qquad (B.26)$$

Instead, in the problem shown in Fig. B.4, the gradients are equally oriented and thus negative Lagrange multipliers are possible.

The situation shown in Fig. B.5 can be extended to the case of more active constraints. Figure B.6 shows the basic idea. The gradient of the performance index has to be directed in such a way as to make possible a reduction of the performance index only through a violation of the constraints.

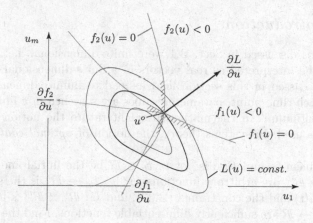

Fig. B.6 Optimization with inequality constraints: two active constraints

Accordingly, the gradients of the active constraints at the optimum define the base vectors, whose directions, weighted with negative constants, allow for the evaluation of the gradient of the performance index,

$$\left.\frac{\partial L}{\partial u}\right|_{u=u^o} = -\lambda_1 \cdot \left.\frac{\partial f_1}{\partial u}\right|_{u=u^o} - \lambda_2 \cdot \left.\frac{\partial f_2}{\partial u}\right|_{u=u^o} \ldots - \lambda_\nu \cdot \left.\frac{\partial f_\nu}{\partial u}\right|_{u=u^o}, \qquad (B.27)$$

where $\nu \leq n$ is the number of active constraints.

Introducing the Hamiltonian

$$H(u,\lambda) = L(u) + \lambda^T \cdot f(u), \qquad (B.28)$$

this condition follows,

$$\frac{\partial H}{\partial u} = \frac{\partial L}{\partial u} + \lambda^T \cdot \frac{\partial f}{\partial u} = 0, \qquad (B.29)$$

where

$$\begin{aligned} \lambda \geq 0 \text{ if } & f(u) = 0 \text{ (active constraint)} \\ \lambda = 0 \text{ if } & f(u) < 0 \text{ (inactive constraint)} \end{aligned} \qquad (B.30)$$

Thus, in the subspace in which the constraints are not active, a stationary point will be sought in these directions. Where the constraints are active, since the sign of the multipliers is given, a special optimization problem with equality constraints will be solved.

It is possible to imagine "pathological" situations in which the gradients of the constraints are linearly dependent and therefore (B.26) cannot be satisfied. Such situations have to be excluded a priori, see [41].

B.2   Optimal Control

B.2.1   Introduction

The problems discussed in Sect. B.1 were finite-dimensional, i.e., their solutions could be interpreted as real vectors in a finite-dimensional space. The problems discussed in this section lead instead to infinite-dimensional solutions. For each time point, parameter values are searched in a finite interval, i.e., in a continuum. In the mathematical literature the notion *calculus of variations* is used. In systems theory, the notion of *optimal control* is more common.

The problem formulation is: let $t_a, t_b \in \mathbb{R}$ be the initial and final time, $u : [t_a, t_b] \to \mathbb{R}^m$ are arbitrary functions, $x : [t_a, t_b] \to \mathbb{R}^n$ is the state vector defined by $u(t)$ and the constraints (B.32), and $L : \mathbb{R}^n \times \mathbb{R}^m \times [t_a, t_b] \to \mathbb{R}$, $\varphi : \mathbb{R}^n \times \mathbb{R} \to \mathbb{R}$ are sufficiently differentiable functions. Find the control law $u(t)$ that minimizes the performance index

$$J(u) = \varphi(x(t_b), t_b) + \int_{t_a}^{t_b} L(x(t), u(t), t)\, dt, \qquad (B.31)$$

while the state $x(t)$ satisfies the differential equation

$$\dot{x} = f(x(t), u(t), t), \quad x(t_a) = x_a. \qquad (B.32)$$

Starting from this formulation, many variants can be derived (final time unspecified, final state partially or completely specified, constraints over the state vector or the control vector, etc.), some of which will be discussed below. The next sections follow closely the text of [41] and the same notation is used. Readers interested in a thorough treatment of the subject are referred to that monograph.

B.2.2 Optimal Control for the Basic Problem

Instead of dealing with (B.31) and (B.32) separately and on the basis of the results of Sect. B.1, it is reasonable to solve a combined optimization problem in which the constraints are integrated in one performance index by means of the Lagrange multipliers,

$$\tilde{J}(u) = \varphi(x(t_b), t_b) + \int_{t_a}^{t_b} \left\{ L(x(t), u(t), t) + \lambda^T(t) \cdot [f(x(t), u(t), t) - \dot{x}(t)] \right\} dt.$$
(B.33)

Note that the Lagrange multipliers are now functions of time. As in Sect. B.1, a Hamiltonian formulation is introduced with the definition

$$H(u(t), x(t), \lambda(t), t) = L(x(t), u(t), t) + \lambda^T(t) \cdot f(x(t), u(t), t).$$
(B.34)

The augmented performance index (B.33) is written as

$$\tilde{J}(u) = \varphi(x(t_b), t_b) + \int_{t_a}^{t_b} H(x(t), u(t), \lambda(t), t)dt - \int_{t_a}^{t_b} \lambda^T(t) \cdot \dot{x}(t) \, dt.$$
(B.35)

An integration by parts of the third term in (B.35) yields the final equation of the optimality condition

$$\tilde{J}(u) = \varphi(x(t_b), t_b) - \lambda^T(t_b) \cdot x(t_b) + \lambda^T(t_a) \cdot x(t_a) + \\ + \int_{t_a}^{t_b} H(x(t), u(t), \lambda(t), t)dt + \int_{t_a}^{t_b} \dot{\lambda}^T(t) \cdot x(t) \, dt.$$
(B.36)

Now the control variable $u(t)$ will be varied by $\delta u(t)$. With (B.37) the resulting variation of the augmented performance index is obtained by means of the chain rule

$$\delta \tilde{J}(u) = \left[\left(\frac{\partial \varphi(x, t)}{\partial x} - \lambda^T(t) \right) \cdot \delta x(\delta u) \right]_{t=t_b} + \lambda^T(t_a) \cdot \delta x_a \\ + \int_{t_a}^{t_b} \left[\frac{\partial H(x, u, \lambda, t)}{\partial u} \cdot \delta u + \left(\frac{\partial H(x, u, \lambda, t)}{\partial x} + \dot{\lambda}^T(t) \right) \cdot \delta x(\delta u) \right] dt$$
(B.37)

In this equation, δx_a is the arbitrary variation of the initial condition, which depends on $\delta u(t)$. Using the constraints (B.32), the variations $\delta x(u)$ are of course related to the variations $\delta u(t)$.

To avoid the calculation of such a dependence, it is possible to use the degrees of freedom $\lambda(t)$, which have not been specified yet. In fact, if the derivative of λ is chosen as

$$\dot{\lambda}(t) = - \left(\frac{\partial H(x, u, \lambda, t)}{\partial x} \right)^T = - \left(\frac{\partial L(x, u, t)}{\partial x} \right)^T - \left(\frac{\partial f(x, u, t)}{\partial x} \right)^T \cdot \lambda(t),$$

$$\lambda(t_b) = \left(\frac{\partial \varphi(x, t)}{\partial x} \right)^T_{t=t_b}$$
(B.38)

the variation $\delta \tilde{J}$ will be independent of δx. Equation (B.38) is a system of ordinary differential equations for the Lagrange multipliers. Unfortunately, the final value rather than the initial value of $\lambda(t)$ is specified. Both the systems (B.32) and (B.38) are coupled by the optimality condition for the control vector.

With the choice of (B.38) the variation of the augmented performance index (B.37) becomes

$$\delta \tilde{J} = \lambda^T(t_a) \cdot \delta x_a + \int_{t_a}^{t_b} \left[\frac{\partial H(x, u, \lambda, t)}{\partial u} \cdot \delta u(t) \right] dt. \qquad (B.39)$$

Two conclusions can be drawn from this equation: (i) if the control vector is kept constant, then $\lambda^T(t_a)$ is the gradient of the performance index with respect to the initial conditions, and (ii) if the initial conditions are kept constant, the optimal control vector has to yield $\delta \tilde{J} = 0$. For arbitrary $\delta u(t)$, the choice

$$\frac{\partial H(x, u, \lambda, t)}{\partial u} = 0, \qquad \forall t \in [t_a, t_b] \qquad (B.40)$$

is the only way to accomplish that. Therefore, all the quantities in the equations of the optimization problem defined above are known. Table B.1 provides a compact summary.

Table B.1 Sufficient conditions for optimal control (Euler–Lagrange equations)

System	$\dot{x} = f(x(t), u(t), t)$
Performance index	$J(u) = \varphi(x(t_b), t_b) + \int_{t_a}^{t_b} L(x(t), u(t), t) \, dt$
Adjoint system	$\dot{\lambda}(t) = -\left(\dfrac{\partial L(x, u, t)}{\partial x} \right)^T - \left(\dfrac{\partial f(x, u, t)}{\partial x} \right)^T \cdot \lambda(t)$
Optimal control	$\dfrac{\partial L(x, u, t)}{\partial u} + \lambda^T(t) \cdot \dfrac{\partial f(x, u, t)}{\partial u} = 0$
Boundary conditions	$x(t_a) = x_a$ and $\lambda(t_b) = \left(\dfrac{\partial \varphi(x, t)}{\partial x} \right)^T_{t=t_b}$

As already mentioned, this $2n$-dimensional system of coupled differential equations is not easily solved, since n boundary conditions are given at the initial time and the other n values at the final time (two-point boundary problem). If the equations are solvable in closed form, this leads to some implicit relationships, which are sometimes solvable (see the example below).

In numerical methods it is necessary to proceed iteratively, since the Lagrange multipliers are not directly related to the system considered, and then no estimation is possible of $\lambda(t_a)$, which strongly complicates the calculation.

Example 3: Optimal "Rendez-Vous" Maneuver. A vehicle in the plane has to be controlled to move in a prefixed time t_b from a given initial state as closely as possible to a desired final state $x(t_b) \approx p$ and $v(t_b) \approx w$. The control signal must be optimal in terms of fuel consumption, which is assumed to be proportional to the square of the input $u(t)$. From Table B.1 the following relationships can be derived:

$$\dot{x}(t) = v(t), \qquad \dot{v}(t) = u(t), \tag{B.41}$$

$$L = \frac{\mu_1}{2} \cdot u^2(t), \qquad \varphi = \frac{\mu_2}{2} \cdot (x(t_b) - p)^2 + \frac{\mu_3}{2} \cdot (v(t_b) - w)^2, \tag{B.42}$$

$$\dot{\lambda}_1(t) = 0, \qquad \dot{\lambda}_2(t) = -\lambda_1(t), \tag{B.43}$$

$$u^o(t) = -\frac{1}{\mu_1} \cdot \lambda_2(t), \tag{B.44}$$

$$x(0) = x_0, \quad v(0) = v_0 \tag{B.45}$$

$$\lambda_1(t_b) = \mu_2 \cdot (x(t_b) - p), \quad \lambda_2(t_b) = \mu_3 \cdot (v(t_b) - w). \tag{B.46}$$

The constants $\mu_{\{1,2,3\}}$ are used as tuning parameters with which the relative importance of the otpimization objectives can be qualified. In this simple case, the equations of the system can be solved analytically. For the Lagrange multipliers

$$\lambda_1(t) = \mu_2 \cdot (x(t_b) - p) = c_0(x(t_b)), \tag{B.47}$$

$$\lambda_2(t) = [\mu_3 \cdot (v(t_b) - w) + \mu_2 \cdot (x(t_b) - p) \cdot t_b] - [\mu_2 \cdot (x(t_b) - p)] \cdot t \tag{B.48}$$

$$= c_1(x(t_b), v(t_b)) - c_0(x(t_b)) \cdot t.$$

With $\lambda_2(t)$ being linear with time, the optimal control law can be determined and the solution of the system follows

$$v(t) = v_0 - \frac{c_1(x(t_b), v(t_b))}{\mu_1} \cdot t + \frac{1}{2} \cdot \frac{c_0(x(t_b))}{\mu_1} \cdot t^2, \tag{B.49}$$

$$x(t) = x_0 + v_0 \cdot t - \frac{1}{2} \cdot \frac{c_1(x(t_b), v(t_b))}{\mu_1} \cdot t^2 + \frac{1}{6} \cdot \frac{c_0(x(t_b))}{\mu_1} \cdot t^3. \tag{B.50}$$

Of course, both of these equations must satisfy the compatibility conditions

$$v(t_b) = v_0 - \frac{c_1(x(t_b), v(t_b))}{\mu_1} \cdot t_b + \frac{1}{2} \cdot \frac{c_0(x(t_b))}{\mu_1} \cdot t_b^2, \tag{B.51}$$

$$x(t_b) = x_0 + v_0 \cdot t_b - \frac{1}{2} \cdot \frac{c_1(x(t_b), v(t_b))}{\mu_1} \cdot t_b^2 + \frac{1}{6} \cdot \frac{c_0(x(t_b))}{\mu_1} \cdot t_b^3, \tag{B.52}$$

which yield implicit equations for $x(t_b)$ and $v(t_b)$. After some manipulation, the explicit solution can be found, as

$$
v(t_b) = \left\{ 12 \cdot \mu_1^2 \cdot v_0 + \mu_2 \cdot \mu_3 \cdot t_b^4 \cdot w + 2 \cdot \mu_1 \cdot t_b \cdot \right.
$$
$$
\left. \cdot [6 \cdot \mu_3 \cdot w - \mu_2 \cdot t_b \cdot (3 \cdot x_0 + t_b \cdot v_0 - 3 \cdot p)] \right\} \cdot \tag{B.53}
$$
$$
\cdot \left(12 \cdot \mu_1^2 + 12 \cdot \mu_1 \cdot \mu_3 \cdot t_b + 4 \cdot \mu_1 \cdot \mu_2 \cdot t_b^3 + \mu_2 \cdot \mu_3 \cdot t_b^4 \right)^{-1},
$$

$$
x(t_b) = \left\{ \mu_2 \cdot \mu_3 \cdot t_b^4 \cdot p + 12 \cdot \mu_1^2 \cdot (t_b \cdot v_0 + x_0) + 2 \cdot \mu_1 \cdot t_b \cdot \right.
$$
$$
\left. \cdot [2 \cdot \mu_2 \cdot p \cdot t_b^2 + 3 \cdot \mu_3 \cdot (t_b \cdot v_0 + t_b \cdot w + 2 \cdot x_0)] \right\} \cdot \tag{B.54}
$$
$$
\cdot \left(12 \cdot \mu_1^2 + 12 \cdot \mu_1 \cdot \mu_3 \cdot t_b + 4 \cdot \mu_1 \cdot \mu_2 \cdot t_b^3 + \mu_2 \cdot \mu_3 \cdot t_b^4 \right)^{-1}.
$$

Figure B.7 shows the system behavior for the case $\mu_1 = \mu_2 = \mu_3 = 1$, $v_0 = x_0 = 0$, $w = 1$, $p = 5$, $t_a = 0$, $t_b = 5$. The behavior computed for the Hamiltonian will be discussed further below.

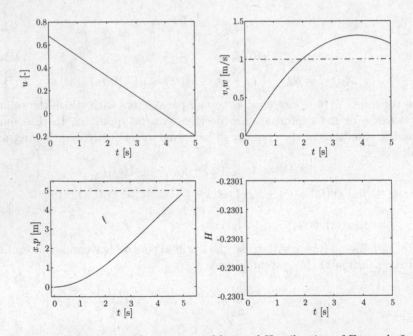

Fig. B.7 Control variable, state variables, and Hamiltonian of Example 3

Figure B.8 shows the sensitivity of the quadratic errors in position and speed

$$
\epsilon_p = (x(t_b) - p)^2, \qquad \epsilon_v = (v(t_b) - w)^2, \tag{B.55}
$$

and of the control effort

$$
\epsilon_u = \int_{t_a}^{t_b} u^2(t) \, dt, \tag{B.56}
$$

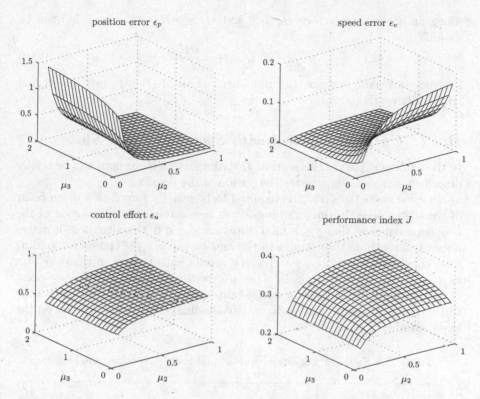

Fig. B.8 Optimization results of Example 3 for different weights μ_2 and μ_3 ($\mu_1 = 1$)

as well as of the performance index with respect to variations of the weights $\mu_2 \in [0.1, 1.0]$, $\mu_3 \in [0.1, 2.0]$.

B.2.3   First Integral of the Hamiltonian

For time-invariant problems, i.e., when both the performance index (B.31) and the system (B.32) are not explicitly functions of time, it can be shown that along the optimal solution

$$\frac{d}{dt}H(x^o, u^o, \lambda^o) = 0, \quad \forall t \in [t_a, t_b], \tag{B.57}$$

that is, the Hamiltonian is constant along an optimal solution. This property does not add any new information, but it is sometimes very useful to test the correctness of a solution or of its implementation (see the example above). The proof is easy to derive. The derivative of H is

$$
\begin{aligned}
\frac{d}{dt}H(x^o, u^o, \lambda^o) &= \frac{\partial H}{\partial t} + \frac{\partial H}{\partial u}\cdot \dot{u} + \frac{\partial H}{\partial x}\cdot \dot{x} + \frac{\partial H}{\partial \lambda}\cdot \dot{\lambda} \\
&= \frac{\partial H}{\partial t} + H_u \cdot \dot{u} + (H_x + \dot{\lambda}^T)\cdot f.
\end{aligned}
\tag{B.58}
$$

Along an optimal trajectory $H_u = 0$ and $\dot{\lambda}^T = -H_x$, whence (B.58) can be simplified as

$$\frac{d}{dt}H(x^\circ, u^\circ, \lambda^\circ) = \frac{\partial H}{\partial t}. \tag{B.59}$$

For time-invariant systems, the right-hand term of (B.59) must vanish and (B.57) follows therefrom.

B.2.4   Optimal Control with Specified Final State

In the remaining part of this section, three important extensions will be briefly described, without showing the derivation of the results.

In some cases the system is required to be exactly forced to a given point of the state space. In the following it will be assumed that a solution of the problem exists and that $q \leq n$ final states are given. In this situation, it makes no sense to include these states in the end cost $\varphi(x(t_b))$. The corresponding final conditions for $\lambda_j(t_b)$, $j = 1, \ldots, q$ are no longer given in this way, but as $x_j(t_b) = x_{bj}$, $j = 1, \ldots, q$. In other words: the problem is still a two-point boundary problem, but now q initial values of the Lagrange multipliers have to be chosen in such a way as to satisfy the final conditions imposed on the state variables.

Table B.2 Sufficient conditions for optimal control, specified final states

System	$\dot{x} = f(x(t), u(t), t)$
Performance index	$J(u) = \varphi(x(t_b), t_b) + \int_{t_a}^{t_b} L(x(t), u(t), t)\, dt$
Adjoint system	$\dot{\lambda}(t) = -\left(\dfrac{\partial L(x, u, t)}{\partial x}\right)^T - \left(\dfrac{\partial f(x, u, t)}{\partial x}\right)^T \cdot \lambda(t)$
Optimal control	$\dfrac{\partial L(x, u, t)}{\partial u} + \lambda^T(t) \cdot \dfrac{\partial f(x, u, t)}{\partial u} = 0$
Boundary conditions	$x(t_a) = x_a$, $x_j(t_b) = x_{bj}$, $j = 1, \ldots, q$
	$\lambda(t_b) = \begin{cases} \nu_j \in \mathbb{R} \text{ arbitrary} & j = 1, \ldots, q \\ \left(\dfrac{\partial \varphi(x, t)}{\partial x_j}\right)^T_{t = t_b} & j = q+1, \ldots, n \end{cases}$

Remark: If a state $x_j(t_b)$ is completely irrelevant for the optimization, i.e., it is neither given nor included in $\varphi(x(t_b))$, the variation of the performance index with respect to this variable vanishes

$$\frac{\partial J}{\partial x_j}\bigg|_{t=t_b} = 0 \tag{B.60}$$

and the final values of the corresponding Lagrange multipliers are thus zero, $\lambda_j(t_b) = 0$.

The same is valid when certain initial states $x_j(t_a)$ are not assigned. The corresponding Lagrange multipliers are thus given as $\lambda_j(t_a) = 0$. Also in this case there is no further information. Now, instead of the optimal initial value of the Lagrange multipliers, the corresponding value of the state variable has to be found that still leads to a two-point boundary problem with n unknowns at $t = t_a$.

B.2.5 Optimal Control with Unspecified Final Time

Optimal control problems with unspecified final time and with fully unspecified or partially given final states can be solved in principle through an iterative procedure starting from a given final time. That final time, to which the smallest performance index corresponds, is the solution of the problem. In other words, the problems with unspecified terminal time have an additional degree of freedom.

It is possible to show that using the conditions given in Tables B.1 and B.2, this requirement is

$$\left(\frac{\partial \varphi(x,t)}{\partial t} + H(x,u,\lambda,t)\right)_{t=t_b} = 0. \tag{B.61}$$

For the important case of minimum time problems ($L(x(t),u(t),t) = 1$), the necessary conditions of Table B.3 apply.

B.2.6 Optimal Control with Bounded Inputs

In almost every real application, the control variables of a system are limited. One possible formulation of this fact is to impose[5]

$$F(u,t) \leq 0. \tag{B.62}$$

The remainder of the problem formulation analyzed below is the same as in Sect. B.2.

This problem has played a great role in the development of the theory of optimal control. While considering only slight variations in the control signal (u and $\dot{u}$ are limited), it is possible to derive a solution (Euler–Lagrange approach) with analogous considerations to those shown in Sect. B.2. However, if large variations are allowed (a case that has a great significance in practice), it is necessary to adopt more advanced concepts of optimal control theory. The result of this analysis is well known as *Pontryagin's maximum principle*:

---

[5] The general case $\tilde{F}(x,u,t) \leq 0$ is clearly more difficult.

Table B.3 Sufficient conditions for minimum time control, fixed final states

System	$\dot{x} = f(x(t), u(t), t)$
Performance index	$J(u) = \varphi(x(t_b), t_b) + t_b - t_a$
Adjoint system	$\dot{\lambda}(t) = -\left(\dfrac{\partial f(x, u, t)}{\partial x}\right)^T \cdot \lambda(t)$
Optimal control	$\lambda^T(t) \cdot \dfrac{\partial f(x, u, t)}{\partial u} = 0$
Boundary conditions	$x(t_a) = x_a,\ x_j(t_b) = x_{bj},\ j = 1, \ldots, q$
	$\lambda(t_b) = \begin{cases} \nu_j \in I\!R \text{ arbitrary} & j = 1, \ldots, q \\ \left(\dfrac{\partial \varphi(x, t)}{\partial x_j}\right)^T_{t=t_b} & j = q+1, \ldots, n \end{cases}$

The Hamiltonian has to be minimized over all the control signals possible.

The key point here is that discontinuities in the control signal $u(t)$ are permitted and that in some cases the minimum is attained at the limits of the range of $u(t)$. The following example will illustrate these ideas.

Example 4: Optimal-Time Rendez-Vous with Limited Control Variable. A simple non-trivial problem consists of transferring a material point in the plane from a known rest position (without loss of generality, chosen as the origin of the coordinate system, i.e., $x(t_a) = 0$, $v(t_a) = 0$) to another rest position $x(t_b) = x_b$, $v(t_b) = 0$, with a limited acceleration. After a proper scaling, the problem may be written as

$$\dot{x}(t) = v(t), \qquad \dot{v}(t) = u(t), \qquad |u(t)| \leq 1, \tag{B.63}$$

$$L = 1, \tag{B.64}$$

$$\dot{\lambda}_1(t) = 0, \qquad \dot{\lambda}_2(t) = -\lambda_1(t). \tag{B.65}$$

The Hamiltonian of this problem is linear in the control variable

$$H(t) = 1 + \lambda_1(t) \cdot v(t) + \lambda_2(t) \cdot u(t), \tag{B.66}$$

thus the minimum $u(t)$ will be reached for a limit value of $u(t)$. According to Pontryagin's minimum principle,

$$u(t) = \begin{cases} +1, \text{ if } & \lambda_2(t) < 0 \\ -1, \text{ if } & \lambda_2(t) > 0 \end{cases}. \tag{B.67}$$

The singular case, i.e., when $\lambda_2(t) \equiv 0$ in a finite interval, can be excluded here. In more complex cases, for instance with various input or state limitations, this situation can actually occur.

The optimal control law found in this way is discontinuous[6] and it is clear that, in order to pass from one rest position to another, *at least* one change of sign of the acceleration has to take place. The Lagrange multipliers vary in the following way

$$\lambda_1(t) = c_1$$
$$\lambda_2(t) = -c_1 \cdot t + c_0. \tag{B.68}$$

In accordance with the control law (B.68) *at most* one switch can take place.

The final time t_b has to be minimized depending on the problem formulation, that is, the additional condition

$$\left(\frac{\partial \varphi}{\partial t} + H \right)_{t=t_b} = H(t_b) = 0 \tag{B.69}$$

has to be satisfied. If the problem is time-invariant, $H(t) \equiv 0, \forall t \in [0, t_b]$. From (B.66), evaluated at $t = 0$,

$$H(0) = 1 + c_1 \cdot v(0) + (-c_1 \cdot 0 + c_0) \cdot u(0) = 1 + c_0 \cdot u(0) \quad \Rightarrow$$
$$\Rightarrow c_0 = \begin{cases} -1 \text{ if } u(0) > 0 \\ +1 \text{ if } u(0) < 0 \end{cases}. \tag{B.70}$$

The initial values of u and λ_2 are also opposite in sign. Inserting the resulting control law in the system equations, e.g., for the case $u(0) = 1$,

$$lllv(t) = t, \qquad\qquad x(t) = \tfrac{1}{2} \cdot t^2, \qquad\qquad \text{for } t < t_s$$

$$v(t) = 2 \cdot t_s - t, \qquad x(t) = -t_s^2 + 2 \cdot t_s \cdot t - \tfrac{1}{2} \cdot t^2, \qquad \text{for } t > t_s. \tag{B.71}$$

The final time and the switch time are

$$t_b = 2 \cdot t_s, \qquad t_s = \sqrt{x(t_b)}, \tag{B.72}$$

and the still unknown constant c_1 is

$$c_1 = \frac{-1}{\sqrt{x(t_b)}}. \tag{B.73}$$

Figure B.9 shows the trajectories of the optimally controlled system in the state space (plane v–x, since time no longer appears explicitly). These trajectories are parabolae. Two of them are particularly important, since they lead to the desired final state without any further switches in u.

---

[6] Often referred to as "bang-bang control."

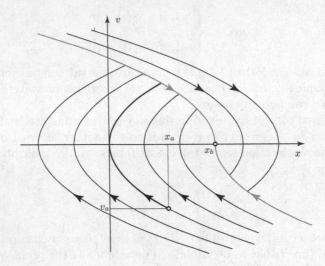

Fig. B.9 State trajectories of Example 4; dark: regular solutions; gray: switching parabolae

Incidentally, the same figure shows what happens if instead of a transfer from one rest point to another, any other initial and final states are considered. By choosing one final state, two parabolae are always defined, one with positive ($u(0) > 1$) and one with negative curvature, which pass through the final point. Only initial points situated on this branch reach the desired final state without any switch. All the other initial conditions lead to one switch.

A further interesting interpretation of Fig. B.9 results from the implementation of the control law found. Instead of calculating the control variable for each point in advance, the two switching curves can be calculated and then stored in memory. During the normal operation, the controller checks on-line whether the state is in the positive ($u(0) > 1$) or in the negative semi-plane, in order to determine the correct input. In this way an open-loop control scheme becomes a closed-loop one.

For any arbitrary initial point the total transfer time to the origin is given by

$$t_{total} = v_0 + 2 \cdot \sqrt{\frac{1}{2} \cdot v_0^2 + x_0}. \tag{B.74}$$

The *cost-to-go* curves (isolines of transfer time) have the shape shown in Fig. B.10. This consideration plays an important role in the extension of this approach, towards a formulation as a feed-back solution.

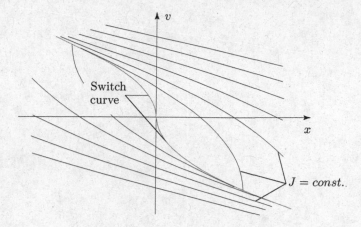

Fig. B.10 Cost-to-go isolines of Example 4

C

Appendix III – Dynamic Programming

This appendix summarizes the main concepts of dynamic programming, its implementation, and some automotive applications. Dynamic programming was developed during the 1950's by Richard Bellman [30] and has ever since been used as a tool to design optimal controllers for systems with constraints on the state variables and the control inputs. The first section explains the theory behind the algorithm and its complexity. The second section identifies some of the problems encountered when implementing dynamic programming algorithms. Finally, the third and last section shows an automotive application of dynamic programming, which illustrates some of the points introduced before. Readers interested in the details of the theory of deterministic and stochastic dynamic programming are referred to the standard text books [30, 33].

C.1 Introduction

For a given system, dynamic programming can be used to find the optimal control input that minimizes a certain cost function. The benefit of dynamic programming compared to standard optimal control theory is its ability to handle multiple complex constraints on both states and inputs and its low computational burden. The main drawback of the dynamic programming approach is that all disturbances[1] (in the case of deterministic dynamic programming), or at least their stochastic properties (in the case of stochastic dynamic programming) have to be known a priori. Therefore, dynamic programming is often not a useful method for the design of real-time control systems. Only in those cases where the disturbances (or their stochastic properties) are known at the outset, deterministic (or stochastic) dynamic programming can be used in real-time control applications. Nevertheless, dynamic programming is a very

---

[1] All exogenous signals are considered to be "disturbances," e.g. reference signals $r(t_k)$ generated by human drivers are considered to be disturbances as well.

useful tool since it can be used to provide an optimal performance benchmark. In the design of a causal real-time controller, this benchmark can then be used to assess the quality of the found controller. Moreover, in some cases the optimal non-realizable solution provides insights in how the suboptimal but realizable control system should be designed.

The dynamic programming theory [31, 30, 33] has been used in automotive applications, using both stochastic dynamic programming [196, 174, 170, 156] and deterministic dynamic programming [291, 108]. Dynamic programming can also be used to optimize the parameters of a power train [312, 313].

In this appendix, the emphasis will be on introducing the main ideas, discussing some implementation issues, and on providing one automotive example illustrating how dynamic programming can be used. The example shown at the end of this chapter deals with the torque split problem in a mild parallel hybrid electric vehicle. In addition to that, two detailed case studies (see Sect. A.2 and Sect. A.8) apply dynamic programming ideas to the problem of gear shifting and the optimization of driving strategies.

C.2   Theory

C.2.1   Problem Definition

This section provides the basic theoretical concept which will be used in the other sections. It is *not* exhaustive and readers interested in a detailed treatment of the subject are referred to the standard textbooks [30, 33]. The nomenclature used is adopted from [33].

In dynamic programming problems the following discrete-time dynamic system is considered

$$x_{k+1} = f_k(x_k, u_k, w_k), \quad k = 0, 1, ..., N - 1. \tag{C.1}$$

The dynamic states $x_k \in X_k \subset I\!R_\delta^n$, the control inputs $u_k \in U_k \subset I\!R_\delta^m$, and the disturbances $w_k \in D_k \subset I\!R_\delta^d$ are discrete variables both in time (index k) and value (thus the subscript $I\!R_\delta$ of the vector spaces). The control inputs u_k are limited to the subset U_k which can depend on the values of the state variables x_k, i.e., $u_k \in U_k(x_k)$. As mentioned above, the disturbance w_k must be known in advance for all $k \in [0, N - 1]$.

A specific control sequence (or "policy") is denoted by $\pi = \{\mu_0, \mu_1, ...\mu_{N-1}\}$, and the cost of using π on the problem (C.1) with the initial condition x_0 is defined by

$$J_\pi(x_0) = g_N(x_N) + \sum_{k=0}^{N-1} g_k(x_k, \mu_k(x_k)). \tag{C.2}$$

With these definitions, the optimal trajectory π^o is the trajectory that minimizes J_π

$$J^o(x_0) = \min_{\pi \in \Pi} J_\pi(x_0). \tag{C.3}$$

C.2.2 Principle of Optimality

The principle of optimality [30], as illustrated in Fig. C.1, provides the main insight in how to solve this problem.

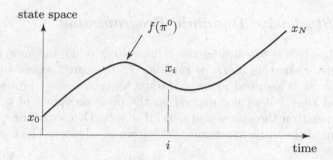

Fig. C.1 Principle of optimality: assume that $\pi^o(x_0)$ is the optimal policy when going from x_0 to x_N and that, when using this policy, a point x_i is reached; then the policy $\pi^o(x_i)$ is the optimal solution of the optimization problem between x_i and x_N

Let $\pi^o = \{\mu_0^o, \mu_1^o, ... \mu_{N-1}^o\}$ be an optimal policy/trajectory for the basic problem (C.1), and assume that when using π^o a given state x_i is reached at time i. Now consider the optimization problem with the initial condition at x_i at time i and the cost-to-go from time i to time N defined by

$$E\left\{ g_N(x_N) + \sum_{k=i}^{N-1} g_k(x_k, \mu_k(x_k), w_k) \right\}. \tag{C.4}$$

Then the truncated policy $\pi^o(x_i) = \{\mu_i^o, \mu_{i+1}^o, ... \mu_{N-1}^o\}$ is optimal for this new problem.

C.2.3 Deterministic Dynamic Programming

For every initial state x_0 the optimal cost $J^o(x_0)$ of the deterministic problem (C.1) is equal to $J_0(x_0)$ where J_0 is given by the last step of the following algorithm, which proceeds *backwards* in time from $N - 1$ to 0:

1. end cost calculation step

$$J_N(x_N) = g_N(x_N), \tag{C.5}$$

2. intermediate calculation step

$$J_k(x_k) = \min_{u_k \in U_k(x_k)} \{g_k(x_k, u_k) + J_{k+1}(f_k(x_k, u_k))\}, \tag{C.6}$$

then, if $u_k^o = \mu_k^o(x_k)$ minimizes the right side of (C.6) for each x_k and k, the policy $\pi^o = \{\mu_0^o, ..., \mu_{N-1}^o\}$ is optimal. This algorithm is referred to as deterministic dynamic programming (DDP) since the disturbance has to be known exactly in advance.

C.2.4 Stochastic Dynamic Programming

Another approach is to consider the same system (C.1), but now to assume that the disturbance $w_k \in D_k$ is random. To be more precise, the random disturbance w_k is assumed to be a Markov process, i.e., its probability distribution at time k does not depend on the previous values of k. However, w_k can depend on the states and control inputs. Of course the probability distribution of the disturbance must be known in advance. The cost criteria

$$J_\pi(x_0) = E_{w_{k_{k=0,1,...,N-1}}} \left\{ g_N(x_N) + \sum_{k=0}^{N-1} g_k(x_k, \mu_k(x_k), w_k) \right\} \quad (C.7)$$

is the *expected* cost of using the control policy $\pi = \{\mu_0, \mu_1, ...\mu_{N-1}\}$ on this stochastic problem with $u_k = \mu(x_k)$ and with the initial condition x_0. The optimal policy π^o is then the policy that minimizes J_π

$$J^o(x_0) = \min_{\pi \in \Pi} J_\pi(x_0). \quad (C.8)$$

When applying the principle of optimality to the problem (C.1), (C.7)-(C.8) it is possible to create the algorithm:

1. end cost calculation step

$$J_N(x_N) = g_N(x_N), \quad (C.9)$$

2. intermediate calculation step

$$J_k(x_k) = \min_{u_k \in U_k(x_k)} E_{w_k} \left\{ g_k(x_k, u_k, w_k) + J_{k+1}(f_k(x_k, u_k, w_k)) \right\}, \quad (C.10)$$

which proceeds backward in time from $N-1$ to 0. For every initial state x_0, the optimal cost $J^o(x_0)$ of the stochastic problem is equal to $J_0(x_0)$ where J_0 is given by the last step of the algorithm (C.9)-(C.10). The expectation is taken with respect to the probability distribution of w_k, which, in general, depends on x_k and u_k. Furthermore, if $u_k^o = \mu_k^o(x_k)$ minimizes the right side of (C.10) for each x_k and k, the policy $\pi^o = \{\mu_0^o, ..., \mu_{N-1}^o\}$ is optimal. The algorithm (C.9)-(C.10) is referred to as the stochastic dynamic programming (SDP) approach.

In some problems a stochastic model of the disturbances can be obtained with relatively low effort a priori, and a *real-time* SDP can be developed using the approach described above. Usually, the DDP approach is used to provide a benchmark, which is then used to assess the quality of other real-time controllers. However, in those cases where the "disturbance" is known a priori,[2] the DDP approach can be used to design real-time controllers as well. Adopting a predictive control strategy [112] represents a compromise between the two extreme cases.

In the following sections the term dynamic programming will refer to both the deterministic and the stochastic approach. Which one is meant will be clear from the context.

C.2.5 Complexity

The objective of the problem (C.1)-(C.3) and (C.7)-(C.8) is to find an optimal solution π^o. The obvious brute force method is to test all possibilities $\pi \in \Pi$. However, this is, in general, not feasible. For instance, to optimize the gear shifting control strategy for a vehicle with 5 gears with a vehicle model sampled with 1 s intervals on a driving cycle lasting 60 s, the number of possible control inputs u_k is 5^{60}. If each evaluation of the model requires 10^{-9} s CPU time, such an approach would require in the order of 10^{25} years. Using the principle of optimality and the dynamic programming algorithm the computation time will be $60 \cdot 5 \cdot 5 \cdot 10^{-9}$ s $= 1.5$ μs. In fact, for this example the DDP algorithm has a number of computations in the order of only $\mathcal{O}(N \cdot p \cdot q)$ where N is the number of time steps, and p and q are the numbers of possible state and input values (value discretization).

In general, the computational burden of all DDP algorithms scales linearly with the problem time N. Unfortunately, all dynamic programming algorithms have a complexity which is exponential in the number of states n and control inputs m

$$\mathcal{O}(N \cdot p^n \cdot q^m). \tag{C.11}$$

This makes the algorithm suitable only for low order systems. The parallel hybrid energy management problem analyzed in this appendix has a long problem duration, but is of low order. Therefore, DDP can be used successfully to derive an optimal control signal.

If the system has a high order (state variables and/or inputs), there are several methods to approximate the optimal solution (approximate dynamic programming) [33]. Obviously, the solution obtained with these methods is not guaranteed to be a global optimum.

---

[2] For instance, public transportation systems always follow fixed and well-known routes. The resulting speed and torque profiles, i.e., the "disturbances," can be assumed to be well known a priori.

C.3   Implementation Issues

There are several issues to consider when implementing the *deterministic* dynamic programming algorithm in general and when using Matlab in particular. The following points will be discussed in this section

- grid selection;
- nearest neighbor or interpolation; and
- scalar or set implementation

The grid selection refers to the selection of the discrete state and input spaces for the algorithm. The second issue deals with the problems a discrete state space introduces to the algorithm. The final issue is specific to the Matlab language and refers to the poor efficiency of some of the built in functions. The first two issues are not dependent on the implementation language and are thus relevant in all implementations.

C.3.1   Grid Selection

All dynamic programming algorithms are based on discrete decision processes and therefore a dynamic system with continuous inputs and states has to be approximated by a discrete-value system. This requires the state space to be limited and discretized as a first step. There are several ways of choosing the grid and the number of elements in it. For some problems this is straightforward. For example, the gear shifting case study of Sect. A.2 inherently has discrete state and input spaces. This is not the case for the parallel hybrid electric vehicle example analyzed below in Sect. C.3.4, where the input signal (the torque split factor) and the state variable (the battery state of charge) are continuous. A large number of grid elements p and q will yield an accurate solution, but will require long computational times and vice versa.

C.3.2   Nearest Neighbor or Interpolation

In all DP algorithms the computation of the last term of (C.6)

$$\ldots + J_{k+1}\big(f_k(x_k, u_k)\big)$$

poses the following problem: since $x_k \in X_k$ is discretized into a finite set of possible states, the term $J_k(x_k)$ also is only defined for these possible states. A problem arises when the value of the new state, calculated using $f_k(x_k, u_k)$, does *not* match any of the possible states in X_k. Figure C.2 shows this problem with $X_k = X_{k+1} \forall k$. All possible inputs, u_{1-3}, have to be evaluated when calculating $J_k(x_i)$, and the cost of using the respective control input has to be added to the cost-to-go from the resulting state

$$J_k(x_i) = \min \begin{bmatrix} g_k(x_i, u_1) + J_{k+1}(f_k(x_i, u_1)) \\ g_k(x_i, u_2) + J_{k+1}(f_k(x_i, u_2)) \\ g_k(x_i, u_3) + J_{k+1}(f_k(x_i, u_3)) \end{bmatrix} \qquad (C.12)$$

The state space X_k, and thus the cost-to-go J_k, is only defined for discrete values of the state x_k. Therefore the term $J_{k+1}(f_k(x_k, u_k))$ in (C.6) must be approximated. There are two main approaches to solve this problem, both of which have some benefits and some drawbacks. The first, very simple solution is to take the closest value and evaluate the cost-to-go using this value. Using the example (C.12) illustrated in Fig. C.2, this method produces the result

$$J_k(x_i) = \min \begin{bmatrix} g_k(x_i, u_1) + J_{k+1}(x^{i+1}) \\ g_k(x_i, u_2) + J_{k+1}(x^i) \\ g_k(x_i, u_3) + J_{k+1}(x^i) \end{bmatrix} \qquad (C.13)$$

The advantage of the nearest neighbor method is its computational speed. Its drawback is a relatively poor accuracy, which can require a very fine grid to achieve the desired precision.

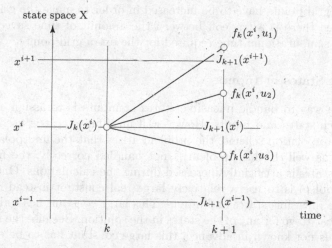

Fig. C.2 Illustration of the calculation of $J_k(x_i)$. The state variables at time $k+1$ are found starting at the time k with x_i and using the possible inputs u_{1-3}.

The second method to solve the problem is to use an interpolated value of the cost-to-go J_{k+1}. Different interpolation methods have been used, but linear interpolation is often sufficient to provide good accuracy while limiting the extra computational cost. For instance, the approximation of (C.12) when

using linear interpolation for the example illustrated in Fig. C.2 yields the
results

$$
J_k(x_i) = \min
\begin{bmatrix}
g_k(x_i, u_1) + (f_k(x_i, u_1) - x^{i+1}) \cdot \frac{J_{k+1}(x^{i+2}) - J_{k+1}(x^{i+1})}{x^{i+2} - x^{i+1}} \\[2ex]
g_k(x_i, u_2) + (f_k(x_i, u_2) - x^{i}) \cdot \frac{J_{k+1}(x^{i+1}) - J_{k+1}(x^{i})}{x^{i+1} - x^{i}} \\[2ex]
g_k(x_i, u_3) + (f_k(x_i, u_3) - x^{i-1}) \cdot \frac{J_{k+1}(x^{i}) - J_{k+1}(x^{i-1})}{x^{i} - x^{i-1}}
\end{bmatrix}
$$

$$(C.14)$$

When implementing the dynamic programming algorithm using interpolation
in Matlab there is an advantage of implementing custom interpolation functions since Matlab's interpolation functions `interp1` and `interp2` are not
optimized for speed and contain rigorous error checks. This makes custom
interpolation functions, optimized for speed, almost a necessity when using
interpolation in any dynamic programming algorithms.

Equally Spaced Grids

When using interpolation it can be beneficial to use grids that are equally
spaced between the elements. By using such grids the interpolation function's
(correctly implemented) computational time does not depend on the size of
the grid which reduces the computational time substantially. However, if the
number of grid points have to be increased in order to make the grid equally
spaced, then there is a tradeoff between the amount of time saved and the
additional computational time required for the extra grid points.

Infeasible States or Inputs

A common way to handle infeasible states or inputs is to assign an infinite
cost to such states and inputs. However, if an infinite cost is used together
with an interpolation scheme, the problem arises that the interpolated value
is infinity as well. If this problem is not handled correctly, the number of
infeasible states is artificially increased during the calculations. One solution
to this problem is to use a relatively large real constant instead of infinity
to penalize infeasible states and inputs. This large constant has to be greater
than the cost-to-go for any of the states in the solution. Because the maximum
cost-to-go is not known in advance, this large constant has to be estimated,
and some iterations may be necessary.

A more detailed description of the problems that occur when using dynamic programming with continuous state variables is found in [310]. The
authors point out that when interpolation is used, the evaluation of the term
$J_{k+1}(f_k(x_k, u_k))$ might rely on a grid-point that is not feasible, i.e., the final
state constraint cannot be met when starting from this point. If this problem
is not handled properly, a large error is introduced into the calculations, and
the found solution deteriorates. The authors propose to calculate the feasible

region prior to the actual dynamic programming iteration. The benefit of this approach is that a much coarser grid can be used for evaluation, which in turn reduces the computational effort. While this method works only for problems with just one state variable, the authors of [93] generalized these ideas for problems with any number of state variables.

C.3.3 Scalar or Set Implementation

The way the model (C.1), together with the dynamic programming algorithm, is implemented can reduce substantially the computational requirements due to certain specific properties of Matlab. A simple implementation of the DDP algorithm has a form similar to the one shown below. This implementation has been adopted for the generic dynamic programming function for MATLAB [311], which can be downloaded from http://www.idsc.ethz.ch/Downloads.

Standard Pseudo Code (Scalar Implementation)

```
for k = N-1 to 1
    forall indices i_x in X_grid
        reset Jk
        forall indices i_u in U_grid
            [x_k+1 g_k] = f(X_grid(i_x),U_grid(i_u),k,...)
            Jk = g_k + J(x_k+1,k+1)
            if Jk < previous Jk
                J_opt = Jk
                u_opt = U_grid(i_u)
            end if
        end forall
        J(i_x,k) = J_opt
        U(i_x,k) = u_opt
    end forall
end for
```

With this standard implementation there is one for-loop for every state and input variable plus one for-loop associated to the time. Unfortunately, Matlab handles for-loops not very efficiently. A much better approach is to utilize Matlab's vector-based algorithms, which requires the for-loops associated with the state and input variables

```
[x_k+1 g_k] = f(X_grid(i_x),U_grid(i_u),k,...)
```

to be replaced by a vector-input vector-output approach

```
[x_k+1 G] = F(X_grid,U_grid,k,...)
```

The form of such a *set implementation* is shown below.

Optimized Pseudo Code (Set Implementation)

```
for k = N-1 to 1
    [x_k+1 G] = F(X_grid,U_grid,k,...)
    Jk = G + V(x_k+1,k+1)
    J(i_x,k) = min Jk
    U(i_x,k) = argmin Jk
end
```

The advantages that can be obtained by using this set implementation are substantial. Fig. C.3 shows the computation times needed to analyze the example discussed below. Obviously, the speedup factor is so large that it is worth implementing the set function approach, even if this entails some extra programming efforts.

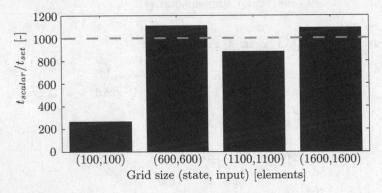

Fig. C.3 Computation time ratio between scalar implementation and set implementation for different state and input grids for the HEV Example (see Fig. C.4). The dashed line shows an average (1000 times) obtained with the three largest grids.

C.3.4 Example: Mild Parallel HEV – Torque Split

For the case of a parallel hybrid electric vehicle, this example shows how the optimal torque control strategy can be found using DDP. An overview of the model and its subsystems is shown in Fig. C.4. The discrete model is very similar to the model of the dual-clutch gearbox system analyzed in the case study of Sect. A.2). In addition, the HEV model includes a torque split device, an electric motor, and a battery. The test cycle is again the MVEG–95, while the gear shifting strategy is the standard one (middle plot in Fig. A.5).

The model has one state x_k, the battery state of charge, and one input u_k, the torque split factor. A time step Δt of 1 s is used. The total torque demanded from the two energy converters is

$$T_{dem} = T_{e0} + T_{m0} + T_g \tag{C.15}$$

where T_{e0} and T_{m0} are the drag torques of the engine and the electric motor. The torque split factor $u \in (-\infty, 1]$ determines the electric motor torque and the internal combustion engine torque according to

$$T_e = (1 - u) \cdot T_{dem} \tag{C.16}$$

and

$$T_m = u \cdot T_{dem} \tag{C.17}$$

In the HEV example considered, the motor and engine speeds are equal to the gearbox speed on the engine side $\omega_e = \omega_m = \omega_g$. The electric motor model is simply represented by an efficiency map, $\eta_m(\omega_m, T_m)$. The electric power provided to/by the battery is

$$P_m = \frac{T_m \cdot \omega_m}{\eta_m(\omega_m, T_m)} \tag{C.18}$$

Using the ideas introduced in Sect. 4.5.2, the battery is modeled as a voltage source (open circuit voltage) in series with a resistance (internal resistance). Both the open circuit voltage U_{oc} and the internal resistance R_i depend on the state of charge of the battery. The battery current is

$$I_b = \frac{U_{oc} - \sqrt{U_{oc}^2 - 4 \cdot R_i \cdot P_m}}{2 \cdot R_i} \tag{C.19}$$

The state of charge is calculated using

$$x_{k+1} = \frac{-I_b \, \eta_b(I_b)}{Q_0} \cdot \Delta t + x_k \tag{C.20}$$

where η_b is the battery charging efficiency and Q_0 is the battery capacity.

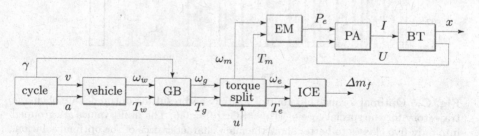

Fig. C.4 Parallel hybrid electric vehicle model components and the signal flow from the drive cycle through the model

DDP is used to calculate the optimal torque split factor. A midsize vehicle with a 1.6 l naturally aspirated engine and an electric motor with a maximum power of 23 kW is analyzed. The cost function is defined by

$$g_k(x_k, u_k) = \Delta m_f \qquad \text{(C.21)}$$

with the final cost

$$g_N(x) = \begin{cases} 0 & x \geq x_0 \\ \infty & x < x_0 \end{cases} \qquad \text{(C.22)}$$

The following discretization of the state- and input spaces is adopted

$$x \in [0.450, 0.452, \ldots, 0.658, 0.660]$$
$$u \in [-2.0, -1.9, \ldots, 0.9, 1.0] \qquad \text{(C.23)}$$

The resulting optimal torque-split control strategy, which depends on the state and on the time, is illustrated in Fig. C.5. This map is then used during a forward simulation of the model to evaluate and collect the various parameters shown in Fig. C.6.

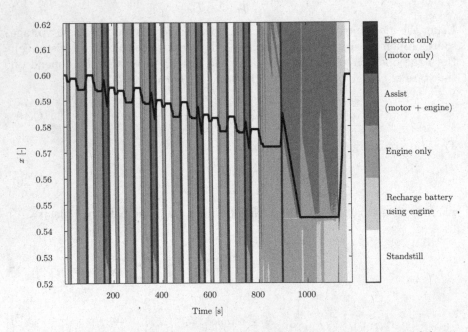

Fig. C.5 Optimal torque-split control strategy and resulting optimal state of charge trajectory for the special case of $x(0) = \mathrm{SOC}(0) = 0.6$. The input values are grouped into only five classes to better show the main characteristics of the optimal solution.

Figure C.6 shows that in this example the electric machine is used only for boosting and during braking maneuvers to recuperate energy. The resulting CO_2 emissions for the hybrid electric vehicle considered are $164\,\mathrm{g/km}$.

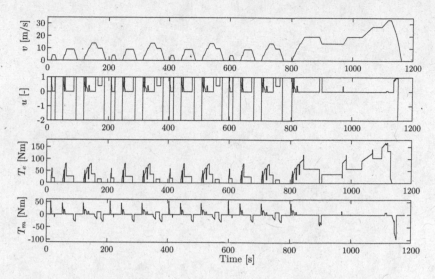

Fig. C.6 MVEG-95 cycle (top), optimal torque split factor (top, middle), engine torque (bottom, middle), electric motor torque (bottom)

References

1. Abe, S., Murata, M.: Development of IMA motor for 2006 Civic hybrid. SAE Paper 2006-01-1505 (2006)
2. Ai, X., Mohr, T., Anderson, S.: An electro-mechanical infinitely variable speed transmission. SAE Paper 2004-01-0354 (2004)
3. Allègre, A.L., Bouscayrol, A., Trigui, R.: Influence of control strategies on battery/supercapacitor hybrid energy storage systems for traction applications. In: Proc. of the IEEE Vehicle Power and Propulsion Conference, Dearborn, MI (2009)
4. Ambühl, D., Sciarretta, A., Onder, C., Guzzella, L., Sterzing, S., Mann, K., Kraft, D., Küsell, M.: A causal operation strategy for hybrid electric vehicles based on optimal control theory. In: Proc. of the 4th Braunschweig Symposium on Hybrid Vehicles and Energy Management, Braunschweig, Germany (2007)
5. Ambühl, D., Guzzella, L.: Predictive reference signal generator for hybrid electric vehicles. IEEE Transactions on Vehicular Technology 58(9), 4730–4740 (2009)
6. Ambühl, D., Sundström, O., Sciarretta, A., Guzzella, L.: Explicit optimal control policy and its practical implementation for hybrid electric powertrains. Control Engineering Practice 18, 1429–1439 (2010)
7. Amine, K.: Low cost components: Advanced high power and high energy battery materials. DOE Vehicle Technologies Program Annual Merit Review (2008)
8. Amphlett, J.C., Baumert, R.M., Mann, R.F., Peppley, B.A.: System analysis of an integrated methanol steam reformer/PEM fuel cell power generating system. In: Proc. of the 27th Intersociety Energy Conversion Engineering Conference, San Diego, CA (1992)
9. Amphlett, J.C., Baumert, R.M., Mann, R.F., Peppley, B.A., Roberge, P.R., Rodrigues, A.: Parametric modelling of the performance of a 5-kW proton-exchange membrane fuel cell stack. Journal of Power Sources 49, 349–356 (1994)
10. Amphlett, J.C., Creber, K.A.M., Davis, J.M., Mann, R.F., Peppley, B.A., Stokes, D.M.: Hydrogen production by steam reforming of methanol for polymer electrolyte fuel cells. International Journal of Hydrogen Energy 19(2), 131–137 (1994)

11. Amphlett, J.C., Mann, R.F., Peppley, B.A., Roberge, P.R., Rodrigues, A.: A model predicting transient responses of proton exchange membrane fuel cells. Journal of Power Sources 61, 183–188 (1996)

12. Amstutz, A., Guzzella, L.: Fuel cells for transportation – an assessment of its potential for CO_2 reduction. In: Proc. of the 2nd Conference on Greenhouse Gas Control, Interlaken, Switzerland (1999)

13. An, F., Vyas, A., Anderson, J., Santini, D.: Evaluating commercial and prototype HEVs. SAE Paper 2001-01-0951 (2001)

14. Anatone, M., Cipollone, R., Donati, A., Sciarretta, A.: Control-oriented modeling and fuel optimal control of a series hybrid bus. SAE Paper 2005-01-1163 (2005)

15. Ao, G.Q., Qiang, J.X., Zhong, H., Mao, X.J., Yang, L., Zhuo, B.: Fuel economy and NOx emission potential investigation and trade-off of a hybrid electric vehicle based on dynamic programming. Proc. of the IMechE, Part D - Journal of Automobile Engineering 222(D10), 1851–1864 (2009)

16. Apter, R., Präthaler, M.: Regeneration of power in hybrid vehicles. In: Proc. of the IEEE 55th Vehicular Technology Conference, Birmingham, AL (2002)

17. Arsie, I., Pianese, C., Rizzo, G., Santoro, M.: A model for the energy management in a parallel hybrid vehicle. In: Proc. of the 3rd International Conference on Control and Diagnostics in Automotive Applications, Genova, Italy (2001)

18. Audi Press Release: Audi Q5 hybrid quattro (2011), online version: http://www.audi-mediaservices.com

19. Audi Press Release: Audi A1 e-tron – electric driving in the city (2010), online version: http://www.audi-mediaservices.com

20. Bach, C., Lämmle, C., Bill, R., Dyntar, D., Onder, C.H., Boulouchos, K., Guzzella, L., Geering, H.P.: Clean engine vehicle – a natural gas-driven Euro-4/SULEV vehicle with 30% reduced CO2 emissions. SAE Paper 2004-01-0645 (2004)

21. Back, M.: Prädiktive Antriebsregelung zum energieoptimalen Betrieb von Hybridfahrzeugen. Diss. Univ. Karlsruhe (2005), online version: http://www.ubka.uni-karlsruhe.de

22. Bailey, K.E., Cikanek, S.R., Sureshbabu, N.: Parallel hybrid electric vehicle torque distribution method. In: Proc. of the American Control Conference, Anchorage, AK (2002)

23. Bailey, K.E., Powell, B.K.: A hybrid electric vehicle powertrain dynamic model. In: Proc. of the American Control Conference, Seattle, WA (1995)

24. Bansal, D., Rajagopalan, S., Choi, T., Guezennec, Y.G., Yurkovich, S.: Pressure and air-fuel ratio control of PEM fuel cell systems for automotive control. In: Proc. of the IEEE Vehicle Power and Propulsion Symposium, Paris, France (2004)

25. Barba, G., Glielmo, L., Perna, V., Vasca, F.: Current sensorless induction motor observer and control for hybrid electric vehicles. In: Proc. of the IEEE 32nd Power Electronics Specialists Conference, Vancouver, Canada (2001)

26. Barbarisi, O., Vasca, F., Glielmo, L.: State of charge Kalman filter estimator for automotive batteries. Control Engineering Practice 14(3), 267–275 (2006)

27. Barsali, S., Miulli, C., Possenti, A.: A control strategy to minimize fuel consumption of series hybrid electric vehicles. IEEE Transactions on Energy Conversion 19(1), 187–195 (2004)

28. Bazaraa, M.S., Sherali, H.D., Shetty, C.M.: Nonlinear programming. Theory and algorithms. John Wiley & Sons, West Sussex (1993)

29. Baumann, B.M., Washington, G., Glenn, B.C., Rizzoni, G.: Mechatronic design and control of hybrid electric vehicle. IEEE/ASME Transactions on Mechatronics 5(1), 58–72 (2000)

30. Bellman, R.E.: Dynamic programming. Princeton University Press, Princeton (1957)

31. Bellman, R.E., Lee, E.: History and development of dynamic programming. IEEE Control Systems Magazine 4(4), 24–28 (1984)

32. Bernard, J., Sciarretta, A., Touzani, Y., Sauvant-Moynot, V.: Advances in electrochemical models for predicting the cycling performance of traction batteries: experimental study on Ni-MH and simulation Oil & Gas Science and Technology – Rev. IFP 65(1), 55–66 (2010)

33. Bertsekas, D.P.: Dynamic programming and optimal control. Athena Scientific, Nashua NH (2007)

34. Biensan, P., Borthomieu, Y.: Saft Li-ion space batteries roadmap. In: Proc. of the NASA Aerospace Battery Workshop, Huntsville, AL (2007)

35. BMW Press Release: The 2010 BMW Activehybrid X6 – American market vision (2009), online version: http://www.press.bmwgroup.com

36. BMW Press Release: The 2010 BMW Activehybrid 7 – American market version (2009), online version: http://www.press.bmwgroup.com

37. Bohn, T.: Active combination of ultracapacitors and batteries for PHEV ESS. DOE Vehicle Technologies Program Annual Merit Review (2009)

38. Bosch Press Release: Start-stop systems reduce consumption and emissions (2007), online version: http://www.bosch-presse.de

39. Bossel, U.: Does a hydrogen economy make sense? Proceedings of the IEEE 94(10), 1826–1837 (2006)

40. Bowles, P., Peng, H., Zhang, X.: Energy management in a parallel hybrid electric vehicle with a continuously variable transmission. In: Proc. of the American Control Conference, Anchorage, AK (2002)

41. Bryson, E., Ho, Y.C.: Applied Optimal Control. Taylor & Francis, New York (1975)

42. Buie, L., Fry, M., Fussey, P., Mitts, C.: An application of cost based power management control strategies to hybrid fuel cell vehicles. SAE Paper 2004-01-1299 (2004)

43. Buller, S., Thele, M., Karden, E., De Doncker, R.W.: Impedance-based nonlinear dynamic battery modeling for automotive applications. Journal of Power Sources 113, 422–430 (2003)

44. Burnham, A., Wang, M., Wu, Y.: Development and Applications of GREET 2.7. Argonne National Laboratory, Report ANL/ESD/06-5 (2006), http://www.transportation.anl.gov/software/GREET/

45. Butler, K.L., Ehsani, M., Kamath, P.: A Matlab-based modeling and simulation package for electric and hybrid electric vehicle design. IEEE Transactions on Vehicular Technology 48(6), 1770–1778 (1999)

46. Ceraolo, M.: New dynamical models of lead–acid batteries. IEEE Transactions on Power Systems 15(4), 1184–1190 (2000)

47. Ceraolo, M., Barsali, S., Lutzenberger, G., Marracci, M.: Comparison of SC and high-power batteries or use in hybrid vehicles. SAE Paper 2009-24-0069 (2009)

48. Chan, C.C., Lo, E.W.C., Weixiang, S.: The available capacity computation model based on artificial neural network for lead–acid batteries in electric vehicles. Journal of Power Sources 87, 201–204 (2000)

384   References

49. Chan, C.C., Wu, J., Zhu, G.L., Chan, T.W.: Digital simulation of PWM inverter-induction motor drive system for electric vehicles. In: Proc. of the 14th Conference of Industrial Electronics Society, Singapore (1988)
50. Chan, H.L., Sutanto, D.: A new battery model for use with battery energy storage systems and electric vehicles power systems. In: Proc. of the IEEE Power Engineering Society Winter Meeting Conference, Singapore (2000)
51. Chan, S.H., Wang, H.M.: Thermodynamic and kinetic modelling of an autothermal methanol reformer. Journal of Power Sources 126, 8–15 (2004)
52. Chang, W.C., Nguyen, M.T.: Investigations of a platinumruthenium/next termcarbon nanotube catalyst formed by a two-step spontaneous deposition method. Journal of Power Sources 196(14), 5811–5816 (2011)
53. Chasse, A., Pognant-Gros, P., Sciarretta, A.: Online implementation of an optimal supervisory control for hybrid powertrains. SAE Paper 2009011868 (2009)
54. Chasse, A., Sciarretta, A.: Supervisory control of hybrid powertrains: An experimental benchmark of offline optimization and online energy management. Control Engineering Practice 19, 1253–1265 (2011)
55. Chau, K.T., Wong, Y.S.: Overview of power management in hybrid electric vehicles. Energy Conversion and Management 43, 1953–1968 (2002)
56. Chiasson, J., Tolbert, L., Lu, Y.: A library of SIMULINK blocks for real-time control of HEV traction drives. SAE Transactions Journal of Engines 2002, 2376–2385 (2002)
57. Christen, T., Carlen, M.W.: Theory of Ragone plots. Journal of Power Sources 91, 210–216 (2000)
58. Christen, T., Ohler, C.: Optimizing energy storage devices using Ragone plots. Journal of Power Sources 110, 107–116 (2002)
59. Christensen, J., Albertus, P., Sanchez-Carrera, R.S., Lohmann, T., Kozinsky, B., Liedtke, R., Ahmed, J., Kojic, A.: A critical review of Li/air batteries. Journal of the Electrochemical Society 159(2), R1–R30 (2012)
60. Chrysler Press Release: New 2009 Chrysler Aspen HEMI Hybrid and Dodge Durango HEMI Hybrid (2009), online version:
http://www.media.chrysler.com
61. Cipollone, R., Sciarretta, A.: Analysis of the potential performance of a combined hybrid vehicle with optimal supervisory control. In: Proc. of the IEEE International Conference on Control Applications, Munich, Germany (2006)
62. Cipollone, R., Sciarretta, A.: The quasi-propagatory model: a new approach for describing transient phenomena in engine manifolds. SAE Paper 2001-01-0579 (2001)
63. Cloudt, R., Willems, F.: Integrated emission management strategy for cost-optimal engine-aftertreatment operation. SAE Paper 2011-01-1310
64. College of the Desert: Hydrogen fuel cell engines and related technologies. Technical Report, Energy Technology Training Center (2001), online version:
http://www.eere.energy.gov
65. Conlon, B.: Comparative analysis of single and combined hybrid electrically variable transmission operating modes. SAE Paper 2005-01-1162 (2005)
66. Continental Press Release: Continental to supply booster module to PSA Peugeot Citroën for its e-HDI start-stop systems (2010), online version:
http://www.conti-online.com
67. Cross, D., Brockbank, C.: Mechanical hybrid system comprising a flywheel and CVT for motorsport and mainstream automotive applications. SAE Paper 2009-01-1312 (2008)

68. Cummins Press Release: Cummins demonstrates solid oxide fuel cell auxiliary power system (2010), online version: http://www.cumminspowerdocs.com
69. Daimler Press Release: ESX3 lowers fuel consumption, emissions and cost (2000), online version: http://www.daimlerchrysler.com
70. Daimler Press Release: The first fuel cell buses with hybrid technology are now in service in Hamburg (2011), online version: http://www.daimler.com
71. DaimlerChrysler Press Release: Sodium borohydride: fuel cell vehicle "Natrium" uses clean, safe technology to provide hydrogen on demand (2002), online version: http://www.daimlerchrysler.com
72. Daimler Press Release: Necar 1 to Necar 5: chemical powerplants on board (2004), online version: http://www.daimlerchrysler.com
73. Daimler Press Release: Mercedes-Benz ML 450 Hybrid: The perfect hybrid powertrain (2009), online version: http://www.media.daimler.com
74. Daimler Press Release: The 2009 Mercedes-Benz S-Class: setting the standard in the luxury class (2009), online version: http://www.media.daimler.com
75. Debal, P., Faid, S., Bervoets, S., Tricoche, L., Pauwels, B.: Development of a post-transmission hybrid powertrain. In: Proc. of the 24th Electric Vehicle Symposium, Stavanger, Norway (2009)
76. Debert, M., Colin, G., Mensler, M., Chamaillard, Y., Guzzella, L.: Li-ion battery models for HEV simulator. In: Proc. of the International Conference on Advances in Hybrid Powertrains, Rueil-Malmaison, France (2008)
77. Delprat, S., Lauber, J., Guerra, T.M., Rimaux, J.: Control of a parallel hybrid powertrain. IEEE Transactions on Vehicular Technology 53(3), 872–881 (2004)
78. Derdiyok, A.: A novel speed estimation algorithm for induction machines. Electric Power Systems Research 64, 73–80 (2003)
79. De Vidts, P., Delgado, J., White, R.E.: Mathematical modeling for the discharge of a metal hydride electrode. Journal of Electrochemical Society 142(12), 4006–4013 (1995)
80. Di Domenico, D., Prada, E., Creff, Y.: An Adaptive Strategy for Li-ion Battery SOC Estimation. In: Proc. of the 18th IFAC World Congress, Milan, Italy (2011)
81. Dietrich, P.: Gesamtenergetische Bewertung verschiedener Betriebsarten eines Parallel-Hybridantriebes mit Schwungradkomponente und stufenlosem Weitbereichsgetriebe für einen Personenwagen. Dissertation ETH no. 12958. Swiss Federal Institute of Technology, Zurich, Switzerland (1999)
82. Dietrich, P., Hörler, H., Eberle, M.K.: The ETH-hybrid car – a concept to minimize consumption and reduce emissions. In: Proc. of the 26th ISATA Conference, Aachen, Germany (1993)
83. Do, K., Guezennec, Y.G., Rizzoni, G.: Dynamic testing and electrical/thermal model validation of supercapacitors for hybrid electric vehicle powertrains. In: Proc. of the IEEE Vehicular Power and Propulsion Symposium, Paris, France (2004)
84. Dones, R., Bauer, C., Bolliger, R., Burger, B., Faist, M., Frischknecht, R., Heck, T., Jungbluth, N., Röder, A.: Sachbilanzen von Energiesystemen: Grundlagen für den ökologischen Vergleich von Energiesystemen und den Einbezug von Energiesystemen in Ökobilanzen für die Schweiz. Final report ecoinvent 2000, no. 6, Paul Scherrer Institut (2004), online version: http://www.ecoinvent.ch
85. Dönitz, C.: Hybrid pneumatic engines. Dissertation ETH no. 18761, Swiss Federal Institute of Technology, Zurich, Switzerland (2009)

86. Dönitz, C., Vasile, I., Onder, C., Guzzella, L.: Modelling and optimizing two- and four-stroke hybrid pneumatic engines. Proc. of the IMechE, Part D – Journal of Automobile Engineering 223(2), 255–280 (2009)

87. Dönitz, C., Voser, C., Vasile, I., Onder, C., Guzzella, L.: Validation of the fuel saving potential of downsized and supercharged hybrid pneumatic engines using vehicle emulation experiments. Journal of Engineering for Gas Turbines and Power 133(9) (2011)

88. Doyle, M., Newmann, J., Gozdz, A.S., Schmutz, C.N., Tarascon, J.M.: Comparison of modeling predictions with experimental data from plastic lithium ion cells. Journal of Electrochemical Society 143(6), 1890–1903 (1996)

89. Eaton technical: New hybrid technology in use (2010), online version: http://www.eaton.com

90. Ebbesen, S., Guzzella, L.: Trade-off between fuel economy and battery life for hybrid electric vehicles. In: Proc. of the 4th Annual Dynamic Systems and Control Conference, Arlington, VA (2011)

91. Economic Commission for Europe of the United Nations (UN/ECE). Regulation no. 101 (2008)

92. EG&G Technical Service, Science Applications International Corporation: Fuel cell handbook. US Department of Energy, National Energy Technology Laboratory, Morgantown WV (2002)

93. Elbert, P., Ebbesen, S., Guzzella, L.: Implementation of dynamic programming for n-dimensional optimal control problems with final state constraints. IEEE Transactions on Control Systems Technology (2012)

94. Ender, M.: Der Taktbetrieb als teillastverbessernde Massnahme bei Ottomotoren. Dissertation ETH no. 11835, Swiss Federal Institute of Technology, Zurich, Switzerland (1996)

95. Engstrom, S.: JCS PHEV system development-USABC. DOE Vehicle Technologies Program Annual Merit Review (2010)

96. Enova Technical Data: Enova systems at a glance (2011), online version: http://www.enovasystems.com

97. Ekdunge, P.: A simplified model of the lead/acid battery. Journal of Power Sources 46, 251–262 (1993)

98. Fan, H., Dawson, G.E., Eastham, T.R.: Model of an electric vehicle induction motor drive system. In: Proc. of the Canadian Conference on Electrical and Computer Engineering, Vancouver, Canada (1993)

99. Fenton, J.: Handbook of automotive powertrain and chassis design. Professional Engineering Publishing, London (1998)

100. Filipi, F., Louca, L., Daran, B., Lin, C.C., Yildir, U., Wu, B., Kokkolaras, M., Assanis, D., Peng, H., Papalambros, P., Stein, J., Szkubiel, D., Chapp, R.: Combined optimisation of design and power management of the hydraulic hybrid propulsion system for the 6 $times$ 6 medium truck. International Journal of Heavy Vehicle Systems 11(3/4), 371–401 (2004)

101. Filipi, Z., Kim, Y.J.: Hydraulic hybrid propulsion for heavy vehicles: combining the simulation and engine-in-the-loop techniques to maximize the fuel economy and emission benefits. Oil & Gas Science and Technology 65(1), 155–178 (2010)

102. Ford Press Release: Ford springs surprise with mighty F-350 Tonka (2002), online version: http://www.ford.com

103. Ford Press Release: Escape Hybrid (2004), online version: http://www.media.ford.com

104. Ford Press Release: Focus FCV: combining fuel cell and hybrid technology for the next generation of the automobile (2004), online version: http://www.media.ford.com

105. Ford Press Release: Ford to expand fuel-saving start-stop technology from hybrids to conventional cars, crossovers (2010), online version: http://www.media.ford.com

106. Ford Technical Data: Ford Focus FCV (2004), online version: http://www.ford.com

107. Fulop, R.: A123 systems. DOE Vehicle Technologies Program Annual Merit Review (2009)

108. Haj-Fraj, A., Pfeiffer, F.: Optimal control of gear shift operations in automatic transmissions. Journal of the Franklin Institute 338, 371–390 (2001)

109. Fuchs, R.D., Hasuda, Y., James, I.B.: Full toroidal IVT variator dynamics. SAE paper 2002-01-0586 (2002)

110. Gagliardi, F., Piccolo, A., Vaccaro, A., Villacci, D.: A fuzzy based control unit for the optimal power flow management in parallel hybrid electric vehicles. In: Proc. of the 19th Electric Vehicles Symposium, Busan, Korea (2002)

111. Galler, D.: Motors. In: Dorf, R. (ed.) The Electrical Engineering Handbook. CRC, Boca Raton (2000)

112. Garcia, J., Prett, J., Morari, M.: Model predictive control: theory and practice. Automatica 25, 335–348 (1989)

113. Gao, L., Liu, S., Dougal, R.A.: Dynamic lithium-ion battery model for system simulation. IEEE Transactions on Components and Packaging Technologies 25(3), 495–505 (2002)

114. Girishkumar, G., McCloskey, B., Luntz, A.C., Swanson, S., Wilcke, W.: Lithium-Air Battery: Promise and Challenges. Journal Phys. Chem. Lett. (2010)

115. GM Press Release: GM Hy-wire: major step forward in reinventing automobile (2003), online version: http://www.gm.com

116. GM Press Release: 2009 Cadillac Escalade Hybrid is the first hybrid in a large luxury SUV (2009), online version: http://www.media.gm.com

117. GM Press Release: Chevrolet Volt's revolutionary Voltec electric drive system (2010), online version: http://www.media.gm.com

118. Gomadam, P.M., Weidner, J.W., Dougal, R.A., White, R.E.: Mathematical modeling of lithium-ion and nickel battery systems. Journal of Power Sources 110, 267–284 (2002)

119. Gonder, J., Pesaran, A., Lustbader, J., Tataria, H.: Fuel economy and performance of mild hybrids with ultracapacitors: simulations and vehicle test results. Presented at the 5th International Symposium on Large EC Capacitor Technology and Application, Long Beach, CA (2009)

120. Gong, Q.M., Li, Y.Y., Peng, Z.R.: Trip-based optimal power management of plug-in hybrid electric vehicles. IEEE Transactions on Vehicular Technology 57(6), 3393–3401 (2008)

121. Gopalakrishnan, S., Namuduri, C., Reynolds, M.: Ultracapacitor based active energy recovery scheme for fuel economy improvement in conventional vehicles. SAE Paper 2011-01-0345 (2011)

122. Grondin, O., Thibault, L., Moulin, P., Chasse, A., Sciarretta, A.: Energy Management Strategy for Diesel Hybrid Electric Vehicle. In: Proc. of the IEEE Vehicle Power and Propulsion Conference, Chicago, IL (2011)

123. Gu, W.B., Yang, C.Y., Li, S.M., Geng, M.M., Liaw, B.Y.: Modeling discharge and charge characteristics of nickel–metal hydride batteries. Electrochemica Acta 44, 4525–4541 (1999)
124. Gu, B., Rizzoni, G.: An adaptive algorithm for hybrid electric vehicle energy management based on driving pattern recognition. In: Proc. of the 2006 ASME International Mechanical Engineering Congress and Exposition, Chicago, IL (2006)
125. Guebeli, M., Micklem, J.D., Burrows, C.R.: Maximum transmission efficiency of a steel belt continuously variable transmission. Transactions of the ASME, Journal of Mechanical Design 115, 1044–1048 (1993)
126. Guzzella, L.: Control oriented modeling of fuel-cell based vehicles. Pres. at NSF Workshop on the Integration of Modeling and Control for Automotive Systems, Santa Barbara CA (1999)
127. Guzzella, L., Amstutz, A.: CAE tools for quasi-static modeling and optimization of hybrid powertrains. IEEE Transactions on Vehicular Technology 48(6), 1762–1769 (1999)
128. Guzzella, L., Onder, C.H.: Introduction to modeling and control of internal combustion engines systems. Springer, Berlin (2004)
129. Guzzella, L., Schmid, A.M.: Feedback linearization of spark-ignition engines with continuously variable transmissions. IEEE Transactions on Control Systems Technology 3(1), 54–60 (1995)
130. Guzzella, L., Sciarretta, A.: Global optimization of rendez-vous maneuvers for passenger cars. In: Proc. of the 5th Stuttgart International Symposium, Stuttgart, Germany (2003)
131. Guzzella, L., Wenger, U., Martin, R.: IC engine downsizing and pressure wave supercharging for fuel-economy. SAE Paper 2000-01-1019 (2000)
132. Guzzella, L., Wittmer, C., Ender, M.: Optimal operation of drivetrains with SI-engines and flywheels. In: Proc. of the 13th IFAC World Congress, San Francisco, CA (1996)
133. Hauer, K.H.: Numerical simulation of two different ultra capacitor hybrid fuel cell vehicles. In: Proc. of the 18th Electric Vehicle Symposium, Berlin, Germany (2001)
134. Hauer, K.H., Moore, R.M., Ramaswamy, S.: A simulation model for an indirect methanol fuel cell vehicle. SAE Paper 2001-01-3083 (2001)
135. Hawkins, L., Murphy, B., Zierer, J., Hayes, R.: Shock and vibration testing of an AMB supported energy storage flywheel. In: Proc. of the 8th International Symposium on Magnetic Bearings, Mito, Japan (2002)
136. Heywood, J.B.: Internal combustion engine fundamentals. McGraw Hill, New York (1988)
137. Higelin, P., Charlet, A., Chamaillard, Y.: Thermodynamic simulation of a hybrid pneumatic-combustion engine concept. Int. Journal of Applied Thermodynamics 5(1), 1–11 (2002)
138. Hiratsuka, H., Sakamoto, T., Arimoto, S., Furuta, H., Yamamoto, N.: Development of Nickel-Manganese-Cobalt based cathode material (PSS) for lithium ion batteries. Matsushita Technical Journal 52, 36–40 (2006)
139. Hissel, D., Péra, M.C., Francois, X., Kauffman, J.M., Baurens, P.: Contribution to the modelling of automotive systems powered by polymer electrolyte fuel cells. In: Proc. of the European Automotive Congress, Bratislava, Slovakia (2001)

140. Honda Press Release: The New Honda IMA System is more efficient than ever (2007), online version: http://www.world.honda.com
141. Honda Press Release: Honda introduces new fuel cell-powered vehicle, FCX-V4 (2001), online version: http://www.world.honda.com
142. Huang, Y.M., Wang, K.J.: A hybrid power driving system with an energy storage flywheel for vehicles. SAE Paper 2007-01-4114 (2007)
143. Hucho, W.H.: Aerodynamics of road vehicles. SAE Publishing, Warrendale (1998)
144. Husted, H.: A comparative study of the production applications of hybrid electric powertrains. SAE Paper 2003-01-2307 (2003)
145. Hyundai Press Release: Hyundai Sonata Hybrid debuts at New York Auto Show (2010), online version: http://www.hyundainews.com
146. Hyundai Press Release: Hyundai unveils Tucson ix hydrogen fuel cell electric vehicle at Fuel Cell and Hydrogen Energy 2011 (2011), online version: http://www.hyundaiusa.com
147. Idaho National Laboratory: Full size electric vehicles (2010), online version: http://avt.inel.gov/fsev.shtml
148. Ide, T., Udagawa, A., Kataoka, R.: Experimental investigation on shift-speed characteristics of a metal V-belt CVT. SAE Paper 9636330 (1996)
149. Inanc, N.: A new slinding mode flux and current observer for direct field oriented induction motor drives. Electric Power Systems Research 63, 113–118 (2002)
150. Infiniti Press Release: The drivers hybrid: low CO2 Infiniti with the power to please (2010), online version: http://www.infinitipress.eu
151. Isidori, A.: Nonlinear control systems. Springer, Berlin (2003)
152. Itagaki, K., Teratani, T., Kuramochi, K., Nakamura, S., Tachibana, T., Nakao, H., Kamijo, Y.: Development of the Toyota mild-hybrid system (THS-M). SAE Paper 2002-01-0990 (2002)
153. Jeanneret, B., Markel, T.: Adaptive energy management strategy for fuel cell hybrid vehicles. SAE Paper 2004-01-1298 (2004)
154. Jeon, S.I., Jo, S.T., Park, Y.I., Lee, J.M.: Multi-mode driving control of a parallel hybrid electric vehicle using driving pattern recognition. Transactions of the ASME, Journal of Dynamic Systems, Measurement, and Control 124, 141–149 (2002)
155. Jeon, S.I., Kim, K.B., Jo, S.T., Lee, J.M.: Driving simulation of a parallel hybrid electric vehicle using receding horizon control. In: Proc. of the IEEE International Symposium on Industrial Electronics, Pusan, Korea (2001)
156. Johannesson, L., Asbogård, M., Egardt, B.: Assessing the potential of predictive control for hybrid vehicle powertrains using stochastic dynamic programming. IEEE Transactions on Intelligent Transportation Systems 8(1), 71–83 (2007)
157. Johannesson, L., Egardt, B.: A novel algorithm for predictive control of parallel hybrid powertrains based on dynamic programming. In: Proc. of the 5th IFAC Symposium on Advances in Automotive Control, Monterey, CA (2007)
158. Johnson, V.H., Wipke, K.B., Rausen, D.J.: HEV control strategy for real-time optimization of fuel economy and emissions. SAE Paper 2000-01-1543 (2000)
159. Johnson, V.H.: Battery performance models in ADVISOR. Journal of Power Sources 110, 321–329 (2002)
160. Kailath, T.: Linear systems. Prentice Hall, Englewood Cliffs (1980)

161. Kargul, J.J.: Affordable advanced technologies. Presented to FACA Meeting (2004), online version: http://www.epa.gov

162. Kassakian, J.G., Miller, J.M., Traub, N.: Automotive electronics power up. IEEE Spectrum, 34–39 (May 2000)

163. Kaushik, R., Mawston, I.G.: Discharge characterization of lead/acid batteries. Journal of Power Sources 28, 161–169 (1989)

164. Kelly, K.J., Zolot, M., Glinsky, G., Hieronymus, A.: Test results and modeling of the Honda Insight using ADVISOR. SAE Paper 2001-01-2537 (2001)

165. Kempton, K., Kubo, T.: Electric-drive vehicles for peak power in Japan. Energy Policy 28, 9–18 (2000)

166. Kessels, J.T.B.A., Koot, M.W.T., van den Bosch, P.P.J., Kok, D.B.: Online energy management for hybrid electric vehicles. IEEE Transactions on Vehicular Technology 57(6), 3428–3440 (2008)

167. Kessels, J.T.B.A., Willems, F., Schoot, W., van den Bosch, P.P.J.: Integrated energy and emission management for hybrid electric truck with scr aftertreatment. In: Proc. of the IEEE Vehicle Power and Propulsion Conference (VPPC), Lille, France (2010)

168. Kiencke, U., Nielsen, L.: Automotive control systems. Springer, Berlin (2000)

169. Kim, K.H., Baik, I.C., Chung, S.K., Youn, M.J.: Robust speed control of brushless DC motor using adaptive input–output linearisation technique. IEE Proceedings on Electric Power Applications 144, 469–475 (1997)

170. Kim, M.J., Peng, H., Lin, C.C., Stamos, E., Tran, D.: Testing, modeling, and control of a fuel cell hybrid vehicle. Control Engineering Practice 13, 41–53 (2005)

171. Kim, N., Cha, S., Peng, H.: Optimal control of hybrid electric vehicles based on Pontryagin's minimum principle. IEEE Transactions on Control Systems Technology 99, 1–9 (2010)

172. Kircher, O., Brunner, T.: Advances in cryo-compressed hydrogen vehicle storage (CcH2). Pres. at the FISITA 2010 World Automotive Congress, Budapest, Hungary (2010)

173. Kleimaier, A., Schröder, D.: An approach for the online optimized control of a hybrid powertrain. In: Proc. of the 7th International Workshop on Advanced Motion Control, Maribor, Slovenia (2002)

174. Kolmanovsky, I., Siverguina, I., Lygoe, B.: Optimization of powertrain operating policy for feasibility assessment and calibration: Stochastic dynamic programming approach. In: Proc. of the American Control Conference, Anchorage, AK (2002)

175. Komiyama, S., Inoue, H., Iwano, H., Yamaguchi, I., Negome, T.: Drive control device for hybrid vehicle. U.S. Patent no. 7055636 B2 (2004)

176. Koo, E.S., Lee, H.D., Sul, S.K., Kim, J.S.: Torque control strategy for a parallel hybrid vehicle using fuzzy logic. In: Proc. of the 1998 IEEE Industry Application Conference, St. Louis, MO (1998)

177. Koot, M., Kessels, J., de Jager, B., Heemels, W., van den Bosch, P., Steinbuch, M.: Energy management strategies for vehicular electric power systems. IEEE Transactions on Vehicular Technology 54(3), 1504–1509 (2005)

178. Kötz, R., Baertschi, M., Buechi, F., Gallay, R., Dietrich, P.: HY.POWER – A fuel cell car boosted with supercapacitors. In: Proc. of the 12th International Seminar on Double Layer Capacitors and Similar Energy Storage Devices, Deerfield Beach, FL (2002)

179. Krause, P., Wasynczuk, O., Sudhoff, S.: Analysis of electric machinery. Wiley-IEEE Press, West Sussex (2002)

180. Kuhn, E., Forgez, C., Lagonotte, P., Friedrich, G.: Modelling Ni-mH battery using Cauer and Foster structures. Journal of Power Sources 158(2), 1490–1497 (2006)

181. Marie-Françoise, J.N., Gualous, H., Outbib, R.: Berthon 42 V power net with supercapacitor and battery for automotive applications. Journal of Power Sources 143(1-2), 275–283 (2005)

182. Kuperman, A., Aharon, I.: Battery–ultracapacitor hybrids for pulsed current loads: A review. Renewable and Sustainable Energy Reviews 15(2), 981–992 (2011)

183. Lajnef, W., Vinassa, J.M., Briat, A.S., Woigard, E.: Characterization methods and modelling of ultracapacitors for use as peak power sources. Journal of Power Sources 168(2), 553–560 (2007)

184. Larsson, V., Johannesson, L., Egardt, B.: Influence of state of charge estimation uncertainty on energy management strategies for hybrid electric vehicles. In: Proc. of the 18th IFAC World Congress, Milan, Italy (2011)

185. Lee, J.H., Lalk, T.R., Appleby, A.J.: Modeling electrochemical performance in large scale proton exchange membrane fuel cell stacks. Jounal of Power Sources 70, 258–268 (1998)

186. Lee, C.K., Kwok, N.M.: Torque ripple reduction in brushless DC motor velocity control systems using a cascade modified model reference compensator. In: Proc. of the 24th IEEE Power Electronics Specialists Conference, Seattle, WA (1993)

187. Lee, C.Y., Zhao, H., Ma, T.: Analysis of a cost effective air hybrid concept. SAE Paper 2009-01-1111 (2009)

188. Lee, H.D., Sul, S.K.: Fuzzy-logic-based torque control strategy for parallel-type hybrid electric vehicle. IEEE Transactions on Industrial Electronics 45(4), 625–632 (1998)

189. Lee, W., Choi, D., Sunwoo, M.: Modelling and simulation of vehicle electric power system. Journal of Power Sources 109, 58–66 (2002)

190. Lee, H.H., Muralidharan, P., Ruffo, R., Mari, C.M., Cui, Y., Kim, D.K.: Ultrathin spinel LiMn2O4 nanowires as high power cathode materials for Li-ion batteries. Nano Lett. 10(10), 3852–3856 (2010)

191. Lescot, J., Sciarretta, A., Chamaillard, Y., Charlet, A.: On the integration of optimal energy management and thermal management of hybrid electric vehicles. In: Proc. of the IEEE Vehicle Power and Propulsion Conference, Lille, France (2010)

192. LG Chem Press Release: LG Chem battery cells to power Chevrolet Volt (2009), online version: http://www.compactpower.com/

193. Lightning Hybrids Technical Data: Carbonweight high pressure bladder accumulators (2011), online version: http://www.lightninghybrids.com

194. Lin, C.C., Kang, J.M., Grizzle, J.W., Peng, H.: Energy management strategy for a parallel hybrid electric truck. In: Proc. of the American Control Conference, Arlington, VA (2001)

195. Lin, C.C., Peng, H., Grizzle, J.W., Kang, J.M.: Power management strategy for a parallel hybrid electric truck. IEEE Transactions on Control Systems Technology 11(6), 839–849 (2003)

196. Lin, C.C., Peng, H., Grizzle, J.W.: A stochastic control strategy for hybrid electric vehicles. In: Proc. of the 2004 American Control Conference, Boston, MA (2004)

197. Lin, F.J., Wai, R.J., Kuo, R.H., Liu, D.C.: A comparative study of sliding mode and model reference adaptive speed observers for induction motor drive. Electric Power Systems Research 44, 163–174 (1998)

198. Liu, H., Wang, Y., Yang, W., Zhou, H.: A large capacity of LiV_3O_8 cathode material for rechargeable lithium based batteries. Electrochemica Acta 56(3), 1392–1398 (2011)

199. Liu, C., Yu, Z., Neff, D., Zhamu, A., Jang, B.Z.: Graphene-Based Supercapacitor with an Ultrahigh Energy Density. Nano Letters 10(12), 4863–4868 (2010)

200. Liu, J.M., Peng, H.E.: Modeling and control of a powersplit hybrid vehicle. IEEE Transactions on Control Systems Technology 16(6), 1242–1251 (2008)

201. Magasinki, A., Dixon, P., Hertzberg, B., Kvit, A., Ayala, J., Yushin, G.: High-performance lithium-ion anodes using a hierarchical bottom-up approach. Nature Materials 9, 353–358 (2010)

202. Maggetto, G., Van Mierlo, J.: Electric vehicles, hybrid vehicles and fuel cell electric vehicles: state of the art and perspectives. Ann. Chim. Sci. Mat. 26(4), 9–26 (2001)

203. Maggetto, G., Van Mierlo, J.: Electric and electric hybrid vehicle technology: a survey. IEE Seminar on Electric, Hybrid and Fuel Cell Vehicles (2000)

204. Mann, R.F., Amphlett, J.C., Hooper, M.A.I., Jensen, H.M., Peppley, B.A., Roberge, P.R.: Development and application of a generalized steady-state electrochemical model for a PEM fuel cell. Journal of Power Sources 86, 173–180 (2000)

205. Marino, R., Peresada, S., Tomei, P.: On-line stator and rotor resistance estimation for induction motors. IEEE Transactions on Control Systems Technology 8, 570–580 (2000)

206. Martellucci, L.: Sipre: a family of new hybrid propulsion systems. In: Proc. of the 27th ISATA Conference, Aachen, Germany (1994)

207. Martin, S., Wörner, A.: On-board reforming of biodiesel and bioethanol for high temperature PEM fuel cells: Comparison of autothermal reforming and steam reforming. Journal of Power Sources 196(6), 3163–3171 (2010)

208. Martinet, S., Pelissier, S.: Batteries for electric and hybrid vehicles: State of the art, modeling, testing, aging. Tutorial in the IEEE Vehicle Power and Propulsion Conference, Lille, France (2010)

209. Maxoulis, C.N., Tsinoglou, D.N., Koltsakis, G.C.: Modeling of automotive fuel cell operation in driving cycles. Energy Conversion and Management 45, 559–573 (2004)

210. Maxwell Press Release: Valeo and Maxwell technologies agree to collaboratively develop ultracapacitor-based energy storage system for hybrid autos (2007), online version: http://www.about.maxwell.com

211. Maxwell Technical Data: 125 Volt transportation modules (2011), online version: http://www.maxwell.com

212. Mayer, J.S., Wasynczuk, O.: Analysis and modeling of a single-phase brushless DC motor drive system. IEEE Transactions on Energy Conversion 4, 473–479 (1989)

213. Mazda Press Release: Mazda develops fuel cell electric vehicle, "Demio FCEV" (1997), online version: http://www.mazda.com

214. McConnell, V.: High-temperature PEM fuel cells: Hotter, simpler, cheaper. Fuel Cells Bulletin 2009(12), 12–16 (2010)
215. Mercedes-Benz Technical Data: The B-Class with fuel cell electric drive technology: The propulsion system of tomorrow (2009), online version: http://www.mbusa.com
216. Miller, J., Everett, M.: An assessment of ultra-capacitors as the power cache in Toyota THS-II, GM Allison AHS-2 and Ford FHS hybrid propulsion systems. In: Proc. of the IEEE Applied Power Electronics Conference and Exposition, Austin, TX (2005)
217. Miller, J., McCleer, P., Everett, M.: Comparative assessment of ultra-capacitors and advanced battery energy storage systems in powersplit electronic-CVT vehicle powertrains. In: Proc. of the IEEE International Conference on Electric Machines and Drives, San Antonio, TX (2005)
218. Miller, J.M., Sartorelli, G.: Battery and ultracapacitor combinations – where should the converter go? In: Proc. of the IEEE Vehicle Power and Propulsion Conference, Lille, France (2010)
219. Miller, M.A., Holmes, A.G., Conlon, B.M., Savagian, P.J.: The GM "Voltec" 4ET50 multi-mode electric transaxle. SAE Paper 2011-01-0887 (2011)
220. Mischker, K., Denger, D.: Requirements for a fully variable valvetrain and realization with the electro-hydraulic valvetrain system EHVS. VDI-Fortschritt-Berichte 12(539) (2003)
221. Mitsubishi Press Release: Mitsubishi Motors to bring new-generation EV i-MiEV to market (2009), online version: http://www.mitsubishi-motors.com
222. Mitsui, K., Nakamura, H., Okamura, M.: Capacitor-Electronic Systems (ECS) for ISG/idle stop applications. In: Proc. of the 19th Electric Vehicle Symposium, Busan, Korea (2002)
223. Mock, P., German, J., Bandivadekar, A., Riemersma, I.: Discrepancies between type-approval and "real-world fuel-consumption and CO2 values. The international council on clean transportation (2012), online version: http://www.theicct.org
224. Moura, S.J., Stein, J.L., Fathy, H.K.: Battery health-conscious power management for plug-in hybrid electric vehicles via stochastic control. In: Proc. of the 2010 ASME Dynamic Systems and Control Conference (2010)
225. Musardo, C., Rizzoni, G., Staccia, B.: A-ECMS: an adaptive algorithm for hybrid electric vehicle energy management. In: Proc. of the 44th IEEE Conf. on Decision and Control, and the 2005 European Control Conf., Seville, Spain (2005)
226. Musser, J., Wang, C.Y.: Heat transfer in a fuel cell engine. In: Proc. of the 34th National Heat Transfer Conference, Pittsburgh, PA (2000)
227. Nesscap Press Release: Launches 48-Volt ultracapacitor modules for heavy-duty transportation and industrial applications (2010), online version: http://www.nesscap.com
228. Nielsen, M.P., Knudsen Kaer, S.: Modeling a PEM fuel cell natural gas reformer. In: Proc. of the 6th International Conference on Efficiency, Costs, Optimization, Simulation and Environmental Impact of Energy Systems, Copenhagen, Denmark (2003)
229. Nissan Press Release: Nissan releases Tino Hybrid (2000), online version: http://www.nissan-global.com

230. Nissan Press Release: Nissan unveils "LEAF" — The world's first electric car designed for affordability and real-world requirements (2009), online version: http://www.nissan-global.com

231. Nissan Press Release: Nissan pioneers fuel-cell Limo (2007), online version: http://www.nissan-global.com

232. Nitz, L., Truckenbrodt, A., Epple, W.: The new two-mode hybrid system from the Global Hybrid Cooperation. In: Proc. of the 27th International Vienna Motor Symposium, Vienna, Austria (2006)

233. Novotnak, R.T., Chiasson, J., Bodson, M.: High-performance motion control of an induction motor with magnetic saturation. IEEE Transactions on Control Systems Technology 7(3), 315–327 (1999)

234. National Renewable Energy Laboratory: Go/No-Go recommendation for sodium borohydride for on-board vehicular hydrogen storage. Report NREL/MP-150-42220 (2007), online version: http://www.nrel.org

235. Ogburn, M.J., Nelson, D.J., Luttrell, W., King, B., Postle, S., Fahrenkrog, R.: Systems integration and performance issues in a fuel cell hybrid electric vehicle. SAE Paper 2000-01-0376 (2000)

236. Ogden, J., Steinbuegler, M., Kreutz, T.: A comparison of hydrogen, methanol and gasoline as fuels for fuel cell vehicles: implications for vehicle design and infrastructure development. Journal of Power Sources 79, 143–168 (1999)

237. Opel Press Release: HydroGen3 fuel cell study moves closer to volume production (2001), online version: http://www.opel.com

238. Opila, D.F., Wang, X., McGee, R., Gillespie, R.B., Cook, J.A., Grizzle, J.W.: An energy management controller to optimally trade off fuel economy and drivability for hybrid vehicles. IEEE Transactions on Control Systems Technology PP(99), 1–16 (2011)

239. Otsu, A., Takeda, T., Suzuki, O., Hatanaka, K.: Controlling apparatus for a hybrid car. U.S. Patent no. 6123163 (2000)

240. Ott, T., Zurbriggen, F., Onder, C., Guzzella, L.: Cycle Averaged Efficiency of Hybrid Electric Vehicles. Accepted for Publication in the Journal of Automobile Engineering (2012)

241. Paganelli, G., Ercole, G., Brahma, A., Guezennec, Y.G., Rizzoni, G.: A general formulation for the instantaneous control of the power split in charge-sustaining hybrid electric vehicles. In: Proc. of the 5th International Symposium on Advanced Vehicle Control, Ann Arbor, MI (2000)

242. Paganelli, G., Guerra, T.M., Delprat, S., Santin, J.J., Combes, E., Delhom, M.: Simulation and assessment of power control strategies for a parallel hybrid car. Journal of Automobile Engineering 214, 705–718 (2000)

243. Paganelli, G., Guezennec, Y., Rizzoni, G.: Optimizing control strategy for hybrid fuel cell vehicle. SAE Paper 2002-01-0102 (2002)

244. Parker press release: Parker Runwise advanced series hybrid drive sets the standard for class 8 refuse vehicles (2009), online version: http://www.parker.com

245. Pathapati, P.R., Xue, X., Tang, J.: A new dynamic model for predicting transient phenomena in a PEM fuel cell system. Renewable Energy 30, 1–22 (2005)

246. Paxton, B., Newman, J.: Modeling of nickel/metal hydride batteries. Journal of Electrochemical Society 144(11), 3818–3831 (1997)

247. Pell, W.G., Conway, B.E., Adams, W.A., De Oliveira, J.: Electrochemical efficiency in multiple discharge/recharge cycling of supercapacitors in hybrid EV applications. Journal of Power Sources 80, 134–141 (1999)

248. Pesaran, A.A.: Battery thermal models for hybrid vehicle simulations. Journal of Power Sources 110, 377–382 (2002)
249. Pfiffner, R., Guzzella, L.: Optimal operation of CVT-based powertrains. International Journal of Robust and Nonlinear Control 11(11), 1003–1021 (2001)
250. Pfiffner, R., Guzzella, L., Onder, C.: A control-oriented CVT model with nonzero belt mass. Transactions of the ASME, Journal of Dynamic Systems, Measurement, and Control 124, 481–484 (2002)
251. Phillips, A.M., Bailey, K.E., Jankovic, M.: Control system and method for a parallel hybrid electric vehicle. U.S. Patent no. 6735502 B2 (2004)
252. Phillips, A.M., McGee, R.A., Lockwood, J.T., Spiteri, R.A., Che, J., Blankenship, J.R., Kuang, M.L.: Control system development for the dual drive hybrid system. SAE Paper 2009010231 (2009)
253. Pillay, P., Krishnan, R.: Modeling of permanent magnet motor drives. IEEE Transactions on Industrial Electronics 35, 537–541 (1998)
254. Piller, S., Perrin, M., Jossen, A.: Methods for state-of-charge determination and their applications. Journal of Power Sources 96, 113–120 (2001)
255. Plett, G.L.: Extended Kalman filtering for battery management systems of LiPb-based HEV battery packs. Part 1. Background. Journal of Power Sources 134, 252–261 (2004)
256. Pop, V., Bergveld, H.J., Notten, P.H.L., Regtien, P.P.L.: State-of-the-art of battery state-of-charge determination. Measurement Science and Technology 16, R93–R110 (2005)
257. Poulikakos, D.: Conduction and heat transfer. Prentice Hall, Englewood Cliffs (1994)
258. Pourmovahed, A., Beachley, N.H., Fronczak, F.J.: Modeling of a hydraulic energy regeneration system – Part I: analytical treatment. Transactions of the ASME, Journal of Dynamic Systems, Measurement, and Control 114, 155–159 (1992)
259. Powell, B.K., Bailey, K.E., Cikanek, S.R.: Dynamic modeling and control of hybrid electric vehicle powertrain systems. IEEE Control Systems Magazine 18, 17–33 (1998)
260. Powell, B.K., Pilutti, T.E.: A range extender hybrid electric vehicle dynamic model. In: Proc. of the 33rd Conference on Decision and Control, Lake Buena Vista, FL (1994)
261. Prada, E., Le Berr, F., Dabadie, J.C., Bernard, J., Mingant, R., Sauvant-Moynot, V.: Impedance-based Li-ion modeling for HEV/PHEV's. In: Proc. of the LMS European Vehicle Conference, Munich, Germany (2011)
262. PSA Peugeot Citroën Press Release: PSA Peugeot Citroën unveils diesel hybrid technology (2006), online version:
    http://www.psa-peugeot-citroen.com/en/
263. PSA Peugeot Citroën Press Release: Fuel savings of 10 to 15% in cities with PSA Peugeot Citroën's new Stop & Start system (2004), online version:
    http://www.psa-peugeot-citroen.com/en/
264. PSA Peugeot Citroën Press Release: Peugeot 3008 HYbrid4 (2010),
    http://www.peugeot-pressepro.com
265. Pukrushpan, J.T., Peng, H., Stefanopoulou, A.G.: Control-oriented modeling and analysis for automotive fuel cell systems. Transactions of the ASME, Journal of Dynamic System, Measurement, and Control 126, 14–24 (2004)

266. Pukrushpan, J.T., Stefanopoulou, A., Peng, H.: Control of fuel cell power systems: principles, modeling, analysis and feedback design. Springer, Berlin (2004)

267. Pukrushpan, J.T., Stefanopoulou, A.G., Peng, H.: Control of fuel cell breathing. IEEE Control System Magazine, 30–46 (April 2004)

268. Rabou, L.: Modelling of a variable-flow methanol reformer for a polymer electrolyte fuel cell. International Journal of Hydrogen Energy 20(10), 845–848 (1995)

269. Rahman, K.M., Fahimi, B., Suresh, G., Rajarathnam, A.V., Ehsani, M.: Advantages of switched reluctance motor applications to EV and HEV: design and control issues. IEEE Transactions on Industry Applications 36(1), 111–121 (2000)

270. Renault Press Release: Renault Fluence Z.E. and Kangoo Express Z.E.: Finalized designs revealed and pre-reservations open (2010), online version: http://www.media.renault.com

271. Ricardo Press Release: Ricardo Kinergy delivers breakthrough technology for effective, ultra-efficient and low cost hybridisation (2009), online version: http://www.ricardo.com

272. Rizzoni, G., Guzzella, L., Baumann, B.: Unified modeling of hybrid electric vehicle drivetrains. IEEE/ASME Transactions on Mechatronics 4(3), 246–257 (1999)

273. Rodatz, P.: Dynamics of the polymer electrolyte fuel cell: experiments and model-based analysis. Dissertation ETH no. 15320, Swiss Federal Institute of Technology, Zurich, Switzerland (2003)

274. Rodatz, P., Garcia, O., Guzzella, L., Büchi, F., Bärtschi, M., Tsukada, A., Dietrich, P.: Performance and operational characteristics of a hybrid vehicle powered by fuel cells and supercapacitors. SAE Paper 2003-01-0418 (2003)

275. Rodatz, P., Paganelli, G., Sciarretta, A., Guzzella, L.: Optimal power management of an experimental fuel cell/supercapacitor powered hybrid vehicle. Control Engineering Practice 13(1), 41–53 (2005)

276. Rodatz, P., Tsukada, A., Mladek, M., Guzzella, L.: Efficiency improvements by pulsed hydrogen supply in PEM fuel cell systems. In: Proc. of the 15th IFAC World Congress on Automatic Control, Barcelona, Spain (2002)

277. Rogers, S.: Advanced power electronics and electric machines (APEEM) R&D program overview. US Department of Energy (2010), online version: http://www1.eere.energy.gov

278. Sadler, M., Stapleton, A., Heath, R., Jackson, N.: Application of modeling techniques to the design and development of fuel cell vehicle systems. SAE Paper 2001-01-0542 (2001)

279. Salameh, Z.M., Casacca, M.A., Lynch, W.A.: A mathematical model for lead–acid batteries. IEEE Transactions on Energy Conversion 7, 93–97 (1992)

280. Salman, M., Chang, M.F., Chen, J.S.: Predictive energy management strategies for hybrid vehicles. In: Proc. of the IEEE Vehicle Power and Propulsion Conference, Chicago, IL (2005)

281. Santarelli, M.G., Torchio, M.F.: Experimental analysis of the effects of the operating variables on the performance of a single PEMFC. Energy Conversion and Management 48(1), 40–51 (2007)

282. Sapkota, P., Kim, H.: Zincair fuel cell, a potential candidate for alternative energy. Journal of Industrial and Engineering Chemistry 15(4), 445–450 (2009)

283. Sasaki, S.: Toyota's newly developed hybrid powertrain. In: Proc. of the 1998 International Symposium on Power Semiconductor Devices & ICs, Kyoto, Japan (1998)

284. Savagian, P.: Hybrid systems and New eAssist powertrain. In: Proc. of the SAE Hybrid Technologies Symposium, Anaheim, CA (2011)

285. Schaefer, A., Victor, D.G.: The future mobility of the world population. Transportation Research A 34(3), 171–205 (2000)

286. Schechter, M.: New cycles for automobile engines. SAE paper 1999-01-0623 (1999)

287. Schmidt, A.P., Bitzer, M., Imre, A.W., Guzzella, L.: Experiment-driven electrochemical modeling and systematic parameterization for a lithium-ion battery cell. Journal of Power Sources 195(15), 5071–5080 (2010)

288. Schouten, N.J., Salman, M.A., Kheir, N.A.: Energy management strategies for parallel hybrid vehicles using fuzzy logic. Control Engineering Practice 11, 171–177 (2003)

289. Schweighofer, B., Raab, K., Brasseur, G.: Modelling of high power automotive batteries by the use of an automated test system. In: Proc. of the IEEE Instrumentation and Measurement Technology Conference, Anchorage, AK (2002)

290. Sciarretta, A., Guzzella, L.: Control of hybrid electric vehicles. Optimal energy-management strategies. IEEE Control Systems Magazine 27(2), 60–70 (2007)

291. Sciarretta, A., Back, M., Guzzella, L.: Optimal control of parallel hybrid electric vehicles. IEEE Transactions on Control System Technology 12(3), 352–363 (2004)

292. Sciarretta, A., Dabadie, J.C., Albrecht, A.: Control-oriented modeling of power split devices in combined hybrid-electric vehicles. SAE Paper 2008-01-1313 (2008)

293. Sciarretta, A., Guzzella, L.: Rule-based and optimal control strategies for energy management in parallel hybrid vehicles. In: Proc. of the 6th International Conference on Engines for Automobiles, Capri, Italy (2003)

294. Sciarretta, A., Guzzella, L.: Fuel-optimal control of rendezvous maneuvers for passenger cars. at-Automatisierungstechnik 53(6), 244–250 (2005)

295. Sciarretta, A., Guzzella, L., Onder, C.H.: On the power split control of parallel hybrid vehicles: from global optimization towards real-time control. at-Automatisierungstechnik 51(5), 195–203 (2003)

296. Sciarretta, A., Guzzella, L., Back, M.: Real-time optimal control strategy for parallel hybrid vehicles with on-board estimation of the control parameters. In: Proc. of the IFAC Symposium on Advances in Automotive Control, Salerno, Italy (2004)

297. Sciarretta, A., Guzzella, L., Van Baalen, J.: Fuel optimal trajectories of a fuel cell vehicle. In: Proc. of the International Mediterranean Modeling Multiconference, Bergeggi, Italy (2004)

298. Serrao, L., Onori, S., Sciarretta, A., Guezennec, Y., Rizzoni, G.: Optimal Energy Management of Hybrid Electric Vehicles Including Battery Aging. In: Proc. of the IEEE American Control Conference, San Francisco, CA (2011)

299. Serrao, L., Onori, S., Rizzoni, G.: ECMS as a Realization of Pontryagin's Minimum Principle for HEV Control. In: Proc. of the 2009 American Control Conference, St. Louis, MO (2009)

300. Setlur, P., Wagner, J.R., Dawson, D.M., Samuels, B.: Nonlinear control of a continuously variable transmission (CVT) for hybrid vehicle powertrains. Control Systems Technology 11(1), 101–108 (2003)

301. Shafai, E., Simons, M., Neff, U., Geering, H.P.: Model of a continuously variable transmission. In: Proc. of the 1st IFAC Workshop on Advances in Automotive Control, Ascona, Switzerland (1995)
302. Shepherd, C.M.: Design of primary and secondary cells II. An equation describing battery discharge. Journal of Electrochemical Society 112, 252–257 (1965)
303. Smedley, S.I., Zhang, X.G.: A regenerative zincair fuel cell. Journal of Power Sources 165(2), 897–904 (2007)
304. Society of Automotive Engineers: SAE Standard J1711 (2006), online version: http://www.sae.org
305. Srivastava, N., Haque, I.: A review on belt and chain continuously variable transmissions (CVT). Mechanism and Machine Theory 44(1), 19–41 (2009)
306. Stern, M.O.: Energy storage in magnetic fields. Energy 13(2), 137–140 (1988)
307. Stevens, P., Toussaint, G., Caillon, G., Viaud, P., Vinatier, P., Cantau, C., Fichet, O., Sarrazin, C., Mallouki, M.: Development of a lithium air rechargeable battery. ECS Transactions 28(32), 1–12 (2010)
308. Stockar, S., Marano, V., Canova, M., Rizzoni, G., Guzzella, L.: Energy-optimal control of plug-in hybrid electric vehicles for real-world driving cycles. IEEE Transactions on Vehicular Technology 60(7), 2949–2962 (2011)
309. Sudhoff, S.D., Krause, P.C.: Average-value model of the brushless DC 120° inverter system. IEEE Transactions on Energy Conversion 5, 553–557 (1990)
310. Sundström, O., Ambühl, D., Guzzella, L.: On implementation of dynamic programming for optimal control problems with final state constraints. Oil & Gas Science and Technology – Rev. IFP 65(1), 55–66 (2010)
311. Sundström, O., Guzzella, L.: A generic dynamic programming Matlab function. In: Proc. of the IEEE International Conference on Control Applications and Intelligent Control, St. Petersburg, Russia (2009)
312. Sundström, O., Stefanopoulou, A.: Optimum battery size for fuel cell hybrid electric vehicle, part 1. ASME Journal of Fuel Cell Science and Technology 4(2), 167–175 (2007)
313. Sundström, O., Stefanopoulou, A.: Optimum battery size for fuel cell hybrid electric vehicle with transient loading consideration, part 2. ASME Journal of Fuel Cell Science and Technology 4(2), 176–184 (2007)
314. Tai, C., Tsao, T.C., Levin, M., Barta, G., Schechter, M.: Using camless valvetrain for air hybrid optimization. SAE paper 2003-01-0038 (2003)
315. Takano, K., Nozaki, K., Saito, Y., Negishi, A., Kato, K., Yamaguchi, Y.: Simulation study of electrical dynamic characteristics of lithium-ion battery. Journal of Power Sources 90, 214–223 (2000)
316. Tamai, G., Hoang, T., Taylor, J., Skaggs, C., Downs, B.: Saturn engine stop–start system with an automatic transmission. SAE Paper 2001-01-0326 (2001)
317. Tate, E.D., Grizzle, J.W., Peng, H.: Shortest path stochastic control for hybrid electric vehicles. International Journal of Robust and Nonlinear Control 18(14), 1409–1429 (2008)
318. Thele, M., Bohlen, O., Sauer, D.U., Karden, E.: Development of a voltage-behavior model for NiMH batteries using an impedance-based modeling concept. Journal of Power Sources 175(1), 635–643 (2008)
319. Thirumalai, D., White, R.E.: Steady-state operation of a compressor for a proton exchange membrane fuel cell system. Journal of Applied Electrochemistry 30, 551–559 (2000)

320. Thomas, C.E., Reardon, J., Lomax, F., Pinyan, J., Kuhn, I.: Distributed hydrogen fueling systems analysis. In: Proc. of the 2001 DOE Hydrogen Program Review (2001)

321. Thorstensen, B.: A parametric study of fuel cell system efficiency under full and part load operation. Journal of Power Sources 92, 9–16 (2001)

322. Torotrak Press Release: Torotrak technology supports Volvo-led flywheel hybrid project (2011), online version: http://www.torotrak.com

323. Toyota Press Release: Toyota develops advanced fuel cell hybrid vehicle (2008), online version: http://www.toyota.co.jp

324. Toyota Motor Corporation: Toyota Hybrid System THS-II (2003), online version: http://www.toyota.co.jp

325. Toyota Press Release: Toyota to display fuel cell hybrid and personal mobility concept vehicles at Tokyo Motor Show (2005), online version: http://www.toyota.co.jp/en/

326. Toyota Press Release: How hybrid vehicles work: gas and electric combine for incredible mileage (2007), online version: http://www.pressroom.toyota.com

327. Toyota Press Release: Lexus Hybrid Drive evolution includes world's first full V8 hybrid and all-wheel drive (2007), online version: http://www.pressroom.toyota.com

328. Trajkovic, S., Tunestal, P., Johansson, B.: Simulation of a pneumatic hybrid powertrain with VVT in GT-Power and comparison with experimental data. SAE Paper 2009-01-1323 (2009)

329. Treffinger, P., Gräf, M., Goedecke, M.: Hybridization of fuel cell powered drive trains. In: Proc. of the VDI-Tagung "Innovative Fahrzeugantriebe", Dresden, Germany (2002)

330. Tulpule, P., Marano, V., Rizzoni, G.: Energy management for plug-in hybrid electric vehicles using equivalent consumption minimization strategy. Int. J. Electric and Hybrid Vehicles 2(4), 329–350 (2010)

331. Turner, J., Pearson, R., Kenchington, S.: Concepts for improved fuel economy from gasoline engines. Int. J. Engine Res. 6(2), 137–157 (2005)

332. US Department of Energy, Office of Hydrogen, Fuel Cells and Infrastructure Technologies. In: Proc. of the Workshop on Compressed and Liquefied Hydrogen Storage, Southfield, MI (2002), online version: http://www.eere.energy.gov

333. van Druten, R.: Transmission design of the zero inertia powertrain. Dissertation, TU Eindhoven, The Netherlands (2001)

334. van Keulen, T., de Jager, B., Steinbuch, M.: An adaptive sub-optimal energy management strategy forhybrid drive-trains. In: Proc. of the 17th IFAC World Congress, Seoul, Korea (2008)

335. Van Mierlo, J., Van den Bossche, P., Maggetto, G.: Models of energy sources for EV and HEV: fuel cells, batteries, ultracapacitors, flywheels and engine-generators. Journal of Power Sources 128(1), 78–89 (2003)

336. Vasile, I., Higelin, P., Charlet, A., Chamaillard, Y.: Downsized engine torque lag compensation by pneumatic hybridization. In: Proc. of the 13th International Conference on Fluid Flow Technologies, Budapest, Hungary (2006)

337. Vasile, I., Dönitz, C., Voser, C., Vetterli, J., Onder, C., Guzzella, L.: Rapid start of hybrid pneumatic engines. In: Proc. of the IFAC Workshop on Engine and Powertrain Control, Simulation, and Modeling, Rueil-Malmaison, France (2009)

338. Veenhuizen, P.A., Bonsen, B., Klaassen, T.W.G.L., Albers, P., Changenet, C., Poncy, S.: Pushbelt CVT efficiency improvement potential of servo-electromechanical actuation and slip control. In: Proc. of the International Continuously Variable and Hybrid Transmission Congress, Davis, CA (2004)

339. Vidyasagar, M.: Nonlinear systems analysis. Prentice-Hall, Englewood Cliffs (1978)

340. Villeneuve, A.: Dual mode electric infinitely variable transmission. In: Proc. of the International Continuously Variable and Hybrid Transmission Congress, Davis, CA (2004)

341. Vinot, E., Scordia, J., Trigui, R., Jeanneret, B., Badin, F.: Model simulation, validation and case study of the 2004 THS of Toyota Prius. Int. Journal of Vehicle Systems Modelling and Testing 3(3), 139–167 (2008)

342. Visteon Press Release: Visteon debuts its compact electric air conditioning compressor for hybrid and electric vehicles (2010), online version: http://www.visteon.com/media

343. Volkswagen Press Release: Touareg Hybrid with new V6 TSI can tow 3.5 metric tons (2010), online version: http://www.volkswagen-media-services.com

344. Volvo Technical Data: Powercell – An environmental friendly power generator for the future (2008), online version: http://www.volvogroup.com

345. Von Raumer, T., Dion, J.M., Dugard, L., Thomas, J.L.: Applied nonlinear control of an induction motor using digital signal processing. IEEE Transactions on Control Systems Technology 2, 327–335 (1994)

346. Vroemen, B.G., Serrarens, A., Veldpaus, F.: Hierarchical control of the zero inertia powertrain. JSAE Review 22, 519–526 (2001)

347. Vroemen, B.G.: Component control for the zero inertia powertrain. Dissertation, TU Eindhoven, The Netherlands (2001)

348. Yoon, H., Lee, S.: An optimized control strategy for parallel hybrid electric vehicle. SAE Paper 2003-01-1329 (2003)

349. Wang, H., Liang, Y., Mirfakhrai, T., Chen, Z., Sanchez-Casalongue, H., Dai, H.: Advanced asymmetrical supercapacitors based on graphene hybrid materials. Nano Research (2011) (published online)

350. Wenger, D., Polifke, W., Schmidt-Ihn, E., Abdel-Baset, T., Maus, S.: Comments on solid state hydrogen storage systems design for fuel cell vehicles. International Journal of Hydrogen Energy 34(15), 6265–6270 (2009)

351. West, J.G.W.: DC, induction, reluctance and PM motors for electric vehicles. Power Engineering Journal 8(2), 77–88 (1994)

352. Wiartalla, A., Pischinger, S., Bornscheuer, W., Fieweger, K., Ogrzwalla, J.: Compressor expander units for a fuel cell systems. SAE Paper 2000-01-0380 (2000)

353. Williams Press Release: Williams Hybrid Power contracts with Porsche AG for 911 GT3 R Hybrid (2010), online version: http://www.attwilliams.com

354. Williams, B.D., Kurania, K.S.: Commercializing light-duty plug-in/plug-out hydrogen-fuel-cell vehicles: "Mobile Electricity" technologies and opportunities. Journal of Power Sources 166(2), 549–566 (2007)

355. Wipke, K.B., Cuddy, M.R., Burch, S.D.: ADVISOR 2.1: a user-friendly advanced powertrain simulation using a combined backward/forward approach. IEEE Transactions on Vehicular Technology 48(6), 1751–1761 (1999)

356. Wittmer, C.: Entwurf und Optimierung der Fahrstrategien für ein Personenwagen-Antriebskonzept mit Schwungradkomponente und stufenlosem Zweibereichsgetriebe. Dissertation ETH no. 11672, Swiss Federal Institute of Technology, Zurich, Switzerland (1996)

357. Wittmer, C., Guzzella, L., Dietrich, P.: Optimized control strategies for a hybrid car with a heavy flywheel. At–Automatisierungstechnik 44(7), 331–337 (1996)

358. Wu, B., Dougal, R., White, R.E.: Resistive companion battery modeling for electric circuit simulations. Journal of Power Sources 93, 186–200 (2001)

359. Wu, B., Lin, C.C., Filipi, Z., Peng, H., Assanis, D.: Optimal power management for a hydraulic hybrid delivery truck. Vehicle System Dynamics 42(1-2), 23–40 (2004)

360. Wu, C.H., Hung, Y.H., Hong, C.W.: On-line supercapacitor dynamic models for energy conversion and management. Journal of Power Sources (2011) (in press)

361. Xue, X., Tang, J., Smirnova, A., England, R., Sammens, N.: System level lumped-parameter dynamic modeling of PEM fuel cell. Journal of Power Sources 133, 188–204 (2004)

362. Zhang, R., Chen, Y.: Control of hybrid dynamical systems for electric vehicles. In: Proc. of the American Control Conference, Arlington, VA (2001)

363. Zhu, Y., Chen, Y., Chen, Q.: Analysis and design of an optimal energy management and control system for hybrid electric vehicles. In: Proc. of the 19th Electric Vehicles Symposium, Busan, Korea (2002)

364. Zhu, Y., Chen, Y., Tian, G., Wu, H., Chen, Q.: A four-step method to design an energy management strategy for hybrid vehicles. In: Proc. of the 2004 American Control Conference, Boston, MA (2004)

Index